SALMON AND ACORNS FEED OUR PEOPLE

NATURE, SOCIETY, AND CULTURE
Scott Frickel, Series Editor

A sophisticated and wide-ranging sociological literature analyzing nature-society-culture interactions has blossomed in recent decades. This book series provides a platform for showcasing the best of that scholarship: carefully crafted empirical studies of socioenvironmental change and the effects such change has on ecosystems, social institutions, historical processes, and cultural practices.

The series aims for topical and theoretical breadth. Anchored in sociological analyses of the environment, Nature, Society, and Culture is home to studies employing a range of disciplinary and interdisciplinary perspectives and investigating the pressing socioenvironmental questions of our time—from environmental inequality and risk, to the science and politics of climate change and serial disaster, to the environmental causes and consequences of urbanization and war making, and beyond.

Available titles in the Nature, Society, and Culture series:

Diane C. Bates, *Superstorm Sandy: The Inevitable Destruction and Reconstruction of the Jersey Shore*
Elizabeth Cherry, *For the Birds: Protecting Wildlife through the Naturalist Gaze*
Cody Ferguson, *This Is Our Land: Grassroots Environmentalism in the Late Twentieth Century*
Aya H. Kimura and Abby Kinchy, *Science by the People: Participation, Power, and the Politics of Environmental Knowledge*
Anthony B. Ladd, ed., *Fractured Communities: Risk, Impacts, and Protest against Hydraulic Fracking in U.S. Shale Regions*
Stefano B. Longo, Rebecca Clausen, and Brett Clark, *The Tragedy of the Commodity: Oceans, Fisheries, and Aquaculture*
Stephanie A. Malin, *The Price of Nuclear Power: Uranium Communities and Environmental Justice*
Kari Marie Norgaard, *Salmon and Acorns Feed Our People: Colonialism, Nature, and Social Action*
Chelsea Schelly, *Dwelling in Resistance: Living with Alternative Technologies in America*
Diane Sicotte, *From Workshop to Waste Magnet: Environmental Inequality in the Philadelphia Region*
Sainath Suryanarayanan and Daniel Lee Kleinman, *Vanishing Bees: Science, Politics, and Honeybee Health*

SALMON AND ACORNS FEED OUR PEOPLE

Colonialism, Nature, and Social Action

KARI MARIE NORGAARD

RUTGERS UNIVERSITY PRESS
New Brunswick, Camden, and Newark, New Jersey, and London

Library of Congress Cataloging-in-Publication Data

Names: Norgaard, Kari Marie, author.
Title: Salmon and acorns feed our people : colonialism, nature, and social action / Kari Marie Norgaard.
Description: New Brunswick : Rutgers University Press, 2019. | Series: Nature, society, and culture
Identifiers: LCCN 2019002457 | ISBN 9780813584195 (paperback) | ISBN 9780813584201 (cloth)
Subjects: LCSH: Human ecology—California. | Environmental degradation—California. | Karuk Tribe. | Imperialism. | Klamath River (Or. and Calif.) | Power (Social sciences)—California. | BISAC: NATURE / Ecosystems & Habitats / Rivers. | TECHNOLOGY & ENGINEERING / Agriculture / Forestry. | NATURE / Ecology. | TECHNOLOGY & ENGINEERING / Fisheries & Aquaculture. | SOCIAL SCIENCE / Regional Studies.
Classification: LCC GF13.3.U6 N67 2019 | DDC 304.209794—dc23
LC record available at https://lccn.loc.gov/2019002457

A British Cataloging-in-Publication record for this book is available from the British Library.

Copyright © 2019 by Kari Marie Norgaard and the Karuk Tribe
All rights reserved

No part of this book may be reproduced or utilized in any form or by any means, electronic or mechanical, or by any information storage and retrieval system, without written permission from the publisher. Please contact Rutgers University Press, 106 Somerset Street, New Brunswick, NJ 08901. The only exception to this prohibition is "fair use" as defined by U.S. copyright law.

∞ The paper used in this publication meets the requirements of the American National Standard for Information Sciences—Permanence of Paper for Printed Library Materials, ANSI Z39.48-1992.

www.rutgersuniversitypress.org

Manufactured in the United States of America

To All My Teachers

Parents, grandparents, son, siblings, partner, colleagues, friends, rivers, earth.

With gratitude

CONTENTS

	Introduction	1
1	Mutual Constructions of Race and Nature on the Klamath	25
2	Ecological Dynamics of Settler-Colonialism: Smokey Bear and Fire Suppression as Colonial Violence	72
3	Research as Resistance: Food, Relationships, and the Links between Environmental and Human Health	129
4	Environmental Decline and Changing Gender Practices: What Happens to Karuk Gender Practices When There Are No Fish or Acorns?	165
5	Emotions of Environmental Decline: Karuk Cosmologies, Emotions, and Environmental Justice	198
	Conclusion: Climate Change as a Strategic Opportunity?	223
	Methodological Appendix	241
	Acknowledgments	245
	Notes	249
	Works Cited	257
	Index	283

SALMON AND ACORNS FEED OUR PEOPLE

INTRODUCTION

> It should not have happened that the great civilizations of the Western Hemisphere, the very evidence of the Western Hemisphere, were wantonly destroyed, the graduate progress of humanity interrupted and set upon a path of greed and destruction. Choices were made that forged that path towards destruction of life itself—the moment in which we now live and die as our planet shrivels, overheated. To learn and know this history is both a necessity and a responsibility to the ancestors and descendants of all parties.
> —Roxanne Dunbar-Ortiz, *An Indigenous People's History of the United States*

To the people who live there, the Klamath Basin is their center of the world. It's not hard to see why. Despite its remoteness the region has been a touchstone for landmark environmental policies, not the least of which is the current process for the removal of the Klamath River dams. The Klamath Basin is a place of profound beauty and paradox. Here significantly intact ecological systems coexist alongside advancing environmental degradation. The region is a high point of California's renowned biological diversity—with numerous endemic amphibians, fish and flowering plants, an abundance of lilies, and some of the highest diversity of conifer species to be found worldwide. A wealth of Indigenous cultural knowledge and ingenuity exists alongside intense disenfranchisement, poverty, substance abuse, and domestic violence. The Klamath Basin is remote—from the heart of Karuk ancestral territory at Ka'tim'iin, it is a two-hour drive to the nearest traffic light. There, cutting-edge and innovative research is being conducted by a handful of tribal scientists and underresourced tribal leaders. Its vast "wilderness areas" (representing 15% of the total wilderness designation in California) are full of people who secure their food and drinking water directly from the forest. Early anthropologists marveled at the enormous abundance of natural resources of the people living on the Klamath River. Karuk people, together with their Yurok, Hupa, and Konomihu neighbors, are considered to have been the wealthiest of all Indian people in California. This wealth was a direct result of their intimate knowledge of the land and their ability to sustain and enhance the

Klamath region's year-round abundance of food resources, particularly salmon, deer, elk, and acorns.

For me, the Klamath Basin and its inhabitants have been a source of profound learning and inspiration. In fact, despite having spent significant portions of my life here over the past decade and a half, I continue to learn something new every day I am "on the river." If the natural world is a potent teacher, the complex ecology and geology of the Klamath mountain region make its teachings exceptionally interesting. But my biggest teachers have been the human inhabitants of the region, Native and non-Native alike. For despite having been educated in what must surely be the most radical public school system in the country in Berkeley, California, a mere six hours to the south, like most settlers of this continent, I was not taught much about my Native neighbors. I was not taught, for example, that so many people in my own state were hunting and fishing for their food, weaving baskets using traditional techniques, carrying out important ceremonies to keep the world intact, and generally engaging in so many of the same types of activities their ancestors had done back into time beyond memory. Nor was I aware that they continue these practices despite the concerted efforts of the state of California and the federal government for over a century and a half to prevent them. I was not aware that engaging in these activities today could still be, for those tribal families, acts of fighting for cultural and physical survival.

In my time here, I have been privileged to learn not only fascinating details of river and fire ecology—details that in many cases have been passed down by word of mouth and shared practice in a direct line across generations from those who developed them, but lessons about community and relationships, lessons about resisting and influencing social power, and lessons about ethics and responsibility, including my own. As anyone who knows me can plainly see, I am still learning in each of these realms.

Take, for example, the day I first got a glimpse into the relationships between forest fire and fish habitat. Karuk cultural biologist Ron Reed and I had been working together for several years on the health and cultural importance of salmon in the context of dam removal, and he was giving a talk to a class of mine at Whitman College. Ron had been covering the usual topics of our work up to that point, speaking about the early death of elders from diabetes and the importance of traditional foods, when suddenly he switched gears and began speaking about the importance of "taking care of the upslope" for the health of the river. I had no idea what this meant. Over the next months, Ron (likely spurred on in part by my blank expression) kept bringing up this connection. He began explaining in detail how burning the forest was essential to maintain adequate stream flow since large amounts of brush took up lots of water. He explained that burning also kept bug populations down so that acorn crops would not be damaged and made good habitat and hunting conditions for deer and elk. Ron explained that these systems not only were about ecological practice but also connected to social responsibili-

FIGURE 1. Karuk elder Laverne Glaze with ferns. Photo credit: Frank Lake.

ties, family and community structure, spiritual practices, and even political sovereignty—what he calls "Karuk social management." This was my first awareness of not only the profound degree of interconnection between many domains that sociologists tend to see as separate but also the degree to which what I had been taught about the natural world was just plain inaccurate. Ecologist Kat Anderson (2005) writes, "Traditional management systems have influenced the size, extent, pattern, structure, and composition of the flora and fauna within a multitude of vegetation types throughout the state. When the first Europeans visited California, therefore, they did not find in many places a pristine, virtually uninhabited wilderness but rather a carefully tended 'garden' that was the result of thousands of years of selective harvesting, tilling, burning, pruning, sowing, weeding, and transplanting" (125–126).[1]

This very different land ethos is based in knowledge that Ron and many others hold about how to enhance individual traditional foods through burning, pruning, digging, and harvesting techniques, as well as how to maintain landscape-level ecological relationships, which comes from a direct line of practitioners that extends back from a Karuk perspective to the beginning of time—at least 12,000 years according to the archaeological record. Much of this information is reflected in ceremonies, in the regalia that are worn, and in the actions that take place. And although it is true that much knowledge about specifics has been lost, a great deal also has been transferred and retained from one generation to the next through both oral history and shared practical experience right down to the present. Western scientists are now beginning to acknowledge not only that there was another form of social organization besides "hunter-gatherer" and "agriculturalist" but also the profound refinement and intricacy of Indigenous ecological knowledge. Anderson notes, "California Indians practiced resource management at four levels of biological organization: the organism, the population, the plant community, and the landscape. They used resource management techniques at each of these levels, or scales, to promote the persistence of individual plants, plant populations, animal populations, plant associations, and habitat relationships in many different vegetation types in California" (Anderson 2005, 135).

Since time before memory, large numbers of salmon have made their way up and down the Klamath River. Other riverine species too, like lamprey, sturgeon, and trout, have been in abundance. In fact, until the 1970s, the Klamath River was the third largest salmon-producing stream in the western United States. Karuk creation stories describe how people have been intimately dependent upon salmon and other riverine foods from the beginning. Engaging in fishing, gathering, hunting, burning, and other forms of "traditional management" is central to spiritual life and cultural practices, and it continues to form a basis of what my colleague, Karuk descendant, traditional practitioner, and United States Forest Service (USFS) ecologist Dr. Frank Lake, calls the "social glue" between generations today (Lake, personal communication; see also Norgaard 2012; Willette, Norgaard, and Reed 2016; Anderson 2005). Salmon hold phenomenally important material and symbolic significance for the community. Historically, salmon accounted for over half of the total calories and protein consumed by Karuk people, and they continue to be an important food source today. Salmon also figure centrally in spiritual practices, ceremonies, stories, and, as will be illustrated here, the social organization of daily life, meaning systems, gender structures, mental and physical health, and the dynamics of tribal and state power. As recently as the 1980s, fishing families ate salmon three times per day in season. Yet as of 2014, the wild salmon populations of the Klamath River have been reduced to roughly 4% of their previous productivity. This gives Karuk people the dubious honor of having one of the most dramatic and recent diet shifts of any people in North America.

Today, Karuk people represent the second largest American Indian Tribe[2] in California with over 4,000 members and descendants. Having survived the brutality of overt genocide during the 1800s and state-sponsored forced assimilation into the 1950s—both of which massively reorganized their economic, political, and social systems—in recent decades, the Karuk Tribe has had their federal recognition reconfirmed, brought back nearly all their ceremonial practices, and developed a new political structure. Their Department of Natural Resources

FIGURE 2. Ron Reed fishing with dipnet at Ishi Pishi Falls. Photo credit: Karuk Tribe.

engages in cutting-edge biological research and policy. The region is home to the largest number of native language speakers and traditional basket weavers in the state of California.

The Klamath River itself has many stories to tell. The Klamath region is considered biologically one of the richest temperate areas in the world, with high levels of species diversity and endemism, and among the highest diversity of conifers and lilies in the world. Despite its remoteness, the Klamath River Basin has already figured centrally in a number of national and internationally significant environmental events from efforts to undo the Endangered Species Act in the early 2000s to, more recently, a highly unexpected collaboration of farmers, Tribes, and commercial fishermen around the removal of the four main-stem dams. Should it occur, this will be the largest dam removal in world history.

When I—a non-Native, white sociologist—first came to Karuk country in 2003, I knew little about any of this. I knew very little about salmon, the traditional use of fire, endemic lamprey, and other species that occur only there or the Klamath

River dams. And despite having grown up in the same state, I knew very little about my Indian neighbors to the north. I had no idea that so many people in my own state engaged in very significant levels of subsistence hunting, fishing, and gathering; were actively fighting against federal agencies to maintain ceremonial practices; or had such intimate and complex relationships with the land. Of course, my lack of awareness was not coincidental. The "disappearance" of American Indian people from "mainstream" awareness was—and for the most part still is—universal across the United States, be it in media, educational curricula, the daily news, or academic knowledge, including my own discipline of sociology. Public school curricula describe the "discovery" of the United States, mention Indians briefly, but skip over the "small matter" of American Indian genocide, and rarely bother to mention Native people or their political or cultural achievements again.

Far from a coincidence, the literal disappearance of Native people was considered a fundamental necessity to the establishment of the United States. The use of military force to kill and relocate people and the signing of treaties to contain them in particular areas were of utmost importance in the founding of this country. Upon its inception in 1850, the state of California set a bounty of five dollars per Native "scalp" and reimbursed bounty hunters for their ammunition. After the period of overt genocide ended came attempts to "disappear" people through state-sponsored projects of forced assimilation through boarding schools, the Dawes Act (Indian General Allotment Act), and termination. It is no wonder that for decades, many people tried to hide Native identities, "blend in" and "become white" to avoid being killed, to survive economically, and to keep their children from being targeted by racism.

Now, that disappearance is discursive. It is enacted through what is and is not represented in popular culture, film, news media, and academic theory. This "disappearance" of Native people is a central feature of the logic of what is now called "settler-colonialism," whereby new arrivals legitimate their ties to land through practices and discourses of erasure (Coulthard 2014; Lefevre 2015; Tuck and Yang 2012; Wolfe 2006; Whyte 2016a). Native peoples and their experiences, perspectives, and forms of knowledge have been "disappeared" from the academy, too. In my own discipline of sociology, this erasure has been particularly extreme (see, e.g., Jacob 2017). Little theory is written by or about Native people, there is no subsection of the American Sociological Association on Native or Indigenous studies, and panels on this topic are normally presented only every other year at our annual meeting. Indeed, few practicing sociologists in the United States identify as Native American.[3] Fortunately, just as American Indians across the country have gained economic and political power in recent decades and just as the numbers of people who claim Native ancestry on the census have begun to rise, so too the time may be coming when the academy as a whole and sociology in particular can make a place for Native experiences and perspectives.

In just the past few years, a number of theorists have argued for major theoretical insurrections within disciplines in the social science and humanities. Mark Fiege's *Republic of Nature* maintains that a wholesale incorporation of the natural environment into the discipline of history is necessary. Roxanne Dunbar-Ortiz's *An Indigenous People's History of the United States* fundamentally transforms the organization and canon of U.S. history. Within sociology, Aldon Morris's *The Scholar Denied* provides a detailed exposé of how anti-black racism sidelined and reordered theoretical possibilities for the field of sociology, while Julian Go opens his 2016 monograph, *Postcolonial Thought and Social Theory*, with the provocative statement that "social theory was born of, in and to some extent for modern empire" (1). These scholars and more not only argue for integration of a few missing concepts but show how a range of modernist assumptions fundamentally reorganizes theories across the entirety of a given discipline.

One aim of this book is to be a part of the reweaving of Native presence, experiences, and cosmologies back into my own discipline—to lend my mind to the efforts to "unsettle" and "decolonize" academic theory, especially sociology.[4] In *The White Possessive: Property, Power, and Indigenous Sovereignty*, Goenpul Aboriginal scholar Aileen Moreton-Robinson calls for the field of Indigenous studies to engage and critique existing academic traditions. Drawing upon the work of Torres Strait Islander scholar Martin Nakata, Moreton-Robinson asserts that Indigenous scholars must engage with traditional disciplines "in order to demonstrate how this knowledge is limited in its ability to understand us" and "to move beyond Indigenous endogenous objectification" (xvi). Although I am a white scholar, my project of challenging sociological theorizing with respect to colonialism, Indigenous peoples, and the so-called natural environment is intended in this vein.[5] In neglecting Native experiences, as well as the ongoing operation and political history of colonialism that underlies work in not only the United States but also academic enterprises globally, many important and powerful theories across the social sciences are misspecified. As historian Roxanne Dunbar-Ortiz (2014) writes of her own discipline, "Awareness of the settler-colonial context of US history writing is essential if one is to avoid the laziness of the default position and the trap of a mythological unconscious believe in manifest destiny. . . . To say that the United States is a colonialist settler-state is not to make an accusation but rather to face historical reality, without which consideration not much in US history makes sense, unless Indigenous peoples are erased" (6–7).

A significant goal of mine is to add to the voices who have recently been indicating how the same centrality of the notion of the United States as a settler-colonial state reshapes and expands theories across both sociology and the social sciences (not to mention the humanities or sciences). I share Go's impassioned critiques of our discipline's amnesia regarding its elitist and colonial origins. But is it postcolonialism or settler-colonialism that we should be theorizing? Can U.S. empire be adequately conceptualized without understanding Native experiences?

Unfortunately, one obvious problem with theories of postcolonialism from an Indigenous standpoint is that they are not actually "post." The danger is that such theories themselves perform colonial acts of erasure. Go closes his powerful 2016 book with the statement, "Colonialism has ended, but the power relations, systems of meaning, and socioeconomic inequalities that it birthed stubbornly endure" (185). While essential, theories of empire and postcolonialism alone do not capture power dynamics with respect to Indigenous peoples, nor do these frameworks account for the ideological and material bases of social power emulating from the natural world. Rather, I hope this case study will illustrate how U.S. sociology and other social sciences developed on this continent have been shaped by settler-colonialism as much as or more than they may have been shaped by postcolonialism. And among the central tenets of colonialism that remain unexamined without this orientation are the importance of land and nature for social life.

But the lessons about health, food, land, power, and colonialism that I have learned from my Karuk colleagues and fifteen years of policy work with the Karuk Tribe are relevant far beyond the social sciences. By tracing relationships between people, a river, and the ongoing use and operation of political power, we come to a fuller understanding of concepts of health, environmental and food justice, and topics of identity, race, gender, and climate change that are of the upmost importance for all peoples today both inside the academy and beyond. *Salmon and Acorns Feed Our People* draws on the experience of one community on the Klamath River to illustrate the rich connections between environmental and human health, gender practices, emotions, and the ongoing power relations of racism and colonialism that are highly relevant for Native and non-Native communities alike in this powerful moment in time.

And the stakes are higher than ever. From globalization and food access to contamination and climate change, the issues Karuk and other Indigenous peoples are facing and responding to are symptomatic of social problems confronted by individuals and communities around the world. Climate change in particular may be the most serious ecological problem our world has encountered. Climate change evokes an urgent need to rethink many aspects of Western social, economic, and political systems from the organization of energy systems around fossil fuels to the sustainability of cultural values of excessive consumption and the relevance of epistemologies that presume a separation of the social and natural worlds. To that end, I have done my best to describe the importance of the varied and creative means of Indigenous insight and resistance at the individual, community, and Tribal levels, of determined and innovative people who engage together in ingenious and effective tactics for social change, and have a profoundly hopeful and healing vision for their communities and the world.

A VIEW FROM THE RIVER: PIKYAV

> The Creator has given me a responsibility. He instructed us how we were to do this from the beginning, and that we were given the promise that the Karuk people would endure forever if you did your part, and if you continue to do what you are instructed to do.
> —Robert Goodwin, former Karuk Tribal Council member

> Now we are being stripped of a lot of our duties as a Karuk person, as a traditional male, and that's just because of regulations... the new regulations they have, rules and regulations, keep us actually from living our traditional way of life... our ceremonies have been, you know, stripping down because of regulations... now we're only allowed to do certain things in our ceremonies, not allowed to do our traditional burns or nothing no more.
> —Kenneth Brink "Binx," traditional practitioner and fisheries technician

Pikyav[6] is "to fix" or "to repair" something in Karuk. The Karuk are known as "Fix the World People" in part because of a set of ceremonies—known as *pikyávish*—that are observed together with neighboring tribes each year to renew the world. The responsibility to fix the world is not just ceremonial. As my longtime research collaborator, former Karuk cultural biologist Ron Reed, tells me, fixing the world means fixing and restoring the intertwined environmental and social degradation that has profound impacts on Karuk people's lives. As outlined in "Practicing *Pikyav*: A Guiding Policy for Research Collaborations with the Karuk Tribe," one fixes the world through fishing for the people, through returning fire to the landscape, engaging in research, carrying out legal actions, developing new policies, implementing Tribal programs, and more:

> In the Karuk language, the verb *pikyav* means "to repair," or "to fix." Another Karuk word is *pikyávish*, which refers to the world-renewal ceremony, a set of ceremonies that the Karuk and neighboring tribes continue to hold every summer. When describing the Karuk culture, tribal members often explain, "We are fix-the-world people." For the Karuk Tribe, the center of the world is *Katimin*, the place where the Klamath River and the Salmon River meet. As part of this philosophy, the Karuk Tribe is working to repair and restore the complex social and ecological systems that make up the Klamath River Basin. This work includes fixing some of the environmental and social damages that continue to have profound impacts on Karuk people and Karuk homelands.

Fixing the world is very much about restoring Karuk ecological management on the ground. Among the far-reaching implications of the discursive disappearance of Indigenous peoples has been that flawed conceptions of "nature" and "environment" permeate public consciousness and underlie the organizational divisions of academic disciplines. Western scientists and social scientists alike follow in the tradition that prior to European contact, our continent was an untouched wilderness. Yet in fact, salmon, acorns, and hundreds of food and cultural use species have been actively managed by Native peoples. For Karuk practitioners on the Klamath, ceremonial practices including the First Salmon Ceremony regulated the timing of fishing to allow for escapement and thus continued prosperous runs. Forests have been burned to increase production of food and medicine species, basket materials, and more. Burning also influences the local hydraulic cycles, increasing seasonal runoff into creeks. The diversity of available food resources provided a safety net should one species fail to produce a significant harvest in a given year. Thus, while salmon have been centrally important, other food resources are consumed fresh and preserved to provide throughout the seasons (Anderson 2005; Karuk Tribe 2010; Lake 2013).

The slogans and phrases "we are all connected" and "there is no away" have been repeated within the environmental movement so often since the 1970s that they have become cliché. Yet still the categories of nature and human remain inscribed as separate in the organizational sensibility of everyday life practices, land management policy, and most academic theory. The full implications of the notion that people have "tended" North American landscapes for a very long time remain difficult for non-Indian academics or natural resource practitioners to grasp. From a practical standpoint, this knowledge makes clear that rather than the concept of an "untouched" wilderness as European settlers professed, many of these landscapes were more akin to carefully tended gardens. What natural scientists have described as "nature" and "natural history" is in fact a human-natural history. For example, fire records in California clearly indicate that Native land management systems have significantly shaped the evolutionary course of plant species and communities for at least the twelve thousand years for which there are records. The abundance of these species was a product of Indigenous knowledge and management in which high-quality seeds were selected, the production of bulbs was enhanced through harvest techniques, and populations of oaks, fish, mushrooms, and huckleberries have been reinforced and carefully managed with prayer and fire. Indigenous knowledge and management generated the abundance in the land that formed the basis of capitalist wealth across North America.

These activities on the landscape continue today, although they are often the site of intense political struggle as we will see. Equally important, interactions with salmon, forest foods, rivers, and rocks organize social activities, individual and group identities, gender constructions, and more. The ongoing ability of Karuk people to engage in what is known as traditional management is important for political sovereignty, subsistence activities, and the mental and physical health of individuals.

What people have described as "traditional management" involves a sophisticated non-Western ecology that includes extensive knowledge of particular species and ecological conditions, as well as the knowledge of how to reproduce them. Rather than doing something *to* the land, ecological systems prosper because humans and nature work together. Working together is part of a pact across species, a pact in which both sides have a sacred responsibility to fulfill. Traditional foods and what the Karuk call "cultural use species" flourish as a result of human activities, and in return, they offer themselves to be consumed. Thus, still today, participation in fishing, burning, gathering, and other aspects of traditional management holds immense personal and spiritual significance for many Karuk people and is central to their identity, as Rabbit, a traditional fisherman in his mid-thirties, describes:

> Salmon is like . . . one of our greatest gifts that Creator has given us, and it's something we focus our ceremonies around—our timing, our traditions, our cultural practices . . . a lot of them really revolve around the Salmon runs. . . . You know, you got people, elders up there on the top of the mountain waiting for you to fish, and it's a really really awesome feeling being able to hand your elders fish, you know that puts warmth in your heart, and it's like definitely culturally and religious, you know, it's fulfilling spiritually.

As Ron Reed describes, participation in these management activities is at the heart of "being Indian": "You can give me all the acorns in the world, you can get me all the fish in the world, you can get me everything for me to be an Indian, but it will not be the same unless I'm going out and processing, going out and harvesting, gathering myself. I think that really needs to be put out in mainstream society, that it's not just a matter of what you eat. It's about the intricate values that are involved in harvesting these resources, how we manage for these resources and when." Unfortunately, the invasion of Karuk territory by non-Native settlers has disrupted these ceremonies and cultural systems. The diets, traditional practices, and daily lives of all peoples and cultures change over time. For Karuk people, however, all these have shifted dramatically in the course of the past generation through what can only be understood as very "unnatural" conditions. Karuk culture and lifeways have been under assault over the past 170 years. This assault occurred first explicitly through interactions with gold miners that led to an unratified treaty, lack of recognition of Karuk land title, state policies of genocide, actions of Christian missionaries, removal of Indian children from their families, legalized indentured servitude, and overt forced assimilation and then implicitly through natural resource policies designed to benefit non-Native people and the resulting degradation of the environment (Norgaard 2014a).[7] Since their inception, the federal government and state of California have implemented land management policies on the Klamath that reflect and privilege non-Native values, economic systems, cultural practices, and cosmologies. Actions by the state, including the failure to recognize Karuk fishing rights, land tenure, and traditional management practices, operate as "racial" and "colonial projects" that move wealth from Native to non-Native social actors (Omi and Winant 2014; Norgaard, Reed, and Van Horn 2011).

Environmental decline is a central feature of colonial violence in the Karuk community today. Forced assimilation continues currently as the above actions of the state degrade the environment and deny Karuk people access to the food resources needed to sustain households and culture. The exclusion of fire began as official policy in the early 1900s with the establishment of the Forest Reserves, which later became known as the National Forests. As such, the U.S. Forest Service became known as the official land manager of the region, and the concept of fire exclusion became its guiding philosophy within a few years. Since the 1960s, dams have blocked access to 90% of the spawning habitat for spring Chinook—historically, the most important salmon run. In the decades following the completion of the lowermost dam, reduced flows, high water temperatures, and algal buildup have drastically reduced the number of salmon and other traditional riverine foods. The absence of traditional foods serves as a mechanism of forced assimilation as people are compelled to replace the traditional subsistence economy with store-bought foods. Forced assimilation happens even more overtly when game wardens arrest people for fishing according to tribal custom rather than state regulation. Testimony of adults and elders about river conditions and foods they ate until recently indicates that very damaging changes to the ecosystem and Karuk lifeways have occurred within the past generation.

For the past several decades, the Klamath River has been consistently plagued by highly impaired water quality. Dams block spawning access for salmon, and the few remaining salmon runs are on the verge of collapse. Through the mismanagement of their ancestral lands by both state and federal agencies, Karuk people are also denied access to sufficient amounts of traditional forest foods, including deer, acorns, and mushrooms, and to participating in many important cross-species relationships of tending and harvesting that they consider their responsibility to uphold. Both environmental decline and non-Native regulations that prohibit burning and reorganize fishing and hunting around non-Native values are threatening the integrity of relationships Karuk people hold with the natural world. The ecological, social, political, psychological, and economic impacts of ecological change are fundamentally interconnected. As Ron Reed explains in the case of fire exclusion,

> You have deer meat, elk, and a lot of times associated with those acorn groves are riparian plants such as hazel, mock orange, or other foods and fibers, materials in there that prefer fire. The use of those materials is dependent upon those prescribed

burns. So when you don't have those prescribed burns, it affects all that in a reciprocal manner. It's a holistic process where one impact has a rippling effect throughout the landscape. We can only have that for a certain amount of time before the place becomes a desert without cultural burns, because the plants are no longer soft and the shoots are no longer food, instead they become these intermediate stages where they are just taking up light and water and tinder for catastrophic fire. So it has an impact not only on the species we are talking about, but how you harvest and manage and hunt those species as well.

It seems impossible for non-Indians to fully grasp the meaning or importance of this complete contrast to the non-Indian perspectives of "food production" and "food consumption." Instead, the significance of American Indian relationships with the natural world is, at best, lost in overglamorized and essentialized characterizations of Noble Savages and, at worst, entirely invisible. To comprehend and acknowledge Native relationships with food and cultural use species would require non-Natives to recognize not only the depth of the human scale of Native American genocide but also the fact that this genocide has been an onslaught against a spiritual order that supported and governed an entire field of ecological, social, and political relationships. If traditional ecological knowledge and management has made the ecology of the Klamath what it is today, racism and cultural genocide are now leading to environmental decline. In short, the view from the river is a view into a world of responsibilities and interconnections, it is a view into hidden stories of structural genocide and ongoing colonialism, and it is a view that reveals profound struggle and creative, sustained resistance.[8]

A VIEW ACROSS THE CONTINENT: NATIVE POLITICAL AND CULTURAL REVIVAL

Just as people fight on the Klamath to retain the cross-species relationships that sustain social life, political structures, and cultural practices, the most important justice struggles in Indian Country concern the fight to maintain and assert political sovereignty, to resist and overcome forced assimilation, and to preserve reservations and reserved treaty rights (Steinman 2012; Tsosie 2003; Whyte 2018a, 2018b; Wilkinson 2005; Wood 1994). As they regain political and economic standing, Native American peoples across the country, including the Karuk, have become increasingly involved in natural resource management. This has been especially true in the West, where Tribes have again become central players in fisheries policy, restoration activities, and climate adaptation. Yet Tribes are disadvantaged in these settings due to both a lack of broader social understanding of their unique cultural perspectives and a lack of acknowledgment of the violent history perpetuated against them—much less the continuing effects of this history. Despite this, Indigenous communities continue to assert political visions

FIGURE 3. Karuk Deputy Director William Tripp at DNR meeting. Photo credit: Will Harling.

for change that combine moral acuity with practical ecological expertise (Whyte, Caldwell, and Schaefer 2018). Climate change poses threats to tribal subsistence, culture, and economy. The disproportionate nature of these impacts results from the ongoing strength and intactness of cross-species connections. Tribes face challenges to knowledge sovereignty in the face of university and government copyright policies. In each case, the effort to maintain relationships with land, traditional foods, and cultural use species, as well as the cultural and spiritual responsibilities to carry out particular activities on their behalf, is at stake. In the face of all this, Tribes repeatedly put forward innovative court cases, integrated educational and health programs, and cutting-edge environmental policies.

One concept from Native studies that is beginning to gain traction in the social sciences and humanities is the need to frame power relations in terms of "settler-colonialism." As with other nonwhite groups, Karuk people experience racism. Yet racism for Native people has been a mechanism of the even more problematic state projects of genocide, assimilation, and colonialism. Colonialism involves the generation of wealth for colonizers through the material separation and alienation of communities from their lands. Unlike concepts such as "internal colonialism" (Blauner 1969), which implies colonialism as a metaphor for a particular geographic region (Byrd 2011), the framework of settler-colonialism describes the logic and operation of power when one group of people arrives on and colonizes lands already inhabited by another with the intention to remain. In the case of settler-colonialism, the so-called metropole and periphery are thus located in the same physical space. This particular social formation of colonialism is characterized by elimination of the original inhabitants; elimination of Indigenous knowledge and political, social, and ecological systems; and their replacement by those of the settler society (Steinmetz 2014; Veracini 2013; Whyte 2016a; Wolfe 2006). Connected to the need for elimination is the quality that rather than an event that occurred in the past, colonization is an ongoing system today that, like racism, is carried out through policies and institutional practices, as well as individual interactions. One ongoing theme throughout this book concerns the relationships between racism, colonialism, capitalism, patriarchy, and environmental degradation—especially as theorized within sociology. Negating the relevance of nature for the social, material, cultural, spiritual, and emotional components of human existence has been central to the discursive legitimation of the new order in North America. Neglect of the natural world as a component of social action within many academic traditions is part of the ongoing system of colonialism.

TURNING THE KALEIDOSCOPE: UNSETTLING ACADEMIA

I first began working on the Klamath in 2003 as a postdoctoral researcher at UC Davis and shortly thereafter as a consultant on behalf of the Karuk Tribe (see Methodological Appendix). For the past decade and a half, this biologically diverse place and its amazing residents have been my teachers. This book applies my understanding of events on the Klamath to extend a number of conversations within sociology, food studies, environmental health, and environmental justice. While these topics may seem both large and disparate (how could one little case study have so much to say?), their connections flow through the landscape, through the political histories of the land, and through the lives of individuals and families. In fact, one of my most important takeaways from working on the Klamath is of the profound interconnection between systems my peers and I have been taught to view as separate. It is only through the alienated cosmology of Western colonial thought that food, health, identity, and environment become separate "topics." To see their profound connections is to begin the project of decolonizing or "unsettling" academia.[9]

Native scholars are among the most underrepresented across all academic disciplines. Even disciplines such as political ecology, where understanding of the relationships between the natural world and political power is most developed, can benefit from further integration of Indigenous experiences and perspectives (see, e.g., Middleton 2015; Hoover 2017). I am writing this at an exciting time. Although still sorely are underrepresented, the existence of enough Indigenous scholars in U.S. universities has begun changing the terms of analyses within ethnic studies (Klopotek 2011), education (Grande 2015), history (Dunbar-Ortiz 2014), English (Brown 2018, Huhndorf 2001, Miranda 2013), philosophy (Whyte 2013a), law (Tsosie 2007), anthropology (TallBear 2013), geography (Middleton

2010, 2015), political science (Stark 2010), and feminist theory (Goeman and Denetdale 2009; Maracle 1996; Teves 2011), not to mention putting fields like Native studies at the center of the most dynamic and important conversations in academia today (see, e.g., Teves, Smith, and Raheja 2015). Within these diverse settings, the disciplinary conversations echo Dunbar-Ortiz's (2014) words with respect to history: "The main challenge for scholars in re-visioning US history in the context of colonialism is not lack of information, nor is it one of methodology.... Rather, the source of the problems has been there refusal or inability of US historians to comprehend the nature of their own history, US history. The fundamental problem is the absence of the colonial framework" (7).

My audience for this project is twofold: primarily non-Native academics across the social sciences and humanities generally and sociology specifically. Second, while Native scholars both inside and outside the academy will find much in these pages to be a (hopefully reasonably articulated version of) commonsense knowledge and reality, I hope my detailed efforts articulating specifics of ecological dynamics and detailed struggles over land management policy will be useful in providing concrete examples for ongoing conversations and discussions on the nature of settler-colonial power and resistance and how settler-colonialism intersects with racism, capitalism, and patriarchy.

UNSETTLING SOCIOLOGY

Today, widespread environmental decline in forms as diverse as decreasing timber harvest levels to ocean acidification affect social phenomena from gender constructions to political movements and processes of racial formation. Yet sociology as a discipline has barely begun to theorize the role of the natural environment on the social. The subfield of environmental sociology has made significant headway over the past forty years, but much of the work is on how society affects the environment (rather than the reverse, or a more mutual exchange), and the subfield remains balkanized such that there is too little cross-pollination between what is known as environmental sociology and the leading theories across the discipline as a whole. This book applies data, insight, and examples of how the natural environment matters for social life from Karuk struggles on the Klamath River to extend four areas of theory specific to sociology: work in race, health, gender, and emotions.

First, this book engages sociological work on race and ethnicity. Within Omi and Winant's powerful theory of racial formation (1994, 2014), there is a surprising absence of sociological attention to the importance of land or nature as a material and symbolic resource for the process of racial formation (Park and Pellow 2004).

And the reverse is also true: while there is a rich body of literature on race and environment within the field of environmental justice (Bullard 2008; Pellow 2002), the majority of this material either is descriptive of the unequal impacts of environmental decline or engages the social movement tactics of communities. Remarkably, few scholars have taken up the 2004 call by Lisa Park and David Pellow to engage the central sociological theories on race to dynamics of environmental discrimination. Yet environmental exposure can be understood to be an instance of racial formation and part of the transmission of political and economic power, as theorists in other social sciences have articulated (e.g., Pulido 2000, 2016; Ducre 2018).

Second, the discipline has focused a great deal of attention on the developmental structuring of racism but is only now beginning to pay attention to colonialism, much less colonialism as an ongoing process in North America. Surprisingly, little sociological or even ethnic studies scholarship has examined the dynamics of colonialism or genocide as ongoing processes in "modern" societies (Fenelon and Trafzer 2014; Pulido 2017b; Smith 2012). The material in these pages will vividly detail how natural resource policies further not only racism but also settler-colonialism. Furthermore, colonialism is not confined to the overt genocide or forced assimilation of the past but very much enacted in the present through state institutional structures that continue to impoverish communities and erode sovereignty in the form of environmental degradation. Not only has the human scale of Native American genocide been of remarkable little sociological focus, but the fact that this genocide has been coupled with a reorganization of the natural world and an assault on a spiritual order that nourished and governed an entire field of ecological relationships also represents a substantial void to present understanding of race and racialization process. Environmental practices have meanings too, as do places. The imbuing of racial meanings to places in the landscape has also been crucial for racial formation. Indeed, while other violent aspects of racialization, including slavery and the imprisonment of Japanese Americans during World War II, have received sociological attention, there are virtually no accounts detailing these processes of Indigenous and white racialization through genocide and displacement.

Third, environmental health has been an important terrain for highlighting linkages across natural and human systems via toxins in material bodies and developed the complex linkages between identity politics, breast cancer movements, and breastmilk. Yet still the field lacks perspective on how colonialism shapes health theories, health social movements, or medical practices. While both the fields of environmental health and environmental justice have a strong tradition of community collaboration and citizen science, relatively few such collaborations occur with Indigenous communities. And while natural resource managers in forestry and fisheries sciences have begun to realize the value of Indigenous ecological knowledge, a parallel recognition of the potential value of Indigenous knowledge for research design or process remains in its infancy within the social or health sciences.

A fourth area where attention to the natural world and the relevance of settler-colonialism can be expanded in sociology and other social sciences is scholarship on emotions. Although psychology has made great strides in theorizing emotional impacts of environmental decline, the few sociological studies that address the environment and mental health tackle exposure to acute contamination events or to disasters. This material is important, but it has been primarily descriptive. There is virtually no interrogation of the concept of emotional harm or the relationships between emotions, environmental change, other features of social structure, and the process of inequality formation. *Salmon and Acorns Feed Our People* extends existing emotions theory by describing the prolonged emotional trauma of longer-term environmental decline and including conditions in the natural environment in a matrix of power and social outcomes that emotions mediate. These pages illustrate how the emotions people experience in the face of environmental decline inscribe gendered and racialized power relations and advance assimilation and genocide.

Last, sociological work on gender and environment are each notable subfields within sociology, yet despite a plethora of work documenting their intersection in the 1980s, this nexus has been surprisingly underdeveloped for the past decade. Feminist sociology has an uneasy relationship with the notion that the natural world would structure gender, yet in the face of today's rapid environmental degradation, the importance of the natural world for social dynamics of power and inequality, as well as gender and racial projects, has become both more visible and more important to understand. This project illustrates how theorizing "the natural" is necessary to understand both the construction of traditional Native masculinities and the operation of gendered colonial violence.

ENVIRONMENTAL JUSTICE

The view from the Klamath is useful to extend several arguments in relation to the interdisciplinary field of environmental justice. From the beginning, the environmental justice movement has challenged the white notion that "the environment" is "out there" or "somewhere else" by shifting focus of the environmental movement away from "the wilderness," into the urban areas, into factories, and, with the focus on toxins, into human bodies. Whereas white conservationists saw environmentalism in terms of defending so-called wilderness areas, environmentalists of color described resistance to toxic exposure in urban areas and the workplace as environmental concerns. Environmental justice scholars began to argue that existing understandings of the environment and the environmental movement were limited to descriptions of the behavior and sensibilities of urban middle-class whites (Bullard 2008; Bryant and Mohai 1992; Taylor 2014). Innovative scholar-activists and legal visionaries such as Robert Bullard, Beverly Wright, Benjamin Chavez, Charles Lee, and Luke Cole began identifying and working with communities of color facing disproportionate siting of toxic facilities and highway redevelopment projects.

While these early self-identified environmental justice efforts included important Indigenous activists, it has taken longer for the centuries-long fact of Indigenous resistance to colonialism to be understood as environmental justice struggles and longer still for Indigenous values, worldviews, or goals to be reflected in broader conceptions of environmental justice. Environmental justice activists and scholars have explicitly identified the state as "neoliberal" and "racist" (Pulido, Kohl, and Cotton 2016; Kurtz 2009) but rarely use the term *colonial*. Karuk and Indigenous perspectives on environmental justice reframe the dominant environmental justice discourse from a focus on "equality" or "rights to clean water or air" to one of caretaking responsibilities that are disrupted by natural resource policies of the settler-colonial state. The notion of "decolonizing environmental justice" is an opportunity to expand understanding of the origins of the environmental and environmental justice movements, assess whether the state is truly a potential ally or explicit foe, and especially interrogate the desired goals and outcomes of social action. Indigenous cosmologies, ethics, and understandings of power can point environmental justice movements toward deeper understandings of sustainability, community, ecological relationships, and the other worlds that are possible beyond the capitalist or colonial imaginations.

Early environmental justice work brought to light the crucial connections of racism and toxic exposure for physical health. A second contribution of this project concerns the integration of emotional experiences as a form of environmental injustice. While a solid disaster literature exists on the negative psychological consequences of environmental degradation and their unequal distribution along the lines of race, class, and gender, mental health impacts are sparsely covered within the environmental framework. Yet when species such as salmon are considered kin, and when the natural world is a stage for social interactions and identity, the grief, anger, shame, and hopelessness associated with environmental decline may become embodied manifestations of racism and colonial violence, and emotions of outrage, hope, and compassion animate resistance.

ENVIRONMENTAL HEALTH

My original work with the Karuk Tribe documented the relationship between environmental decline, the loss of traditional foods, and diet-related diseases (Norgaard 2004, 2005). In 2004, the Karuk Tribe submitted a report to the Federal Energy Regulatory Commission (FERC) entitled "The Effects of Altered Diet on the Health of Karuk People." This represented the first time that a Native American tribe had claimed in a federal process that the presence of a dam had led to an elevated rate of diabetes and other diet-related diseases for their people. Our report made frontpage news in the *Washington Post* as well as on many other newspapers across the West. This research was innovative in the policy world for the links that it made between environmental and human health. This approach contributes to the field of environmental health by situating negative physical health conditions of diabetes, strokes, and heart disease in a context of social power, racial formation, and environmental degradation. Examining how social structure and environmental degradation work together in the production of negative health outcomes can extend understanding of the relationships between human and environmental health.

NATIVE STUDIES

The field of Native studies has been an enormous source of inspiration and innovation for this project. Within the United States, sociologically trained Native studies scholar Duane Champagne has long been a key academic voice for Indigenous peoples. Despite his sociological training, Champagne has been most notable for his role in developing the field of Indigenous studies. Scholars from Champagne, Alfred, Goeman, Goldstein, Whyte, Tsosie, Klopotek, Simpson, and many others at the cusp of Native studies and other fields have provided concepts, without which I could not have crafted these arguments. Native studies within the United States has developed particularly strong cultural critiques. Now Aboriginal scholar Aileen Moreton-Robinson calls for the field of Indigenous studies to "expand its mode of inquiry to a range of intellectual projects that "structure inquiry around the logics of race, colonialism, capitalism, gender and sexuality" (2015, xvii). In particular, Moreton-Robinson emphasizes how Indigenous studies scholarship brings important yet currently absent attention to the importance of land. Whereas studies of whiteness produced by white scholars have emphasized migration, slavery, and the logic of capital, Indigenous studies specifically bring attention to the importance of land and land ownership to the construction of whiteness. Moreton-Robinson underscores how "the existence of white supremacy as hegemony, ideology, epistemology, and ontology requires the possession of Indigenous lands as its proprietary anchor within capitalist economies such as the United States" (xix). It is my hope that my elaboration of the ecological details of settler-colonialism will aid in underscoring how settler-colonialism is ongoing today through land management policies. For just as colonialism is not a single event of the past, we must think beyond the notion that "land theft" and land dispossession are single events of the past. Instead, colonialism is an ongoing process that takes place through the alteration of land, the alteration of species composition and ecological structures, and the alteration of relationships between people and the more-than-human entities known as nature.

FOOD STUDIES

Recently, there has been an explosion of academic and popular interest in food studies, complete with a new vocabulary of terms like *slow food, local food, food security, food deserts, food sovereignty,* and *food justice.* While food holds significant individual and cultural meanings for all peoples, the depth of the Indigenous importance of "food" to identity, community, spiritualty, colonialism, and its resistance is of another order that remains almost entirely invisible within this new field. In contrast to the generally commodified understandings of "food" where there are concerns about inequalities in the "production" and "consumption" of "food," Karuk and other Indigenous people speak of the foods they eat as relations. People speak of a sacred responsibility to tend to their relations in the forest and in the rivers through ceremonies, prayers, songs, formulas, and specific management practices such as the use of fire.

Rather than doing something *to* the land, ecological systems prosper because humans and nature work together. Traditional foods and what Karuk people now call "cultural use species" flourish as a result of human activities, and in return, they offer themselves to be consumed. Notions of food justice or even so-called food sovereignty that circulate in most public conversations today do not even begin to capture the intimate moral dimensions of these relationships or their significance for Indigenous self-determination, environmental reproduction (see Hoover 2018), or ecological systems themselves (Whyte, Caldwell, and Schaefer 2018).

The erasure of Indigenous knowledge, presence, and leadership within the food sovereignty movement is among the most glaring given that the origins of the term and movement tactics come from the Indigenous movement Via Campesina, as well as the fact that in the United States, American Indian sovereignty is such a potentially potent political force. Nonetheless, the term *food sovereignty* is essentially co-opted by food studies literature with minimal understanding of its larger political meaning, or of what Indigenous food sovereignty movements are fighting for or against, and excluding the Native perspective from its dialogue (and, intentional or not, diminishing the strength of the term's legal definition and meaning as a result).[10]

AN OVERVIEW OF THE BOOK

How have physical changes in the land supported and legitimized the emergence of racial categories? To what extent does racialization continue to take place through environmental degradation and environmental policies and movements today? In the first chapter, I use the example of racialization in the mid–Klamath Basin to illustrate the fundamental importance of the natural environment to the development of the categories of White and Native and to the success of a series of racial projects undertaken by the state that led to the reorganization of wealth over the past century and a half. Discussions concerning the relationships between racism, colonialism, patriarchy, and imperialism are exploding across Native studies, ethnic studies, and the disciplinary social sciences (Byrd 2011; Coulthard

2014; Klopotek 2011; Moreton-Robinson 2015; Pulido 2017b). I begin by engaging race and racism not because I believe that race and racism supersede colonialism but because the conversation in the United States and within my own discipline of sociology is most developed with regards to race. I aim to build on this more familiar theoretical terrain both to expand its utility and explanatory power through incorporation of the environment and later to show its limitations. While my primary goal in this chapter is to illustrate the fundamental importance of the natural world for constructions of race and racism, by necessity, my arguments engage the interrelated tangle of colonialism, patriarchy, capitalism, and U.S. imperialism within which racism is bound. Hence, this chapter alludes to the limitations in the framework of race that are taking off in the rich interdisciplinary discussions regarding the relationships between colonialism, patriarchy, racism, and capitalism. These conversations and the notion of racial-colonial formation are running themes throughout the book. In all cases, my primary goal is to center the fundamental importance of the natural environment within this theoretical mix.

I begin chapter 1 by examining how state actions, including genocide, the failure to recognize Karuk land tenure and traditional management practices, and forced assimilation, operate as "racial projects" that move wealth from Native to non-Native social actors. I argue that racialization is not only about "the elaboration of racial meanings to particular relationships, social practices or groups" (Omi and Winant 1994, 91) but also to particular environmental practices and places in the landscape. Indeed, it is not only the movement of wealth but also the creation and even conceptions of the nature of wealth that are fundamental to the process of race making. By examining not only the movement of wealth within society but also the process of manipulating relationships in the natural and social world to create different types of wealth, we come to a fuller and richer analysis of power in general and racism specifically. Moreover, the intensification of environmental degradation in recent decades highlights the interrelationship between racial dominance and environmental degradation.

But although most sociological accounts of Native American experience use the lens of race, the dynamics of power operating in the Karuk community today cannot ultimately be conceptualized within this framework. Building upon these themes regarding the importance of the natural environment to the operation of social power, chapter 2 illustrates the relevance of settler-colonialism for sociological theory using a discussion of fire policy as colonial violence. Whereas theories of race have yet to engage how power operates through the manipulation of the natural world, settler-colonialism puts land as central, making it an especially valuable theoretical lens for environmental sociology in particular. This chapter underscores how colonialism is ongoing through ecological alteration. While Native studies scholars have emphasized that colonialism is a structure, not an event, and that this structure plays out ongoing ways today, less detail has been paid to the particulars of the ecological dynamics through which settler-colonialism is carried forward in the present. The chapter begins with descriptions of Karuk forest management and the use of fire and then details how fire exclusion has become a vehicle for historical and ongoing colonial dispossession through altered ecology. This alteration of ecological relationships in turn becomes a mechanism for forced assimilation, the disruption of knowledge and cultural systems, the deterioration of physical and mental health (the central focus of chapters 3 and 5), forced assimilation, and economic and political dispossession.

Chapter 3 examines food and the links between environmental and human health, focusing on federal and state policies that interfere with traditional management. This chapter situates negative physical health conditions of diabetes, strokes, and heart disease in a context of social power, racial-colonial formation, and environmental degradation. Karuk people have experienced one of the most recent and drastic diet shifts of any tribe in North America. Whereas much work in medical sociological takes an individualistic approach, the Karuk Tribe successfully framed their high rates of diabetes and heart conditions as an artificial consequence of federal land management policies that denied people access to healthy foods. I describe the contents and process of this research, showing how success was a function of merging community-based research with Indigenous traditional ecological knowledge and Western science. This chapter describes the process of our research collaboration in some detail to illustrate the power of Indigenous cosmologies and the value of building on both traditional ecological knowledge and Western science for environmental justice and environmental health research. I then elaborate on the theoretical contributions our work offers for the fields of environmental health, environmental justice, and food studies, none of which have centered Indigenous perspectives in their approach, framing, or tactics.

Changing environmental conditions can affect gender practices as well. Within the Karuk community, tribal fishermen are responsible for providing salmon for food, celebrations, and ceremonies. The "right" to fish at specific sites is an honor that is passed down through families. With the degradation of the river, traditional fishermen are no longer able to fulfill these traditional practices or are only able to do so in a reduced capacity. Chapter 4 asks, what happens to Karuk masculinity when there are no fish? What happens to Karuk femininity when there are no acorns? How has their inability to fish or gather affected the sense of identity and self-esteem of Karuk men and women? This chapter describes how the natural environment is a central influence on gender constructions, including the internalization of identity, social roles, power structures, and resistance to racism and genocide. People's efforts to perform gender are interwoven with racialized colonial ecological violence. At the same time, some people are able to construct new gender identities by transferring cultural responsibilities to fish, and community and "collective continuance" to new settings as activists and fisheries scientists. Such gendered disruptions in the face of environmental decline are a critical but heretofore invisible component of masculinity theory, yet theorizing the natural is necessary to understand both the construction of traditional Native masculinity and femininity and the operation of gendered and racialized colonial ecological violence.[11]

Chapter 5 engages the emotional dynamics of environmental decline. The natural environment upon which humans exist for our survival is being rapidly degraded worldwide. How do communities with intimate ties to landscapes and nonhuman species experience this degradation? How do the emotions associated with environmental change inscribe racialized power relations or advance assimilation and genocide? This chapter extends analyses of environmental influences on social action through an analysis of the emotions experienced by Karuk Tribal members in the face of environmental decline. Among other things, I argue here that these mental health impacts in the face of environmental decline are a critical but heretofore invisible component of environmental injustice.

From globalization and food access to contamination and climate change, the issues faced by the Karuk Tribe are symptomatic of social problems faced by individuals and communities around the world. Karuk people are struggling to retain their traditional subsistence activities and culturally embedded foods at the same time as communities around the United States and globe are energized by efforts to regain local foods, get local foods into schools, and start Community Supported Agriculture ventures. Issues of identity, the search for meaning and community, and questions of what constitutes the good life faced by Karuk people pervade collective response to climate change across the United States. The conclusion examines some of the sources of Indigenous political power and emphasizes the vision that people have moving forward. Karuk activism and resistance to environmental degradation and related social problems are a touchstone and opportunity for lessons for non-Native communities and social movements engaged in similar struggles for food justice, local sustainability, and community resilience.

1 • MUTUAL CONSTRUCTIONS OF RACE AND NATURE ON THE KLAMATH

No country in the world was as well supplied by Nature, with food for man, as California, when first discovered by the Spaniards. Every one of its early visitors has left records to this effect—they all found its hills, valleys and plains filled with elk, deer, hares, rabbits, quail and other animals fit for food; its rivers and lakes swarming with salmon, trout, and other fish, their beds and banks covered with mussels, clams, and other edible mollusca; the rocks on its sea shores crowded with seal and otter; and its forests full of trees and plants, bearing acorns, nuts, seeds and berries.
—Titus Fey Cronise, *The Natural Wealth of California*[1]

Racism is inextricably tied to the theft and appropriation of Indigenous lands in the first world. In fact, its existence in the United States, Canada, Australia, Hawai'i, and New Zealand was dependent on this happening. The dehumanizing impulses of colonization are successfully acted upon because racisms in these countries are predicated on the logic of possession.
—Aileen Moreton-Robinson (2015, xiii)

Early non-Native travelers to the mid-Klamath region, including settlers and anthropologists, marveled at the immense abundance of natural resources to which the Karuk people had access. What these observers recognized as "wealth" in the form of abundant food, profound health, intricate cultural activities, and time for relaxation was a direct result of the success of generations of tribal environmental knowledge and management. Over the past century and a half, traditional Karuk land management practices that generated profound ecological abundance in the Klamath Basin have become criminalized and largely replaced with successive extractive waves of removal of gold, fish, and timber as commodities for sale.[2] Knowledge systems that emphasized relationships, responsibility, and interconnection at multiple scales have been replaced by the atomized worldview of Western science that has in turn led to single-species resource management. Today, in stark contrast to that earlier wealth, Karuk inhabitants along the Klamath River are among the hungriest and—by settler and capitalist reckoning—the poorest people in the region we now call California. The percentage of families living in poverty in Karuk Aboriginal Territory and homelands is nearly three times that of the United States as a whole. At the same time as most Karuk people in the area have become hungrier, the region generated profound wealth for the newly settled, largely white population and for a developing California—wealth that in capitalist terms would eventually elevate California to economic, political, and social prominence on a global scale.

Sociologists understand these kinds of dramatic reversals in economic circumstances and the radical shifts in resources from one group of people to another in terms of the rise of capitalism, institutional racism, and the sociohistorical process of racial formation (HoSang, LaBennett, and Pulido 2012; Omi and Winant 2014; Winant 2004). I begin first with racism. Institutional racism indicates that racial disadvantage is built into the social structures—here directly into the structural development of the state of California. The concept of racial formation details how economic wealth and political resources are moved from one racial group to another through the process of "race making" that has both ideological justifications and material outcomes. Indeed, at the core of the theory of racial formation is the notion not only that "race" is a social construction but also that what we think of as "race" is constructed *in order to* justify such transfers of wealth. In other words, what it means to be black, white, Asian American, American Indian, and so forth is different in different places and times, but if we study how the content and boundaries of these racial categories have been formed through unique histories, we can understand considerably more about the nature of the world today. Race is an irreducible political construct that is used as an organizing principle for group position and the distribution of "resources," broadly construed. By examining unique racialized environmental histories, we can begin to see more clearly the imprints of the past events and struggles that continue to structure present-day ideas of race and environmental policy alike.

If the idea of race is merely a social construct developed to justify white power, how has it become so real in the minds of people? Racial formation takes place through a process called racialization—defined by Omi and Winant (2014) as "the extension of racial meaning to a previously unclassified relationship, social practice or group" (13). Race comes to be real in the minds of a community of people through specific activities that the authors call "racial projects." These racial projects are particular moments when notions of whiteness, blackness, and so forth are made solid through a combination of ideological discourses and institutional actions. These racial projects have been likened to the building blocks of the concept of race (Bonilla-Silva 2012). Thus, racially exclusive immigration laws, changes in the definitions of white on the U.S. Census, or exclusion of black children from schools can all be understood as racial projects—these are specific instances in which the concept of being black, white, Native, and so forth is asserted as real and socially relevant, on one hand, and serves to provide, transfer, or deny material resources, on the other.

In the past thirty years since it was first crafted, the theory of racial formation has become the central and most important explanation for race, racism, and racial outcomes in the discipline of sociology (Sapterstein, Penner, and Light 2013). Yet within this powerfully generative framework, there is a surprising absence of sociological attention to the importance of land or nature as a material and symbolic resource for the process of racial formation (Park and Pellow 2004). By "nature," I mean both the larger complex of plants, animals, rocks, minerals, and other beyond human entities with whom humans share our world—some of which flow in and out of human bodies, as well as their historically contingent social constructions. Instead, emphasis within recent theory has been on the vitally important but more strictly *social* aspects of race making that occur through immigration and incarceration policy (Delgado 2012), discourses of colorblind racism (Garcia 2012), affirmative action debates, and other forms of political marginalization (Carbado and Harris 2012). If racial formation is the sociohistorical process by which racial identities are "created, lived out, transformed and destroyed" (Carbado and Harris 2012, 109), then what might we gain by adding in attention to the natural environment as a key part of this process?

This chapter will detail the importance of the natural world for material and symbolic resources in the process of race making. Next, chapter 2 will engage this history together with current examples of land management practices in the Klamath River Basin to further illustrate why we must understand that the state *is both racial and colonial*. As scholars from Quijano (2000) to Fenelon (2016), Casas (2014), Klopotek (2011), and many more have emphasized, colonialism is the context and raison d'état within which the North American racial formation processes are occurring. Colonialism and racialization continue to operate together in a variety of ways, and as Lakota scholar James Fenelon highlights, the fact that the dominant society still operates in terms of race underscores its continued relevance (2017). Thus I will also use the term *racial-colonial formation*.

I begin the first chapter with race not to signal that race is any way *primary* in relation to colonialism or capitalism but rather because the larger public and sociological conversations to date center on race. Yet within sociology, these discussions fail to incorporate the significance of the natural world. Connections between race and colonialism are particularly important to flag because discussions of race and racism that do not incorporate colonialism extend the discourse of Indigenous erasure that itself is a central mechanism of colonialism.[3] From this jumping-off point, I then elaborate the need for engaging settler-colonial theory in chapter 2.

While some readers may wonder that I begin with race rather than colonialism, others may question why I have not more centrally emphasized capitalism.

This choice too is an artifact of the existing intellectual terrain: this field is essentially better developed. A host of Marxist scholars and environmental sociologists from John Bellamy Foster, Brett Clark, and Silvia Federici to David Harvey, John Foran, Jason Moore, and many others have done crucial work in theorizing the role of the environment for the simultaneous production of wealth and inequality. I hope this discussion will extend their efforts to center how much the natural environment matters across additional important themes in our discipline, from the construction of race and gender to the operation of emotions and social power. My understanding of settler-colonialism is congruent with Marxist critiques of capitalism. Indeed, in an interview with Naomi Klein, Michi Saagiig Nishnaabeg scholar Leanne Simpson (2017) emphasizes that she cannot "think of a system that is more counter to Nishnaabeg thought than capitalism" (77), and

> extraction and assimilation go together. Colonialism and capitalism are based on extracting and assimilating. My land is seen as a resource. My relatives in the plant and animal words are seen as resources. My culture and knowledge is a resource. My body is a resource and my children are a resource because they are the potential to grow, maintain, and uphold the extraction = assimilation system. The act of extraction removes all of the relationships that give whatever is being extracted meaning. Extraction is taking. Actually extracting is stealing—It is taking without consent, without thought, care or even knowledge of the impacts that extraction has on the other living things in that environment. That's always been a part of colonialism and conquest. Colonialism has always extracted the indigenous—extraction of indigenous knowledge, indigenous women, indigenous peoples. (75)

Yellowknives Dene scholar Glen Coulthard (2014) discusses the relationships between Marxist and Indigenous anticapitalism, noting that Marx developed his understanding of primitive accumulation in colonial contexts and that many of his insights regarding the nature of capitalism are foundational for anticolonial resistance. At the same time, Indigenous scholars have critiqued Marx's assumption that the colonization of North America was a necessary step and find the notion of controlling the means of production an insufficient response to capitalism. Coulthard states,

> The history and experience of dispossession, not proletarianization, has been the dominant background structure shaping the character of the historical relationship between Indigenous peoples and the Canadian state. . . . The theory and practice of Indigenous anti-colonialism, including Indigenous anti-capitalism, is best understood as a struggle primarily inspired by and oriented around *the question of land*—a struggle not only *for* land in the material sense, but also deeply informed by what the land *as system of reciprocal relations and obligations* can teach us about

living our lives in relation to one another and the natural world in non-dominating and non exploitative terms—and less around our emergent status as "rightless proletarians." (13)

See Coulthard (2014) as well as Fenelon (2016), Dunbar-Ortiz (2014), and Simpson (2017) for further discussion of the complex relationships between the operation of genocide, racism, capitalism, and colonialism. Theorizing the mutual structures of racism, colonialism, capitalism, patriarchy, and the natural environment are ongoing themes throughout the book.

In this chapter, I aim to illustrate that "nature" often matters a great deal in the project of race making both because it is the ultimate source of all material wealth and because the notion of "nature" or "the natural" is one of the most potent ideological resources available for making claims about what is "real," "inevitable," and "just the way things are." In making such claims, I draw upon a rich literature across the social sciences that includes work from Donna Haraway (1989, 1991), Val Plumwood (1993), and others who have written about how dualisms between nature and culture structure Western thought; scholars like Charles Mills (1997/2014, 2007), Laura Pulido (2000, 2016), Julie Guthman (2008), Noël Sturgeon (1997, 2009), Jake Kosek (2006), and Moore, Kosek, and Pandian (2003), who examine environmental privilege and how racialized discourses structure environmental and natural resource politics; and theorists in the field of new materialism, such as Stacey Alaimo (2010) and Elizabeth Grosz (2010), who are seeking to reweave conceptual divides between nature and culture back into political theory. Similarly, I write alongside scholars in the humanities like Sarah Wald (2016), who traces the dual constructions of nature and race through discourses of immigration and citizenship in California agriculture; Sarah Jacquette Ray's (2013) in-depth analysis of exclusion within the environmental movement; Priscilla Ybarra's (2016) notion of writing the "goodlife"; and Paul Outka (2016), John Claborn (2014), and Kimberly Ruffin's (2010) important thinking regarding cultural productions of race and nature in relation to blackness. All of these are key contributions to a growing interdisciplinary literature in the vein of what David Pellow (2017) calls "critical environmental justice." While these scholars and more across the social sciences and humanities have challenged modernist dictates regarding the irrelevance of the natural world and radically reworked early texts, questions of environment in my own discipline are still considered the terrain of the subfield of "environmental sociology," not woven into theories across the discipline. Indeed, a central theme of this book is that sociologists bring valuable theoretical frameworks to larger interdisciplinary discussions with concepts such as racial-colonial formation, yet at the same time we have undertheorized the role of the natural world in our work. While many of these dynamics have been studied in some detail by academics in anthropology and geography—especially by those who work in the global South, sociologists have been slow to engage the importance of the natural environment for social processes in general. Just as historian Mark Fiege's *Republic of Nature* illustrates how the discipline of U.S. history *as a whole* can be better theorized with attention to the environment, the natural world matters fundamentally for the field of sociology. Sociological concerns, from the trajectory and form of social movements to the rise of political parties or dynamics of youth identity development, are shaped by the general "field" that is the natural world, even though they may not yet have been analyzed that way. And even within disciplines such as anthropology, geography, and ethnic studies, more can be done to center Indigenous experience and vision (see, e.g., Pulido 2017b; Middleton 2015; Klopotek 2011; Smith 2012).

Omi and Winant argue that to understand the racism that has shaped the geography of American life, we must understand the sociohistorical context in which it has emerged. As Dunbar-Ortiz (2014) emphasizes, "Everything in US history is about the land—who oversaw and cultivated it, fished its waters, maintained its wildlife; who invaded and stole it; how it became a commodity ('real estate') broken in to pieces to be bought and sold on the market" (1). Not infrequently, that "context" in which racism has emerged may involve the relationships that racial groups have with the natural environment and, in particular, their ability to manipulate the natural environment in different ways. For as Laura Pulido (2017b) eloquently notes, "Land is thoroughly saturated with racism. There are at least two primary land processes to consider: appropriation and access. Appropriation refers to the diverse ways that land was taken from native people.... Once land was severed from native peoples and commodified, the question of access arose, which is deeply racialized" (528). Former Sierra Club president Carl Anthony describes how notions of blacks as dirty legitimated the placement of contamination in their communities (see also Mills 2001; Pellow 2016, 2018). Access to and control over Ranchero lands in California during the 1800s was key to the process through which Latinx people achieved the racial categorization of white (Almaguer 2008). Furthermore, race and racism are not static. Both changing environments and the changing ways that groups interact with natural environments shape the production of race and racism. In light of today's rapid and widespread environmental degradation, the importance of the natural world for processes of racial domination, racial-colonial projects, and racialization is more important than ever.[4]

This chapter builds on Park and Pellow's (2004) provocative claim that racial formation and institutional racism are a "complex set of practices supported by the linked exploitation of people and natural resources" (403) to develop a framework for identifying the importance of the natural environment to racial formation. For despite substantial literatures on both racial formation and environmental justice, and notwithstanding the relevance of these literatures for one another, few racial formation theorists draw upon the natural environment, and as Pulido notes within the field of environmental justice, there is "minimal attention to racial formation" (2017, 2). Drawing on these and other key theorists from outside sociology (e.g., Mills 1997/2014; Pulido 2000, 2016), I first outline a range of mechanisms through which the natural environment can be understood as a centrally important material and ideological player in the process of racial formation. I then use historical and interview data from Karuk experiences in the Klamath Basin over the past 160 years to develop this claim in more detail. I will illustrate how present-day poverty and hunger in the Karuk Tribe have been produced through state processes that have centrally involved control and manipulation of the natural environment at the same time as they involved constructions of what it meant to be "Indian" and "white."[5] Racialization via environmental policy is actively ongoing today.[6] While many overtly dramatic events occurred over a century ago, such as the failure to sign treaties and the disputed transfer of land ownership to the U.S. Forest Service, the continuing consequences of such events are played out every day through ongoing cultural assumptions that Native people are absent or not to be taken seriously as land mangers or knowledge holders, their input or approval unnecessary, as well as "legal" (by U.S. systems of law) and criminal enforcement of racialized notions of how the land and resources should be used and for/by whom. Last, I provide some discussion of how the intensification of environmental degradation in recent decades makes attention to the relationship between racial dominance and environmental degradation so critically important today.

Incorporating Indigenous perspectives with scholarship on race and capitalism allows us to link how the phase of capitalist development often identified as "primitive accumulation" is also a process of Indigenous land dispossession and erasure. Leanne Simpson emphasizes the centrality of naming capitalism in the process, as well as moving beyond critiques from non-Indigenous standpoints: "Capital in our reality isn't capital. We have no such thing as capital. We have relatives. We have clans, We have treaty partners. We do not have resources or capital. Resources and capital, in fact, are fundamental mistakes within Nishnaabeg thought, as Glenna Beaucage points out, and ones that come with serious consequences—not in the colonial superstitious way, but in the way we have already seen: the collapse of local ecosystems, the loss of prairies and wild rice, the loss of salmon, eels, caribou, the loss of our weather" (77). Yet when it comes to explaining what has happened on the Klamath River, accounts of the importance of the natural world are not the only limitation in existing sociological theories of race or capitalism. While critical race theory has placed much attention on the notion of the racial state (Kurtz 2009; Goldberg 2002), Native scholars across a range of disciplines remind us that the state is also colonial (Byrd 2011; Coulthard 2014; Fenelon 2015, 2016; Woolford, Benvenuto, and Hinton 2014). Colonization and racialization operate together in multiple and dynamic ways. Chapter 2 will examine how relationships with the environment have been and continue to be central to colonial projects as well. Although it is only now beginning to be addressed as such by sociologists here in the United States (Champagne 2006; Fenelon 2015; Glenn 2015; Steinman 2012, 2016; Steinmetz 2014), North American colonialism is an ongoing process today. On the Klamath, natural resource policies related to forest management, water quality, and dams constitute key dimensions of the enactment of colonialism. In chapter 2, the active and ongoing nature of racial-colonial formation will be examined through a detailed study of wildfire policy over the past century. This next chapter will engage theories of settler-colonialism and empire to illustrate their particular relevance for theorizing social dynamics in North America, as well as their ability to center the importance of the natural world to ongoing dynamics of social and economic power. Feminist critiques of gender relations and Marxist analyses of capitalism are fundamental to this project.

Together, these first two chapters describe how Karuk people experience and resist both racism and colonialism as ongoing processes through environmental degradation and (failing) land management practices. As we will see, these land management practices based on single-species management, commodity production, and fire suppression create profound problems for Native and non-Native people alike in the form of large-scale high-intensity wildfires—wildfires that are on another order of magnitude than those that occurred under Native management. At the larger level, the environmental problem of climate change itself can be analyzed as an outgrowth of colonialism (see Whyte 2018c, forthcoming; Wildcat 2010). This will be discussed in the Conclusion.

LAND AND NATURE IN INSTITUTIONAL RACISM, RACIAL FORMATION, AND RACIAL PROJECTS

> Discourses of race and nature provide the resources to express truths, forge identities, and justify inequalities.... Nature is not merely the material environment, nor is race merely a problem of social relations. Race and nature are both material and symbolic. They reach across this imagined divide, acting at once through bodies and metaphors. Natural character is written into discourse and expression but is also worked into flesh and landscape. Racialized discourses mark both living beings and geographical territories with the force of their distinctions.
> —Moore, Kosek, and Pandian (2003, 1–2)

Early theories of racial inequality, including the work of W.E.B. Du Bois (1903, 1935) and Manning Marable (1983), explicitly include the importance of land as a source of wealth (and its absence as a source of poverty).[7] Yet rather than continuing to incorporate the materialist dimension of the physical environment in racial analyses, attention to institutional racism in contemporary race theory is more often conceived as a function of disproportionate access to *social* resources such as educational opportunity or other forms of cultural, economic, and political capital (e.g., Stretesky and Hogan 1998; Garcia 2012).[8] As such, the "wealth" in question appears sui generis, or "out of nowhere." Even environmental justice scholarship—which provides powerful descriptions of exposure to environmental contaminants as a form of racism—has often taken a more descriptive approach, leaving room for further theoretical explication of the role of the environment within racial formation and institutional racism per se (see, e.g., Pellow 2017). Work by Taylor (2009, 2014, 2016), Pulido (2000, 2016, 2017a), Pulido, Kohn and Cotton (2016), Kurtz (2009), and other key scholars provides vivid racial histories of the conversation and environmental justice movements, as well as details the importance of understanding the state as racialized and the indispensable concept of racial capitalism (Pulido 2016; Robinson 2000). These scholars have successfully illustrated (if perhaps only implicitly) the role of the biophysical environment in the racial formation process. Yet tracing back the process through which that wealth comes into social existence yields useful and important understanding concerning both the meaning of racial categories and the process of racial formation. Indeed, both the movement and creation of wealth—and even conceptions of the nature of wealth—are fundamental not only for the development of capitalism but also to the process of race making. By taking a step back and examining not only the movement of wealth within society but also the process of manipulating relationships in the natural and social world to create different types of wealth, we come to a fuller and richer analysis of power in general and racism specifically.[9] Leanne Simpson (2017) explains, "We don't have this idea of private property or 'the commons.' We practice life over a territory with boundaries that were overlapping areas of increased international Indigenous presence, maintained by more intense ceremonial and diplomatic relationship, not necessarily by police, armies, and violence, although under great threat we mobilized to project what was meaningful to us" (78).

If we examine the central components of racial formation with these ideas in mind, it becomes evident that the creation, inhabitation, transformation, resistance, and destruction of racial categories often rest upon (and in some cases may even require) the ability of groups and individual actors to manipulate the natural environment. In her work on the origins of the conservation movement in the United States, Taylor (2016) provides vivid details of multiple racial groups' relationships to land during the 1800s. Du Bois (1935), Taylor (2016), and others show how the ability to manipulate the landscape was critical for the generation of both wealth and the categorical construction of whiteness in the 1800s. As Moore, Kosek, and Pandian (2003) write, "Race and nature work as a terrain of power" (1). For example, just after the Civil War, a time Omi and Winant (2014) describe as being when "whiteness was cast into doubt" (76), former plantation lands were distributed to freed blacks. Whites strongly opposed this land redistribution and lobbied hard for the 1866 revision of the Homestead Act, which not only revoked the land transfers, returning many lands to former slave owners, but also denied blacks from receiving 160-acre homesteads in the West (see, e.g., Finney 2014). At stake was not only access to specific lands as a means of economic benefit but also the resources to legitimate both unique white identities and the system of racial hierarchy. In this moment of potential racial identity crisis, regaining access to land was part of how whites not only amassed the wealth that supported their assertions of racial difference but also enabled their constructions of whiteness as superiority.

Racial formations also often draw upon nature as a symbolic resource, as when anti–Japanese American sentiments were drummed up during World War II using images of Japanese Americans as monkeys as a means to elicit mistrust.[10] As Moore, Kosek, and Pandian (2003) note, "Time and again, race and nature stake claims to commonsensical truth.... Their sense of universality both makes race and nature continually available for naïve rediscovery and continually obscures the historical conditions that make and remake them. Because race and nature always seem to precede history, they can be taken again and again as the very substrate on which myriad social truths are built" (4).

In their study of Silicon Valley, California, Park and Pellow (2004) write, "Racial formation in the United States has always been characterized by an underlying link between ecological and racial domination" (408) and emphasize that attention to ecological degradation enhances our understanding of race and racism in important ways. Park and Pellow note that the state manipulates both racial categories and the environment to generate profits and describe how, "following racial formation theory, we find that the 'content and importance of racial categories' is closely associated with social, political, and economic practices that relegate people of color to the least desirable living and working environments. This racial 'ecology' determines which populations are viewed as 'fit' for particular environments and specific places. Such a system also constructs populations of color as means to profitable ends (i.e., labor) and/or as barriers to progress (e.g., Native peoples occupying resource rich lands)" (404–405). Indeed, following Omi and Winant's (2014) framework that "racial formation is the process through which racial categories are created, lived out, transformed and destroyed" (109), we can see that the environment matters for the construction of race and racism in both symbolic and material ways (see Table 1).

First, racial categories may be constructed in order to justify access to the natural environment and the right for a given group to manipulate it according to their worldview and interests. Indeed, Omi and Winant's discussion of the creation of racial categories during "conquest" is the only reference to the importance of the natural environment for racial formation. Certainly, the conquest period is a clear moment where the importance of land can be identified. Fenelon and Trafzer (2014) underscore that "Euro-Americans constructed the American Indian legally, socially, and racially to the benefit of Euro-American

TABLE 1   Natural Environment as Basis for Racial Formation

	Material outcome	Symbolic justification
Created	Access to land as impetus for creation of racial categories	Notion of race as real legitimated with references to ideas of "natural"
		Constructions of Native people as savages and Whites as civilized justified taking land for "progress" (Doctrine of Discovery, Manifest Destiny)
Lived out	Relationships with the natural world structure racialized experiences (e.g., Latinos and farmworker exposure to toxins, experiences of environmental racism and privilege) (Pulido 1996)	Notions of white innocence and purity normalize white privilege via connections to pure nature
	Notions of whites as civilized upheld through material practice of agricultural production	Concepts of people of color as dirty and polluted naturalize environmental racism (e.g., Mills's Black trash thesis); constructions of African Americans as animals justifying hard physical labor
	Indigenous land management used as basis for denying people access to allotments	
Transformed, resisted	Meaning of racial categories transformed via alteration of natural world (e.g., forced Indigenous assimilated to White through reorganization of land use in the Dawes Act)	Manifest Destiny, inevitability of vanishing Native, last frontier, savage and civilized
		"Cleansing," "purifying" metaphors? "Becoming American?"
	Access to and ability to manipulate natural world enables resistance to racial categorization	Strategic essentialism, Native pride, traditional values (Sturgeon 1997, Vasquez, and Wetzel 2009)
		Latinx resistance of categorization as people of color via land ownership (rancheros), Native resistance to assimilation to whiteness via traditional environmental practices, Karuk knowledge of land enables resistance to genocide
Destroyed	Destruction of natural world as a means of eliminating ethnic groups (e.g., effect of buffalo massacre, pesticide poisoning, and diabetes on Klamath)	Symbolic examples: narrative of erasure, language of settlers as First peoples, notion of land as inert static, "wilderness"

colonizers, and the United States invoked rationalizations and ideologies of invasion and destruction intentionally to steal Native American lands and control for the first nations of the Americas" (11) and "racial formation during this period was over 'Indians' to destroy their nation, enslave them, and take their land, to build their new colonies, often committing genocide, wiping out the Indigenous population base and minimally eliminating their societies and original sovereignty" (6).

As Almaguer (2008) details for the development of white supremacy and California agriculture, it is often through controlling such relationships with the environment that race is constructed and wealth is moved from one racial category to another. White settlers wanted land occupied by Native peoples. Thus, alongside direct genocide, Native people were moved away from areas whites desired, and definitions of the category were set up in such a way as to "disappear" people across time (see, e.g., Chang 2010). Garrouttee (2001) writes, "These theories of race articulated closely with political goals characteristic of the dominant American society. The original stated intention of blood quantum distinctions was to determine the point at which the various responsibilities of that dominant society to Indian peoples ended" (225).

This emphasis on a minimum blood quantum as qualification for the category of indigeneity facilitated the disappearance of Indigenous people through assimilation and thereby White access to lands set aside via treaties—in contrast to the "one drop rule" for blacks that justified the use of their bodies as labor. Such legal definitions are explicit moments whereby the powerful linkage between the social construction of race by a colonial state and land dispossession is visible.

Discourses of Native people as savages or closer to nature and whites as civilized and therefore rightful leaders and decision makers justified direct genocide in the Klamath Basin and elsewhere, and they continue to justify land management decisions by non-Native agencies today in subtle and not so subtle ways (e.g., the notion that tribal consultation need not take place for research on Native land). The creation of these racial categories and their particular contents thus paved the way for white entry into the region and the generation of white wealth through further manipulation of the land via hydraulic gold mining, the taking of land for farms and urban areas, and modern forestry practices.

Second, the content of racial categories often underlies lived-out racialized experiences of the natural world. Some of these racialized experiences are of environmental exposure (pesticide poisoning of Latinx farmworkers, the exposure of Vietnamese American women and men to toxins in nail salons), while others, such as the disproportionate ability of whites to access and enjoy "pristine" wilderness areas for recreation, are of environmental privilege. In other words, one importance of racial categories is that they justify the current situation whereby some people have environmental privilege and others extreme negative environmental exposure. Moore, Kosek, and Pandian (2003) describe how the view that

Indigenous people were naturally tougher was used to justify lower pay scales and longer working hours for Indigenous miners in the Peruvian Andes. Similarly, the supposed stealth of Indian men justified their placement on the front lines in warfare (e.g., World War II, Korean War, Vietnam War), at the same time as Indians' purported lack of the fear of heights justified their exemption from safety precautions in the building of skyscrapers. In agricultural communities, notions of white people as pure and innocent work alongside beliefs about Latinos as uneducated to blame those who have been poisoned by pesticides for their own exposure. At the same time, notions that Latinx people are expendable deflect the moral implications of their experiences (Marquez 2014; Viramontes 1996). In each case, the likelihood that one's body contains particular chemicals or will be found relaxing on the beach becomes part of what it may mean to be Latinx, black, white, or Native in the world today (see Ray 2013 for rich discussion).

Note that it is also through references to particular notions of *nature* that these racial categories may be justified—as in the way that the notion of nature as wild or dirty enables the concept of a Native "savage," whereas very different constructions of nature as pure and pristine underlie notions of white purity and innocence. Furthermore, these notions of nature may then be imposed back onto actual landscapes, as, for example, the concept of nature as "pristine" and apart from humans is then imposed on wilderness areas. Here particular European romantic notions of nature as pure and pristine underlie notions of whiteness as pristine. And this dualism, in turn, marks wilderness areas as places that are coded white—places where white people go to construct their identities at the same time as they are landscapes "for" white people.[11] Through these mutual constructions of race and nature, white supremacy is enacted on both human bodies and the land. Park and Pellow (2004) note that

> this racial hierarchy also has an environmental component in that one's position in society is correlated with (if not determinative, or reflective, of) one's relationship to the environment. Those who are placed at the lower rungs within this hierarchy are generally perceived as 'closer' to nature because of their allegedly 'primitive' cultures. This form of racialization functions to diminish the rights of people of color and immigrants based upon the 'inferior' racial categories they occupy. This racist logic, which gives ethnic minorities an 'animalistic' quality, justifies the concentration of people of color in jobs and residential spaces that are particularly 'dirty' or hazardous (405).

Third, as Omi and Winant (2014) emphasize, "race and racial meanings are neither stable nor consistent" (2). Indeed, the meaning of racial categories or even the categories themselves may be transformed through transformation of the natural world, as will be described in this chapter, whereby the alternation of forests to commodity production or land transformation and subsequent urban migration via the Dawes Act facilitated the development of a pan-Indian identity (Nagel 1994). Justification for the unique racial categories of American Indian and white comes from the association of Native people with the natural world. At the same time, these categories are materially solidified through the unequal wealth outcomes that result from different kinds of relationships with material nature.

Fourth, it may be through relationships to the natural environment that people resist racial and ethnic categorization. In Northern California during the time of outright frontier genocide, the fact that Karuk people lived further inland where they had access to mountains meant that at least some were able to hide from vigilantes and militia—a fact reflected in the greater portions of the Karuk population who survived into the 1880s as compared with the coastal Wiyot people (Raphael and House 2007; Secrest 2003). During this time, the intimate knowledge that people had of their land facilitated their survival. Furthermore, this chapter and the next will draw upon voices of people who seek to resist forced assimilation to whiteness and instead remain "Karuk" today through engagement with the land through fishing, hunting, and traditional burning.

Last, the destruction of the natural environment may be fundamental to the attempted elimination of racial categories as has also been the case for Native Americans. The widespread massacre of American bison during the mid-1800s (or, as Lakota people call them, *Tatanka*) is a classic example (Hubbard 2014). Chapter 2 will detail the relationships between shifting state and Indigenous power via ecological alteration on the Klamath. On the symbolic level, the discourse of the vanishing Indian was popularized by photographer Edward Curtis. Eva Marie Garroutte (2001) writes that for Native Americans, "The ultimate and explicit federal intention was to use the blood quantum standard as a means to liquidate tribal lands and to eliminate government trust responsibility to tribes along with entitlement programs, treaty rights, and reservations. Indians would eventually, through intermarriage combined with the mechanism of blood quantum calculations, become citizens indistinguishable from all other citizens" (225).

In each of these cases, "nature" is not just another "inert" site for the enactment of power, but it is through multidimensional relationships with the material natural world that state power is enacted, negotiated, and resisted (see, e.g., Scott 1998, 2008). In other words, the natural world itself is a "tool of structuration" (Giddens 1991). Again, all this is so in part because nature is the source of material human existence and wealth in the form of food, water, minerals, and more, *and* because the natural world holds profound symbolic significance. "Race provides a critical medium through which ideas of nature operate, even as racialized forces rework the ground of nature itself. Working together, race and nature legitimate particular forms of political representation, reproduce social hierarchies, and authorize violent exclusions—often transforming contingent relations into eternal necessities" (Moore, Kosek, and Pandian, 2003, 2).

RADICAL FORMATION OF THE KLAMATH

In this next section, I use the example of racialization in the mid–Klamath Basin to illustrate the fundamental importance of the natural environment to the development of the categories of white and Native and to the success of a series of racial projects undertaken by the state that led to the reorganization of wealth over the past century and a half.[12] In Northern California at this time, people came from a wide variety of nationalities and ethnicities into this "new" land to become white.

As we follow the emergence of the idea of whiteness and Native in the Klamath region, we can understand the history of this next 160 years as a transition from one cultural logic of relating to the natural world to another. The dynamics I describe in what would soon become known as California had at this point been shaped by several hundred years of material actions and ideological constructions of race in North America, in turn resulting from global political and economic events. Roxanne Dunbar-Ortiz situates California genocide as the second of four "distinct periods" of genocide "on the part of US administrations"—the California genocide follows the Jacksonian era of forced removal.

While whites and Natives were the major players at the time, the presence and absence of blacks and Chinese require account. Between 1840 and 1870, China experienced a series of environmental and political events, including flooding, drought, famine, and military defeat (Leung 2001). Trade with the West increased after this point, and beginning around 1851, several thousand Chinese people, mostly men, came to the Klamath region. Siskiyou County was considered the "second Mother Lode as distinguished from the Mother Lode in the Serra area" (Leung 2001, 30). Chinese people initially worked in the gold mines but, over time, also lived in Arcata, Eureka, and other newly forming towns where they were farmers, small business owners, and servants (Saunders 1998; LaLande 1985, 1981; Leung 2001). "When placer mines were depleted, American miners abandoned them. Chinese miners bought the old claims and worked the abandoned tailings. Some Chinese teamed up to have developed a 'chain pump' method for larger mining operation Later, when hydraulic mining methods were used, many Chinese laborers were brought back to mining fields. By 1870 almost on third of the miners in California were Chinese" (Leung 2001, 29). Oral accounts and physical evidence indicate that Karuk and Chinese people shared beads, coins, and traditional knowledge (Peters and Ortiz 2016). Chinese miners brought and spread medical plant species to the Klamath region that were then brought into use by Karuk practitioners (Leung 2001). Karuk-Chinese relations were mixed. While some Karuk people describe significant conflicts with Chinese miners, one Karuk elder interviewed by Richards and Creasy in 1996 noted, "We always got along with the Chinese. We knew they were having a tough time, too" (369).

White anti-Chinese racism was intense. The Chinese Exclusion Laws of 1882 made immigration illegal and denied Chinese the possibility of naturalization. In 1885, the 480 Chinese people living in Eureka were forcibly "expulsed" from Humboldt County practically overnight. Although the period of time in the region was relatively short, their population size and impacts on Karuk ecology were significant: four thousand Chinese men worked on the railroad line over the Siskiyou mountains (just to the east of Karuk territory), and multiple communities along the Salmon River housed over 200 to 300 Chinese gold miners each (Leung 2001).

That there were few black settlers on the West Coast is equally a function of global economic and political exploitation, racism, slavery, and the racial formation projects happening in the eastern states. African Americans, however, have had a presence throughout California as far back as early Spanish colonization efforts (Almaguer 2008).[13] While the California Constitution officially prohibited slavery, some people were nonetheless there as slaves. Blacks were present working in gold mines, but there are few accounts of African Americans in or near Karuk country at any point in the 1800s.

In the Introduction and elsewhere, I provided some descriptions of the sophisticated Indigenous management systems Karuk people have developed to care for the land and its inhabitants. These practices reflect an ecological, cultural, political, and spiritual order that shaped the mid–Klamath Basin and Karuk culture for generations upon generations. Information about key components of Karuk management systems is reflected in ceremonial regalia and practices. For example, burning is a key cultural and ecological practice, and rolling a burning log down the hill is part of the World Renewal Ceremony in the fall. This act started the fall burning season in which fire was used to enhance a wide variety of bulbs, basketry materials, and acorn trees. As Karuk Eco-Cultural Restoration Specialist and cultural practitioner William Tripp describes, "They used to roll logs off the top of Offield Mountain as part of the World Renewal Ceremony in September, right in the height of fire season so that whole mountain was in a condition to where it wouldn't burn hot. It would burn around to some rocky areas and go out. It would burn slow. Creep down the hill over a matter of days until it just finally went out. When it rained it would go out and that's what we wanted it to do." As Tripp notes, the fact that the log did not start a large fire even in September indicates that the landscape was burned regularly. In contrast to either the fetishization of traditional Indigenous knowledge by non-Native environmentalists or the myths that Indigenous practices exploit the land, Karuk stewardship practices reflect a very real "alternative" to the capitalist and colonial mode of extractive production as means of relating to the land. Their effectiveness is evident in the landscape and increasingly recognized by the natural science community. The Karuk Tribe Eco-Cultural Management Plan notes, "Fire caused by natural and human ignitions affects the distribution, abundance, composition, structure and morphology of trees, shrubs, forbs, and grasses" (Karuk Tribe 2010, 4). People burn to enhance the quality of forest food species like elk, deer, acorns, mushrooms, and lilies, as well as basketry materials such as hazel and willow, but also to keep travel routes open (Lake 2013). Smoke from these fires cools the river temperature and decreases evapotranspiration by plants, thereby increasing river flow (David, Asarian, and Lake 2018). Together, these two factors cool the river enough to trigger salmon to enter and begin their upstream migration.

This sophisticated Karuk system of ecological management centers on the responsibilities within and between families, as well as between humans and other nonhuman species. These labor-intensive practices require a large extended family to care for the environment, leading to mutual relationships between social and ecological reproduction, as well as furthering vital bonding experiences within social groups and between Native peoples and their homelands. Traditional fisherman Kenneth Brink describes how each family has specific responsibilities, which together ensure community-level management: "One family living up on this creek, one family living up on that watershed, and this family over here did all of its work . . . if this family over here didn't do their work, see it all grew back, and sucked up what little water that fish could have . . . so it's very important that everyone has to do their management part. Every little piece counts." Karuk family activities are a conduit of cultural transmission and maintain important social networks necessary for a functioning landscape management system. Former Karuk tribal council member Robert Goodwin captures this as he describes Karuk people as "created for [their] environment," stating that "we were, or have been instructed how to keep that environment intact. What we do each year to make it new again is so that the next year we would have everything we need to sustain life." Goodwin emphasizes the social aspects of ecological renewal: "the primary reason that we have the religious ceremonies that we do" is "to renew who we are as a people." Karuk traditional management practices created profound ecological abundance and enhanced the availability of foods and numerous other cultural use species. Glen Coulthard (2014) describes this system as grounded normativity: "by which I mean the modalities of Indigenous land-connected practices and longstanding experiential knowledge that inform and structure our ethical engagements with the world and our relationships with human and nonhuman others over time" (13).

Over the past century and a half, Karuk management practices that were oriented around species complexity and long-term sustainability were forcibly replaced by extractive management activities that were geared toward the withdrawal of "commodities" (gold, conifer trees, fish; see Diver et al. 2010). Through this period of what is known as capitalist primitive accumulation, these commodities became the basis of monetary wealth for non-Native people through the material reorganization of bodies, organisms, rivers and trees and the meaning systems, cultural values, and spiritual practices that sustained them. Hydraulic mining cannons were used to wash away so much of the hillsides that many of the original village sites along the river are simply gone. Between 1915 and 1928, canneries established at the mouth of the Klamath removed 52,000 salmon annually (about 725,000 pounds) (McEvoy 1986; Snyder 1931). Lamprey, once constituting the largest volume of living organisms in the Klamath River, are now rarely seen. Instead, a dominant biological feature of the river—at least in summer months—is the toxic algal bloom that results from upstream fertilizer input and warm water from dams. As this occurred, the landscape itself has been reimagined to reflect emergent racial constructions in the practice of race making. Today, different species thrive—the species with whom Karuk people have kinship relations have become less plentiful, while species of commercial value to non-Indian whites such as Douglas fir have become significantly more abundant. Leaf and Lisa Hillman note that "today the species composition in the Klamath Basin has been altered dramatically in support of resource management that is as non-Native as it gets: single species management. Replacing those species holding a high cultural value to Native people with those of commercial value to non-Indian whites such as Douglas fir, colonialists have further reduced access to Native foods and preimposed the idea that single-species management is somehow more intelligent and productive than the holistic traditional ecological knowledge and management system" (personal communication). The acorn trees that provided nearly half the calories and protein to the Karuk diet for tens of thousands of years came to be seen as a nuisance (Bowcutt 2011, 2015). As the Forest Service placed value on the sale of timber, these slow- and irregularly growing trees could not match the comparatively quick-growing and high market value of conifers. Species like porcupine—important for regalia and, from a Karuk perspective, highly beneficial in keeping conifers in check—were almost entirely eradicated through intensive programs conducted in the 1950 to 1960s, with hunting and baiting with strychnine-salt bait blocks (Anthony et al. 1986). Today, the porcupine are, as on the title of a presentation on the topic, "An Increasingly Rare Sight in California Mid-Elevation Mixed Conifer Forests."[14] These efforts to eradicate competition for the sole benefit of conifer species—in particular, the old growth and highly valued redwood—further underscore the profound difference between management practices, cultural values, and harvesting etiquette.

Today, maps of the mid-Klamath region display place names written in English and reflect settlers' narratives of history and worldview rather than those of the Karuk and their Indigenous neighbors who have lived here since time immemorial. These maps not only serve to highlight the boundaries of Forest Service ranger districts rather than the territorial boundaries between places tended and cared for by Karuk, Yurok, or Konumihu people but also have the function of erasing traces of Indigenous presence. These maps do not list any of the geographic names for places, creeks, land formations, and so on that reflect Karuk history and worldview—save for, perhaps and interestingly relevant here, Swillup Creek, whose name may be derived from the Karuk word for pine tree or Pine Flat, *ish-vírip*. In fact, no other Karuk names or derivatives for geographic instances have been recognized or adopted by the dominant non-Native government.[15]

Further disruption to the worldview of the local tribes came when the federal government began to place high value on so-called wilderness areas that, according to the 1964 Wilderness Act, are ironically defined as "places untrammeled by man, where man himself is a visitor who does not remain," and "land retaining its primeval character and influence, without permanent improvements or human habitation" and which appear "to have been affected primarily by the forces of nature, with the imprint of man's work substantially unnoticeable."[16] These designations illustrate the ignorance of white land managers and the ongoing erasure of Indigenous worldview given not only the ongoing presence of Karuk people in the region but also the fact that human interaction is what shaped these very lands. Yet through such language and land management policies, the erasure of Native presence and the logics of white hegemony have manifested in the material reality of both the people and the land. To use Rob Nixon's (2010) term, they create "unimagined" communities of "uninhabitants." Again, while I describe these dynamics in the language of racialization, these practices of ecological erasure are very much about colonialism. Kyle Whyte (2018c), in particular, emphasizes how "settlement was inflicting rapid and harmful environmental changes on our peoples, which offset the flourishing moral relationships that supported Anishinaabe resilience. The history of Canadian and U.S. colonialism can be read as the establishment of the conditions for their own resilience in North America at the expense of Indigenous peoples' resilience" (2). Whyte and coauthors (2018) underscore that "for settlers, the presence of Indigenous ecologies—from the human activities themselves to their physical manifestations as particular ecosystems and ecological flows—delegitimizes settlers' claims to have honorable and credible religious 'missions,' universal property rights, and exclusive political and cultural sovereignty. To remove all markers or physical manifestations that challenge their moral legitimacy, power, and self-determination, settlers systematically seek to erase the ecologies required for Indigenous governance systems, such as Indigenous seasonal rounds" (159). Key to racial formation theory is the notion that racial constructions are distinctive, and this circumstance is especially important in the case of Native Americans.[17] Eva Marie Garroutte (2001, 2003), one of the few scholars to engage racial formation of American Indian people in depth, notes, "The specific elements of the racial formation process for Indian people make Native Americans' experience unique among those of modern-day U.S. racial groups" (2001, 234). Fenelon and Trafzer (2014) emphasize that "Indigenous peoples represent the most complex social analytical issues in the world today, including invasion by foreign groups, outright genocide, culturicide and multiple forms of coercive assimilation, and ranging over half a millennium of modern colonization histories covering the Americas and globally" (3). A few key points with respect to the racial formation process for Native American people are worth emphasizing here. First, the logic behind the creation of the category of American Indian or Native American has been its elimination. This elimination was first desired in order for whites to access Indigenous lands. Desire for lands still underscores the logic of eliminating (or ignoring) Native presence today, but as elimination was only partially successful, treaties were negotiated with Tribes across the continent. At this point, treaties, tribal status, and the federal trust responsibility to Indian people were barriers to settler access to land. Thus, elimination of Native people next took the form not of direct genocide but elimination of their status as Indian people through attempts at forced assimilation (Hoxie 1984) and, as Deloria (1998) articulates, the need to erase Native presence to obliterate the unpleasant history of genocide itself and thereby uphold the foundational myths of American innocence and democracy.

Garroutte (2001) also underscores a second unique element in the racial formation process for Native people—namely, the tricky relationships between specific tribal affiliations and one's "race" as American Indian. While the more general racial distinction of "Indian" may be the expected focus in racial formation theory, in the case of Native people, it is not being a generic "Indian" but one's specific tribal affiliation that matters both for the state and for tribal people themselves. At the same time, this affiliation with a specific tribe is a cultural construct that has its origins in the Euro-American colonization and assimilation policies. Intermarriage between villages of "people," as Indigenous peoples almost invariably named themselves in their own languages, was not only common but also sought after: marriage between closely related individuals was considered taboo, with numerous origin stories guiding this closely regulated tribal code. However, with a barrage of policies and federal acts defining who was "Indian" or "enough Indian" to limit the established eligible criteria for federal and state benefits like education and health care, those Indigenous groups who could prove their status and become federally recognized sought to appease the federal government agencies by defining their membership through blood quantum (Schmidt 2011). Thus, like Quijano (2010), Fenelon and Trafzer (2014) emphasize that not only is the racial category of American Indian a construct of the colonizer, but the category of race was also a tool of the colonization process: "This was the first, and perhaps most important, racial construction in the history of the world, and would be applied to all subsequent colonization by European powers, first within the Americas, and later around the world, calling the Indigenous peoples Natives or aborigines, without recognizing their social constructions, instead referring to them as undifferentiated 'tribes' within the broader race construct associated with being "uncivilized" and less developed or evolved" (10). Tribes are culturally and linguistically unique, with their own social and political structures prior to European contact and colonization, but these distinctions go beyond differences between race and ethnicity. Tribal affiliations also have paramount political significance—the state has negotiated different relationships, interests, and levels of responsibility with tribes that are federally recognized and those that are not. In Northern California, it matters whether one is Karuk, Yurok, or Winnemum Wintu for cultural, economic, and political reasons. In this settler-colonial state, one can claim American Indian ancestry by checking a box on the U.S. Census, but from Indigenous standpoints, it remains a social fact that to be American Indian in a cultural, social, or political sense means having a connection with a specific Tribe.[18]

How did Karuk cultural, epistemological, and religious systems that had evolved and endured over tens of thousands of years come to be replaced by another cultural logic and sensibility in such short order? How do a people whose religious texts stipulate that "thou shalt not kill" come to tolerate and enact direct genocide in the Klamath Basin and beyond? How do a people that never recognized a leader other than the "head man" of their village site become boxed together into a tribal designation that fails to recognize those members who have intermarried speaking another—even tribal—language? If the idea of race is a social construction, how do concepts of white or Native come to seem real and inevitable in the minds of a people? Omi and Winant describe how race comes to have a stability or permanence as a result of specific "racial projects" that work to make race real through a combination of ideological justifications and legal mandates. At the same time as legal actions by the state reorganize and redistribute resources on the basis of race, ideological justifications provide interpretations of events and racial dynamics (Omi and Winant 1994, 56). HoSang and coauthors (2012) describe how "racial projects link structures and representations within specific historical contexts; they perform the ideological labor known as racialization—the extension or elaboration of racial meaning to particular relationships, social practices or groups" (91). In this way, "racial projects modify and rearticulate meanings of race" (93).

On the Klamath, we can trace state actions that enabled the solidification of racial hierarchies and the reorganization of wealth across three particularly significant racial projects: outright genocide, lack of recognition of land occupancy and title, and forced assimilation. Each of these actions damaged the ecosystem, disrupted Karuk cultural management systems, and led to the reorganization of wealth in the Klamath Basin over the past century and a half. Each set of actions achieved these outcomes through legal and discursive assertions that race was real and socially meaningful—that Natives were savages with no rights, that whites were justified in killing people and taking their lands. Thus, each set of actions was part of the process of racial formation: in each circumstance, the state's economic, political, and military actions were legitimated via the judicial system and justified by racialized rhetoric (Secrest 2003, 207–213). While outright legal genocide ended by 1873, the racial projects of lack of recognition of land occupancy and title and forced assimilation continue actively today through land management policies and research programs set forth by both the state of California and the federal government. I will argue therefore that *racialization is not only about "the elaboration of racial meanings to particular relationships, social practices or groups" (Omi and Winant 1994, 91) but also particular environmental practices and places in the landscape.* While I will attend to the legal, material, and discursive particulars of race making at the local and regional levels, these dynamics take place in the context of global dynamics of capitalist accumulation and a project of race making that had been ongoing across the continent for well over a century by the time it reached the West Coast of North America.

"FRONTIER GENOCIDE" AND FORCED RELOCATION

> If environmental racism is the unequal burden of ecological hazards imposed on people of color and their surroundings, then the European conquest was the continental embodiment of this process.
> —Park and Pellow (2004, 410)

The first racial project that used the reorganization of the natural world to establish notions of "whiteness" and American Indians as real racial categories on the Klamath was that of frontier genocide (Fenelon and Trafzer 2014; Fenelon 2015; Norton 1979, 2014; Lindsay 2012, 2014; Madley 2016). Fenelon and Trafzer (2014) describe how "racial formation during this period [frontier genocide] was over 'Indians' to destroy their nation, enslave them, and take their land, to build their new colonies, often committing genocide, wiping out the Indigenous population base and minimally eliminating their societies and original sovereignty" (6). Both the ideological and legal basis for genocide comes from the Marshall Doctrine or Doctrine of Discovery, which explicitly legitimated the perspective that Indian lands were available for the exploitation of whites on the basis that whites were civilized and Native people were "savages" (Newcomb 2008). The doctrine also justified the forced removal of Native people from their lands. As Robert Miller (2005) notes, "The Doctrine had its genesis in medieval, feudal, ethnocentric, religious, and even racial theories. Amazingly, perhaps, the Doctrine is still an active part of American law today. [The] Discovery was applied by European-Americans to legally infringe on the real property and sovereign rights of the American Indian nations and their people, without their knowledge or consent, and it continues to adversely affect Indian tribes and people today" (2). Miller details that although it was not adopted by the Supreme Court until 1823, it was strategically used from the beginning of North American colonization to advance settler gains: "In fact, the deed to almost all real estate in the United States originates from a federal title that itself came from an Indian title. Moreover, the Doctrine is still being actively applied today against American Indian people, tribes and their lands, and still restricts their property, governmental, and self-determination rights" (3–4).

While seeds of the racial category of Indigenous people clearly come from the Doctrine of Discovery, the language therein sets up categories of "primitive" and "modern" rather than "white" and any other groupings per se. Fenelon and Trafzer (2014) describe how "the formation of the dominant group was not entirely racial yet, with Christian and European nations gradually supplanted with whites denoting the original 'civilized versus savage' dualism passing into racial groups. The French and English were quick to follow up on these systems of colonization for land and labor exploitation with racial constructs that further institutionalized the stratification by race" (6). Thus, conceptions of whiteness were forged into existence through the project of North American colonization as settlers from diverse cultural and political backgrounds united in the project of settlement and capitalist primitive accumulation. Here, notions of how the whites would use the land for commodity production directly legitimated the creation and establishment of the racial categories of both white and Native. By the time these racial formation events took place in California, settlers were drawing upon 150 years of legal actions and discursive racial constructions. Land ownership via the Homestead Act, for example, required the modification of lands into "productive" forms—and whites were considered the ones qualified to do it. Through such actions, land was redeemed from "wilderness" and people redeemed from "savagery"—both simultaneously elevated to a virtuous status. Sociologist Richard Widick (2009) quotes E. H. Howard, a New York settler credited with discovery of Humboldt Bay: "[Humboldt] was a land abounded to savages, with its untold treasures waiting for the coming race, for the Argonauts of '49 and '50 to break the seal under which they had lain for ages. And here we stand—we can go no further . . . we must acknowledge that the plucky and energetic Anglo-American, in settlement and civilization, has eclipsed every other people in the elevation of his race and the grandeur of his territorial possessions" (153). These same notions of a forest as "wasted" and people as uncivilized if not engaged in commodity production simultaneously situate the concept of human and landscape "productivity" as proper and virtuous. Such sensibilities extended explicitly the logic of the Dawes Act and continue to be relevant in land use decisions today (e.g., in the Klamath National Forest, the 2016 "salvage" timber sale after the 2014 July Complex fire that was opposed by the Karuk Tribe was justified on the basis that timber would be "wasted").

Although there was certainly prior interchange between Karuk and non-Indian people, violent dislocation to the kinship relationships across species and the radical reorganization of wealth and ecology began with the entry of miners to the Klamath region in the gold rush in 1850 and 1851 (Lowry 1999; Madley 2016; Norton 1979, 2014; Raphael and House 2007; Diver et al. 2010). This time of radical restructuring of ecological and human systems led to the development and solidification of new concepts of land ownership and the emergence of commodity extraction, both of which were key to the emergence of capitalism in the newly forming state of California.

The Constitutional Convention of 1849 set the stage for these events. Discussions at the convention gave ideological justification regarding who the land was rightly "for." Notions of race were built into the foundation of the newly forming state in part through debates on whether Native people should be allowed to vote. A review of the 1849 Constitutional Convention by the California Research Bureau described how, "in the end, the majority prevailed and the Convention agreed to the following constitutional provisions regarding suffrage and California Indians" (Johnston-Dodds 2002, 3). Note the fluidity of the social construction of citizenship and nationality as well as the intersections with gender, patriarchy, and race. The language in the Constitution read that

> every white male citizen of the United States, and every white male citizen of Mexico, who shall have elected to become a citizen of the United States, under the treaty of peace exchanged and ratified at Queretaro, on the 30th day of May, 1848, of the age of twenty-one years, who shall have been a resident of the State six months ... shall be entitled to vote at all elections which are now or hereafter may be authorized by law: Provided, that nothing herein contained shall be construed to prevent the Legislature, by a two thirds concurrent vote, from admitting to the right of suffrage, Indians or the descendants of Indians, in such special cases as such a proportion of the legislative body may deem just and proper. (Cited in Johnston-Dodds 2002, 3)

However, no such legislation allowing Native people the right to vote was ever passed, and Native Americans were not granted the right to vote until 1924. Note here how the language illustrates and reveals the intersections of gender and race, as well as the fluidity and social construction of citizenship and nationality. In 1850 and 1851, the legislature passed additional laws that gave specific legal definitions of race. The 1850 statute defined an Indian as having one-half Indian blood. The following year, however, another statute defined an Indian as "having one fourth or more of Indian blood" (Johnston-Dodds 2002, 6). The first session of the California State Legislature also passed the "1850 Act for the Government and Protection of Indians," thereby creating further legal structures solidifying the new white and Native relationships with land: "The 1850 Act and subsequent amendments facilitated removing California Indians from their traditional lands, separating at least a generation of children and adults from their families, languages, and cultures (1850 to 1865), and indenturing Indian children and adults to Whites" (Johnston-Dodds 2002, 5). Thus, blood quantum as a definition of race was specifically used as a means to deny Native people access to land. From 1851 through early 1852, U.S. Indian Commissioners negotiated eighteen treaties with California Indian tribes that would have reserved roughly 7.5 million acres of land for Native use (about 7.5% of California). Among these were treaties with the Karuk people (Hurtado 1990; Heizer 1972). Gold had just been discovered, however, and as Hurtado (1990) described, "Treaties that conflicted with agriculture and mining interests had little hope of finding support in California's state government," which "did everything possible to thwart them" (139–140). Indeed, "at the beginning of the 1852 California legislative session, the Legislature recognized the value of the land represented in the treaties and appointed committees to prepare joint resolutions and committee reports to recommend how California's U.S. Senators should proceed regarding the ratification of the treaties" (Johnston-Dodds 2002, 23). The Special Committee on the Disposal of Public Land summed up the views opposing ratification of the treaties in its report on the public domain:

> Your memorialists feel assured, from all the facts which are daily transpiring, and the state of public feeling throughout the mines, that if those treaties are ratified, without any sufficient amendments to alter their permanent disposition of the public domain, it will be utterly impossible to prevent the continued collisions between the miners and the Indians. It will not be owing to any objection of the former to the mining of the Indians in the placers; but it will be caused by the exclusive privileges attempted to be secured for Indians, to the mines always heretofore open to the labors of the white man. (Johnston-Dodds 2002, 23)

Note in this passage the interplay between the emerging state structure and the actions of new settlers on the ground in structuring the politics, definition, and solidity of the newly forming concepts of race. On July 8, 1852, due to pressure from the governor of California that was directly a function of the desire for land and gold by the new immigrants, Congress refused to ratify this and other California treaties, leaving eighteen tribes, including the Karuk, without any of the protections, land, or rights reserved in their treaties:

> In mid-March 1852, the California Assembly (35 to 6) and Senate (19 to 4) voted to submit resolutions opposing the ratification of the treaties to California's U.S. Senators. The President submitted the treaties to the U.S. Senate on June 1, 1852. On June 7, the Senate read the President's message, and referred the treaties to the Committee on Indian Affairs. The treaties were then considered and rejected by the U.S. Senate in secret session. The treaties did not reappear in the public record until January 18, 1905, after an injunction of secrecy was removed. (Johnston-Dodds 2002, 24)

Access to land was critical in the unfolding of these events: "When American agents negotiated 18 treaties, creating 18 reservations, the California delegation made sure that Senate of the United States met in a secret session and voted against ratifying any of the negotiated treaties. Thus, California Indians had no 'legal' ownership to traditional lands, giving the newcomers time to steal as much Indian lands as possible before federal officials recognized Indian reservations and Indian nations could claim a small portion of their vast former holdings taken by newcomers through 'legal' means established by non-Indians in the state of California" (Fenelon and Trafzer 2014, 17). Instead of ratifying treaties, the disruption of Karuk economic, political, and cultural systems was carried out by created state-sponsored programs promoting the killing of Native people under a succession of the governors of California (Hurtado 1990; Madley 2016). Largely due to state-sponsored genocide, the Karuk population went from about 2,700 people in 1850 to about 800 people by around 1880 (Krober 1925, cited in McEvoy 1986).

During this period of explicit genocide, desire for access to first gold and then land as material environmental resources drove these racialized actions and their justifications. The outright killing of about two-thirds of Karuk people, relocation of villages, and attempts to move people onto reservations were explicitly supported by ideological conceptions of Native people as savages who were closer to nature and white people as both civilized and inevitably destined to overtake Karuk lands (Lowry 1999; Norton 1979; Raphael and House 2007). For example, in a message to the California State Legislature in 1851, California Governor Burnett expressed that "a war of extermination will continue to be waged between the races, until the Indian race becomes extinct, must be expected. While we cannot anticipate this result but with painful regret, the inevitable destiny of the race is beyond the power or wisdom of man to avert." The state of California provided infrastructure and funding for the racial project of genocide, but local militias were largely composed of recent immigrants who traded their labor for funding, and the reward of lands largely carried it out on the ground: "From the state archival record, it is impossible to determine exactly the total number of units and men engaged in attacks against the California Indians. However, during the period of 1850 to 1859, the official record does verify that the governors of California called out the militia on 'Expeditions against the Indians' on a number of occasions at considerable expense" (Johnston-Dodds 2002, 16). Thus, the racial project of genocide proceeded via state and federal actions, on one hand, and settlers on the ground, on the other: "The United States sought total domination of Native Americans, and federal and state officials allowed pioneers to murder, rape, kidnap, steal, and destroy Native Americans, creating systems for superordinating settlers, militia soldiers, and government officials to subordinate Indians, thereby developing caste-like social systems fully alienating Indigenes, usually on their own lands" (Fenelon and Trafzer 2014, 13). One document with the title "General Recapitulation of the Expenditures incurred by the State of California For the Subsistence and Pay of the Troops" details "military" expeditions totaling $843,573.48, ordered by the governor, during 1850, 1851, and 1852, "For the Protection of the

Lives and Property of her Citizens, and for the Suppression of Indian Hostilities within her Borders" (Fenelon and Trafzer 2014). Furthermore, "Muster Rolls" in the California State Archives indicate the participation of approximately 35,000 men—a total of 303 units. Most California counties had a unit (Fenelon and Trafzer 2014).[19] One table in this document records a value of $14,987.00 being disbursed to the "Siskiyou Volunteer Rangers" whose territory would have included both Karuk and Konomihu as well as other Shasta people. These figures represent state expenditures supporting the project of Karuk genocide, which was carried out by settlers on the state's behalf: "Small groups of pioneers met in democratic meetings to discuss Indian policies and decided to murder men, women, and children. They used rumors and theft of livestock as their overt justification for killing Indians, but they also harbored deep-seeded racism and fear of Indigenous, non-Christian peoples the newcomers considered uncivilized savages" (Fenelon and Trafzer 2014, 8).

Not only was access to the natural world of material importance for the racial projects of genocide and forced relocation, but statements by the first three governors of California also illustrate the symbolic importance of "nature" in the construction of the racist ideologies underlying these actions. For example, racial ideologies are central justifications for state violence in an 1852 letter by California Governor John Bigler, asking for assistance from the federal government in protecting white settlers in Northern California from Indians: "The acts of these Savages are sometimes signalized by a ferocity worthy of the cannibals of the South Sea. They seem to cherish an instinctive hatred towards the white race, and this is a principle of their nature which neither time nor vicissitude can impair. This principle of hatred is hereditary . . . Whites and Indians cannot live in close proximity in peace" (quoted in Heizer 1974, 189). Both the characterizations of people as savage and the language "instincts" in this passage naturalizing Native people as "primitive," as well as Bigler's references to "Savages" with a capital "S," ideologically link Native people to a violent and wild nature as the means to justify frontier genocide and colonization. David Pellow notes here how racism and racialization "travel" through comparisons and linkages with other oppressed peoples globally (personal communication). Lisa Hillman reflects, "Whites just didn't even think Indians could be included in the race of men, but were a part of the wild themselves. The whites were the 'first,' because the Indians didn't count. I think it was more like studied and deliberate ignorance of this rather ugly (for them) reality" (personal communication). Leaf Hillman emphasizes "taming the wilderness and exterminating the Indians were viewed as the same thing and not different from one another" (personal communication). Similarly, Anderson (2005) highlights the words of Father Geronimo Boscana of the San Juan Capistrano Mission from 1812 to 1826. Boscana uses words such as *brutes* and refers to the conceptual scheme that simultaneously reflects human hierarchy with respect to nature, progressive cultural evolution, and capitalist notions of productivity as morality that was overt at the time but still dominant today: "No doubt these Indians passed a miserable life, ever idle, and more like the brutes than rational beings. They neither cultivated the ground nor planted any kind of grain, but lived upon the wild seeds of the field, the fruits of the forest, and upon the abundance of game" (73). Note as well Boscana's profound ignorance of the elaborate land management activities in which California Indians engaged. The senses that genocide and white supremacy were inevitable were key justifications that flowed from the logic of "progress." While newspapers of the time detail many of the events in these racial projects, newspaper editorial pages offer important accounts of the ideological justifications at play in this time and place. In 1854, the *Humboldt Times*—the main newspaper of the California north-coast region comprising the western edge of Karuk territory—declared that "this vast forest will be made to yield to the enterprising sprit of the *universal Yankee nation*" (cited in Raphael and House 2007, 143). And upon reviewing many of the settler entries in reference to the genocide of the time, Widick (2009) notes that "both poles of the region's editorial spectrum—the sympathetic and liberal Harte and the murderous conservative Wiley used common terms of nation, race, god, gender and property to present the native as other and obstacle to the total white project" (140).

While none of the treaties were ratified, and all California Indian people suffered direct genocide, this was not the only settler-colonial tactic used at this time to access lands whites desired. The state also employed the tactic of forced relocation to move Karuk and other people onto "reservations." Here too narratives of forced assimilation and cultricide were explicitly at play. From 1853 to 1864, the federal government authorized a series of acts removing Indians to reservations of various sizes and configurations that "must not be located on lands occupied by citizens of California" and were as specified in the 1864 act to be placed "as remote from White settlement as practicable" (Raphael and House 2007, 160). Edward F. Beale, California's first Indian superintendent, developed the California reservation system. In what could be called some of the first and most enduring acts of environmental racism in California, it is reported that "Beale took pride in finding the worst possible tracts, which no white man could possibly covet" (Raphael and House 2007, 160).[20]

In 1864, the Hoopa Valley Indian Reservation was established, and many Karuk people were ordered to leave their ancestral lands along the mid-Klamath and lower Salmon rivers and relocate to this reservation.[21] Many people did so. This overt displacement, together with later relocation programs, the search for employment, and the absence of recognized territory, contributed largely to the fact that still today many Karuk people continue to live on the Hoopa reservation, in cities on the coast, the Interstate 5 corridor, and are spread across California and Oregon. Yet, because Karuk territory is largely mountainous, and the lands along the river corridor were the most desired by whites, many people were also able to escape and find refuge in the mountains at higher elevations until they could return. Thus, not only the reasons and ideological justification behind this racial project of genocide related to nature, but the ability of Karuk people to resist both genocide and forced relocation was also enabled by their relationships with the natural world. The mountainous terrain and people's knowledge of and relationships with their lands made it possible for families and individuals to escape both direct genocide and later to resist the efforts of Indian agents to relocate them. The higher percentage of Karuk survivorship of frontier genocide as opposed to coastal tribes is a direct reflection of these factors. As David Pellow noted, this outcome can be understood as an example of the agency of the natural world and human-nonhuman collaboration in the resistance against colonialism and genocide.

To the extent that they were successful, these state acts of forced removal gave successive waves of white settlers the possibility of imagining much of the Klamath region as being an empty "wilderness."[22] The fact that by the late 1880s there were fewer Karuk people present in the river corridor limited the direct contact between settlers and the people whose lands they were occupying, thereby affirming the myths of an "empty" land and Native disappearance, in turn allowing the possibility of innocence on the part of white settlers. Thus, Raphael and House (2007) note that "the many men and few women who came to Humboldt Bay in the early 1850s thought of themselves as the area's 'first' citizens, and everything they did in their new surrounds was also labeled a 'first.' Elliott's classic History of Humboldt County, published in 1882 lists 52 'firsts' in the table of contents: 'First White person in Humboldt County' . . . 'First Child Born in Ferndale'" (142). Jean O'Brien (2010) centers this discourse of "firstling and lasting" as a key mechanism of Indigenous erasure. Together, the triad of inevitability, ignorance, and innocence became part of the fundamental quality of whiteness and work to support notions of white supremacy. This sense of inevitability served to justify actions in the minds of individual settlers and participants in direct genocide who rationalized their actions with the reasoning that if they did not settle the land, someone else would (a justification that is still commonly heard today by whites who purchase Karuk lands; this supports the idea that an Indian did not count as "someone"). Next, the portrayal of lands as unoccupied, unused, or vacant wilderness allowed settlers to claim ignorance of this process, as well as the corresponding reality that the new immigrants were settling in someone else's recently vacated home. Then (and now), willful ignorance of Indigenous ecological management further supported the notion of empty lands. The supposed innocence of settlers in this process was upheld by claims of inevitability, the discourse of progress and human evolution, and racialized notions of Native inferiority. Qualities of the newly developing racial category of "white" were thus dependent upon racialized notions of Native. Brian Klopotek (2011) describes how "white supremacy in particular has been a central racial project shaping relations among racial and ethno-national groups in the United States. . . . It is an ideology that understands white people and their ancestors as morally, intellectually, politically and spiritually superior to nonwhites, and therefore entitled to various forms of privilege, power and property. It wields power even in the absence of explicit legal protections because it so often goes unnoticed, making social and political inequalities seem 'natural'" (6). The racial projects of genocide and forced relocation undermined Karuk economic, political, and social structure and moved significant amounts of wealth into white hands. In Marxist terms, we understand this as primitive accumulation. The juxtaposition of this transfer of wealth across the newly created racial categories is underscored by Widick (2009), who notes, "While Indians driven into the hills and onto corrupt reservations starved, Humboldt produced 80,000 pounds of butter, 2,000 pounds of cheese, and 3,604 cattle in 1856, an increase from 1,812 head in 1854. . . . During this period, industrial timber extraction became Humboldt's number one commodity for export. In 1854 native forests yielded 27 million board feet of lumber to the hand saw and oxen with 187 boatloads shipped into the world system through the mouth of the bay" (134). Indeed, back in 1848, just before the height of direct genocide, a mining party led by Major Pierson B. Redding removed $80,000 in gold in his first trip up Trinity River—a significant figure at the time. Lands nearer to coast, to the west of Karuk territory in what would become Humboldt County, were both forested with beautiful redwoods and relatively accessible via coastal ports and roads. By 1854, during the period in which the state of California was bankrolling and organizing indigene genocide, some 200 new immigrants were already working as loggers in and near Karuk territory, and another 130 were employed in sawmills. Raphael and House (2002) report that "in 1860 local mills sawed 9.6 million feet of lumber, by 1875 the output had ballooned eightfold to 75 million feet" (196). Already just three years into statehood, 143 ships departed Humboldt Bay to San Francisco, each one carrying lumber into the hands of the new white immigrants (148). Recounting conditions farther south in the state, Anderson (2005) describes how "California Indians provided much of the labor needed to build the new economies of the colonizers during each historical period, and many prominent non-Indian men in California's history built their fortunes on the backs of cheap Indian laborers. But the California Indians gave the newcomers far more than their labor. The success of the mission economies and of the many fur trading, market hunting, gold mining, logging, ranching, and agricultural enterprises of early California rested on a land that was productive and healthy ecologically because of the careful stewardship of many generations of California Indians" (64). Both the human and ecological impacts of this period are devastating to the point of being difficult to comprehend. Throughout the racial project of overt genocide and displacement, white supremacy was enacted onto people, places, and the species known as nature. Severing Indigenous kinship relationships with places and species, as well as reconstructing human-nature relationships around commodity production, was crucial to this first racial project because white settlers and gold explorers wanted the land for themselves. As Karuk people were killed and survivors forced to relocate, Karuk practices of tending the land to ensure productivity and abundance were replaced by extractive technologies from commodity fishing and forestry to hydraulic placer mining that were enormously environmentally destructive. Anderson (2005) writes,

> Significant changes in the environment had occurred during Spanish and Mexican rule, but these were small and restricted compared to the changes wrought in the land during and immediately following the Gold Rush. Mining, logging, and agriculture caused tremendous damage to many of California's rivers and creeks, especially in the foothills of the Sierra Nevada and in the watersheds of the Salmon, Trinity, and Klamath Rivers of northern California, altering their ecologies and forever changing their hydrological functioning. The numbers of cattle, sheep, and pigs escalated, and exotic grasses and forbs spread to the farthest reaches of the state, drastically transforming the species compositions of grasslands and woodlands. (83)

These actions have obvious and lasting impacts on key species dependent upon plant, animal, and insect foods found in these types of habitats. Widick (2009) describes how, "after just three years of white colonization, wild game was already becoming scarce and cattle populations on what had been the Indigenous commons. By 1853 systematic destruction of the region's elk, deer and bear populations by settlers forced the state to recognize that if it wanted to keep the peace, it would have to supply the Natives with meat. That year California Superintendent of Indian affairs Edward Beale reported to the U.S. Secretary of the Interior that white encroachment on Indian-occupied lands and corruption of 'the system of beef delivery to Indians' had led to starvation" (133–134). Not only do these actions also directly impact key traditional Karuk riverine foods such as salmon, steelhead, sturgeon, and lamprey, but they have also caused irreparable damage to the very earth upon which the Karuk built their homes. Following placer mining came hydraulic placer mining that washed away forests, hillsides, and numerous river terraces that had been Karuk villages as highly pressurized water flushed an estimated 12 billion tons of mud and soil into California rivers across the state as a whole (Merchant 1998). Raphael and House describe how, "by the 1860s, placer mining was on the downswing. But the hydraulic mining that replaced it was more destructive yet. Any location that had been the site of a hydraulic operation could never be farmed, for rock tailings replaced all topsoil. While Indians in other areas took up farming to replace or supplement traditional food supplies, agricultural opportunities for the Karuk were extremely limited. Meanwhile, their fisheries had been seriously impacted by placer miners, who churned up the riverbed, and hydraulic miners who tore down the hillsides and deposited unfathomable amounts of silt into the channels" (282–283). Most non-Indians can identify ecological degradation in the form of severe manipulations of the rivers from hydraulic placer mining or in the forest from clear cutting. Some can even identify forest manipulations from the imposition of new fire regimes. What seems quite beyond comprehension, especially for non-Indians, is the scale of ecological damage occurring from the disruption to Native cultural management. As an ecologist, Anderson (2005) addresses some of these changes: "Growing alien food plants in extensive cropping systems, cutting down trees for fuel and construction materials, hunting native animals for the sale of their meat and fur, and grazing large numbers of non-native ungulates all impacted the supplies of salmon, deer, acorns, seeds and grains, greens, and edible bulbs and tubers on which the native people depended for food. Native gathering sites in the rich valley grasslands, coastal prairies, mountain meadows, oak savannas, and forests, kept fertile and open by burning and visited regularly for many generations, became rangelands, timberlands, and farms" (63). Not only has the human scale of Native American genocide been of remarkable little sociological focus, the fact that this genocide has been coupled with a reorganization of the natural world and an assault on a spiritual order that nourished and governed an entire field of ecological relationships represents a substantial void to present understanding of race and the racialization process. Environmental practices have meanings too, as do places. The imbuing of racial meanings to places has also been crucial for racial formation. Indeed, while other violent aspects of racialization, including slavery and the mass incarceration of Japanese Americans, have received sociological attention, there are virtually no sociological accounts detailing these processes of Indigenous and white racialization through genocide and displacement.

LACK OF RECOGNITION OF LAND OCCUPANCY AND TITLE

Karuk people recognize over a million acres of biologically diverse mountains and rivers as their ancestral territory and homelands. Yet as of 2016, Karuk-owned lands consist of only 901 acres of trust lands and another 777 acres of fee lands.[23] Instead, 98% of the lands that Karuk recognize as ancestral territory and homelands, lands and species that have been cared for and tended to by Karuk people since time immemorial, are officially under the management of the U.S. Forest Service. The state's refusal to recognize Karuk land title originates with the Doctrine of Discovery and Princes Right to Conquest (itself illegal) and was further legitimated by the abovementioned failure of the U.S. Congress to ratify the 1851 treaty during the gold rush and the subsequent claiming of Karuk lands as public domain. While the failure to ratify treaties led the way for genocide and forced relocation as described in the last section, the lack of recognition of land occupancy and title underscores the development of the racial categories of white and Native in a variety of ways. The natural world has been equally central to the next set of state actions that solidified the meanings of these racial categories on the Klamath. Like the project of direct genocide, the failure to recognize land occupancy and title has in turn fortified racialized conceptions of Indigenous people as inferior found in the doctrine of discovery, occurring here in their own distinctive American manifestation. Here in the language of racial capitalism, the process of primitive accumulation as land dispossession continues, but as Coulthard (2014) underscores, "The history and experience of dispossession, not proletarianization, has been the dominant background structure shaping the character of the historical relationship between Indigenous peoples and the [Canadian] state" (13).

Overt genocide in California and the Klamath Basin is generally viewed as over by 1873 (see Madley 2016). The last such overt act in the Klamath Basin occurred during the events at Captain Jack's Stronghold. But state-sanctioned and militia violence ceased somewhat earlier in Karuk country farther downriver. As it became safer to do so, Karuk people began returning from hiding in the high country to their traditional homes along the river corridor. In their history of Humboldt County, Raphael and House (2007) describe, "In the early 1850s when several thousand gold miners flocked to the Klamath and Salmon Rivers, many Karuks were forced to leave their villages on benches near the rivers and flee into the nearby mountains. As each claim was worked and then abandoned, Karuk individuals and families who survived the early years of massacres, disease and dislocation gradually returned to the river, but not always to their traditional villages, some of which had been turned into piles of rocks" (282). The remoteness of the region afforded Karuk people some ability to recover and survive, especially as compared with their neighbors on the coast. Thus, Raphael and House note, "When Stephen Powers traveled through Karuk country in 1871, he observed that the people and the culture had not been completely destroyed. Although Karuks were donning manufactured cloths, they seemed to have maintained most of their traditional customs. Enough people spoke English to explain to him their beliefs and tell him their stories, yet the Karuk language was still the standard method of discourse" (282–283). And "in 1908 when two white women came to the Klamath to teach school, they reported that only a handful of Karuks knew how to speak English. And when the anthropologist John P Harrington came through in the 1920s he was able to document highly detailed information about the traditional cultivation and use of native tobacco, including customs that had been handed down continuously for generations" (283). As people returned from the high country to the river corridor, however, some discovered that their lands had been claimed by the U.S. government. The failure to recognize Karuk land occupancy and title is the next racial project that asserted white supremacy and disrupted Karuk relationships with the land. Here too, differences between racialized European and Karuk conceptions of land, appropriate land use, and land "ownership" underlie and in turn became a vehicle justifying the lack of recognition of Karuk land title—namely, that whites were needed to come civilize the West, that land ownership in the capitalist sense and agriculture were the highest moral expression and elevation of human evolution, and that agriculture and civilization were the inevitable destiny of the region.

Scott Quinn, director of Karuk Department of Tribal Lands Management, notes that "prior to the infusion of Europeans into the Upper Klamath River in 1850, ownership of land by individuals was not recognized. But the tribes, and individual people did own rights to hunt, fish, gather and manage particular portions of the surrounding landscape" (Quinn 2007). Federal monies were authorized in 1860 to teach Indians to farm (Norton 1979, 153), and in 1862, the passage of the Homestead Act provided that small parcels of land were allotted to families who were citizens (Native people were not).

These racialized notions of relating to the land in the form of agriculture versus Indigenous practices of tending, gathering, and hunting were explicitly behind the Dawes Act, which was designed to break up tribal land and divide Native land among individuals: "It was hoped that initiating Indians to the concept of private land ownership would aid in integrating them into white society" (Delaney 1981, 2). Through this racial project, resources were diverted from Indian to non-Indian hands, and land management practices shifted from activities geared toward food production and key to binding Indians together and performing socialization frameworks to those that would achieve profits under capitalism (timber and farming).

Here, racialized notions of how the land should be used were legally enforced upon both people and the landscape. The Dawes Act specified that "all applications for allotments under the provisions of this section shall be submitted to the Secretary of Agriculture, who shall determine whether the lands applied for are more valuable for agricultural or grazing purposes than for the timber found thereon; and if it be found that the lands applied for are more valuable for agricultural or grazing purposes, then the Secretary of the Interior shall cause allotment to be made as herein provided." Because the Karuk did not have a reservation and were then living on lands claimed by the U.S. Forest Service, the 1910 amendment of the act to include forest lands was particularly significant (Delaney 1981).[24]

Attempts to alter the relationships people held with land were explicitly used as a means to alter racial constructions, as evident in this discussion: "The Indians who are now residing on private lands, with the consent of the owners, or engaged in cultivating their soil, should not be disturbed in their position. They are already in the best school of civilization. . . . The adoption of this plan would obviate the contemplated permanent disposal of a large portion of our mineral and arable land [to the Indians]" (Johnston-Dodds, 2002, 24). And, non-Indian conceptions of land ownership are codified into laws. Together with racialized rhetoric and ideology, these laws become a vehicle for the transfer of land from Indian into white hands. This process was very overt. Karuk families who applied for allotments had to demonstrate that they would use the lands for agricultural purposes—it was not possible to acquire an allotment to carry out traditional management. Furthermore, demonstrating any agricultural purpose was difficult to do both due to the forested condition of the mountainous region of the mid-Klamath and the fact that the few previously occurring flat areas of bottomland had been mostly washed away by placer and hydraulic mining. Leaf and Lisa Hillman note that villages were remote, and many Indian families had no idea this was going on. When people came to kick them off, they were surprised. There is also the fact that few could read and write, so even if they knew that they would have to apply for an allotment, they might not have had the ability themselves to do so or knew someone who could. Furthermore, some Indians refused allotment because to do so would be an acknowledgment that the land was not theirs. Allotments made in Karuk territory are both fewer and smaller than allotments elsewhere in California and Oregon due to the landscape and history. Central to this racialized notion of land were differences in Karuk and white notions of ownership and rights versus responsibility. Leaf Hillman, Karuk Tishuniik ceremonial leader and DNR director, describes this relationship and its associated responsibilities with reference to the importance of World Renewal Ceremonies: "The rocks and the trees and the water and the air, the responsibility that I have, those are real relations.... We have not forgotten that we are related and that we have a responsibility. And at the same time we give thanks to those other spirit people for helping to subsist us, and reminding them that we haven't forgot that we owe them something too. So the renewal is renewing the bonds that exist." Controlling access to land via allotments was one of the more overtly identifiable mechanisms used by the state in undermining Karuk land occupancy and title.

Leaf Hillman notes further that there was in fact a direct correlation between the implementation of allotment, changing fire regimes (as will be discussed in next chapter), and the appraisal of land: "Folks from the Forest Reserve were some of the first established government officials here, and they created a tie with the Indian Service: Quid pro quo—we'll help the Indians with their allotment process if the Indian Service agency folks would help the forest service by getting the Indians to stop burning." While individuals were thus coerced into engaging in commodity production, the forest was deemed valuable for timber production, and the state began to enact large-scale land management practices through the development of forestry agencies beginning in 1876 with successive iterations in 1881, 1891, and 1901 until reaching its present organizational form and title as the U.S. Forest Service in 1905. At this point, the new agency's land management practices began to further undermine Karuk occupancy. Wildfire, in particular, was seen as a threat to timber by the white settlers, and fire suppression has profoundly disrupted the existing relationships across species that underlie Karuk economic, ecological, and food systems as will be further elaborated in chapter 2.

Later in the 1950s with the widening of State Highway 96, the Bureau of Indian Affairs (BIA) transferred additional Karuk lands to the state of California through right of way, and in the process, many Karuk Public Domain Indian Allotments were divided and decreased further in size to accommodate the modern two-lane highway. In some cases, undivided interest in an approved allotment was allowed to be sold by the BIA to nontribal buyers. This has created ambiguity in the true ownership of the land, as a portion of the property remains in trust. If the new owner does not pay taxes on the small undivided fee interest, the county puts the property up for auction and the new buyer assumes he or she owns the entire property. This has led to a lot of confusion over the true ownership of some allotments. Today, only thirty-five of the ninety original Karuk allotments remain in the ownership of Karuk families. Today, because very little land within the Karuk Aboriginal Territory is in private ownership (approximately 97% of Karuk land became what is now known as national forest in 1908), land that does come onto the market remains too expensive for most Indian families or even often for the Karuk Tribe to purchase. Today, the new "gold rush" (aka marijuana cultivation) has artificially inflated real estate prices astronomically.

The state's lack of recognition of Karuk land title and occupancy operates as a project of racial formation with respect to not only land "ownership" but also other aspects of Indigenous relationships to species. Because treaties outline territories as well as rights to usual and accustomed areas, recognition of land title is coupled with recognition of fishing rights. Shortly after statehood, the same Governor John Bigler who called Indians savages signed into law the first California Fish and Game Act. This act, passed in 1852, closed seasons in many parts of the state for Karuk cultural foods, including deer and elk, as well as other species that were being overhunted by white settlers, including quail, partridge, mallard and wood ducks, and antelope. The state appointed game wardens to enforce hunting and fishing regulations in 1871, thereby authorizing the official harassment of Karuk people for engaging in hunting and fishing practices. Karuk fishing rights have yet to be acknowledged by the U.S. government, although tribal members still fish at one "ceremonial fishery." During the 1970s, the federal government stepped up enforcement and forcibly denied Karuk people the right to continue their traditional fishing practices by arresting them and incarcerating them (Norton 1979). Access to food and notions of how land should be used may be contested, but the state holds the ability to assert its version.

While events such as the failure to sign treaties and the establishment of the U.S. Forest Service happened over a century ago, the continuing consequences of such events are played out every day through the ongoing legal and criminal enforcement of racialized notions of how the land should be used and for whom. The management of Karuk cultural resources by non-Indian agencies, such as the Bureau of Reclamation, U.S. Forest Service, or U.S. Fish and Wildlife Service, and the fact that Karuk cultural management is mostly illegal are also part of the next racial project underlying the movement of wealth out of Karuk hands: that of forced assimilation.

FORCED ASSIMILATION

> Our way of lives has been taken away from us. We can no longer gather the food that we gathered. We have pretty much lost the ability to gather those foods and to manage the land the way our ancestors managed the land.
> —David Arwood

Forced assimilation is the third racial project that shaped the reorganization of human bodies, other species, and the production of wealth in the Klamath Basin over the past century and a half. Here too, the natural world has been crucial to the extension of racial meanings and categories, as well as the ability of Karuk people to resist them. When Native people could not be eliminated by direct genocide, the next attempt was made to eliminate the category of Native itself by subsuming them into whiteness through forced assimilation (Hoxie 1984).[25] While the notion of forced assimilation is most commonly discussed in the context of boarding schools, language suppression (Child 1998), and the outlawing of religious practices such as the ghost dance, it continues today via a wide range of state-sponsored programs, practices, and land management policies. Boarding schools were key institutions where the most explicit attempts were made to force Karuk and other Native people into the dominant culture. Like youth from tribes throughout Canada and the United States, Karuk children were separated from their families at young ages and taken to government boarding schools in Oregon, California, Arizona, Kansas, and Pennsylvania, where the project was, in the words of their founder, Richard Pratt, "to kill the Indian and save the man." Those who were able to eventually return home were forever changed; many, however, were never seen again. In a letter addressed to the Sacramento Indian Agency, Rose Sunderland, a Karuk Indian from Happy Camp, wrote in 1931,

> There were six children in my family ... George, Wingate, Edward, Jackson, Ida and I went to the Hoopa [Indian Boarding] School.[26] From Hoopa Wingate was sent to Carlisle [Indian Boarding School] and graduated in 1901 or 1902 and then went back to California and stopped in San Francisco where he thought he might find employment. I have never seen or heard of him since so presume he is dead. Edward was sent to Chemawa Indian School in Oregon and died at that school. Jackson was sent to Phoenix Indian School in Arizona where he graduated and then started for Happy Camp and died before he reached Happy Camp. Ida was sent to Phoenix Indian School and has lived in Arizona ever since because she had consumption when she was in school and was told she could not live anywhere else. I was sent to Carlisle and graduated in 1905 and stayed in the East because I could do housework and it was the only means I had of earning a "living." (What a life it has been).[27]

The racial projects of lack of land recognition and title have often been coupled with forced assimilation both historically and into the present. Native family structures were explicitly targeted through forced assimilation and via the intentional removal of children and families from the land because the dismantling of native kinship systems was needed to transfer lands into non-Native hands (Anderson and Ball 2011; Stremlau 2005). The theme of forced assimilation is especially key for settler-colonialism and will also be discussed at length in the next chapter.

While there is no overt policy designed to change how Karuk people view and use the land parallel to the ways that boarding schools explicitly enforced non-Native "white" behaviors onto Indian people, forced assimilation occurs today because significant Karuk cultural practices, including fishing, hunting, and the traditional modes of using fire, are illegal. With the establishment of the U.S. Forest Service (1905), U.S. Fish and Wildlife Service (1871), and California Department of Fish and Game (progressively from 1871 to present form in 1951), Karuk land management practices from burning to the collection of mushrooms came officially under the jurisdiction of non-Indian agencies. Hunting and gathering regulations are set by the state of California according to "white man's" law rather than tribal law and managed with the goal of promoting commodity production and recreational outcomes rather than subsistence foods. Even regulations regarding food species such as deer and elk are written with recreational rather than subsistence hunting in mind.

Differences between tribal and nontribal management systems are significant. Karuk tribal management emphasizes the need for sufficient habitat for species to flourish, rather than simply focusing on limiting the number that could be harvested or killed. As Ron Reed notes, "The Karuk way is to enhance whatever you utilize." In contrast, state fish and game regulations focus on how many deer can be harvested and when. Furthermore, these agencies organize regulations under the assumptions that individual hunters rather than extended families and communities will be the consumers. In order for Karuk people to legally harvest a deer in the Western legal system, they must be of a specific age and first have to pass the hunter safety course, qualify for a license, and then buy tags, which requires proof of meeting the California hunter education requirements, all of which are costly in terms of time and money. For California residents over the age of sixteen,[28] a hunting license costs $47.01. The first deer tag costs an additional $31.06 and the second costs $38.62. There is no option for getting a third tag. According to the state of California, the hunting season in Karuk Aboriginal Territory extends just over a month (from mid-September to mid-October). Two deer yield insufficient meat to feed a family for one year, especially when shared with extended family and elders and served at ceremonies. If a Karuk person successfully hunts as prescribed by traditional tribal codes or harvests a deer without purchasing a hunting license and tag from the state, he or she is considered a poacher. Thus, Leaf Hillman describes how, in order to maintain a traditional Karuk lifestyle today, you need to be an outlaw, a criminal, and you had better be a good one or you'll likely end up spending a great portion of your life in prison. The fact of the matter is that it is a criminal act to practice a traditional lifestyle and to maintain traditional cultural practices necessary to manage important food resources or even to practice our religion. If we as Karuk people obey the "laws of nature" and the mandates of our Creator, we are necessarily in violation of the white man's laws. It is a criminal act to be a Karuk Indian in the twenty-first century.

Getting caught for "poaching" has a variety of consequences depending on the circumstances and if it is a repeated offense. Karuk tribal member Jesse Goodwin

TABLE 2 Karuk and Non-Native Management Practices and Cosmologies

	Karuk	Non-Native
Tending	Responsibility to care for land in order to harvest	Invisible within the paradigm in which nature is "static" and humans are separate from it. Conservation management.
Burning	Responsibility to use fire to enhance food, fiber, and medicinal plant resources. Fire is medicine.	Duty to suppress fire to protect human life, property, and capitalist timber resources. Fire is dangerous and "bad."
Fishing	Responsibility to improve habitat and share with community. Fishing can occur when resources are tended and populations are supported by high-quality habitat.	Duty to provide recreational opportunities; maximum of two salmon per day as per Cal Fish and Wildlife regulation; fishing season is regulated to achieve a minimum escapement
Hunting	Responsibility to improve habitat and share with community; hunting can occur when resources are tended and populations are supported by high-quality habitat	Duty to provide recreational opportunities; maximum two deer per season without any requirement for habitat management or enhancement
Gathering	Responsibility to improve habitat, sustainably harvest, and share with community; mushroom patches are "owned" and tended by families	Desire to maximize commercial harvest opportunities, regulate and "cap" free or personal use; until recently, limited awareness of Indigenous uses[1]

[1] Recently this has changed to "free use" or authority under Farm Bill/Coop. Heritage and Cultural Authority 32A for forest products for traditional and cultural purposes.

explains that "usually, they just take our gun rights away from us, try to see if there's any way of us never being able to do it again, and then after that they send you to jail."

Mushroom regulations too have been a source of tension. Matsutake mushrooms (*xayviish* in Karuk) are an important traditional food but commercially highly valuable for the Asian and American markets. Karuk people have particular patches that are tended in a variety of ways. Harvesting techniques, including only taking mature mushrooms and replacing leaf litter, ensure long-term survival of the organism (Richards and Creasy 1996). Resulting in part from decline in Japanese forests, a multimillion-dollar commercial mushroom market developed in the Pacific Northwest in the 1990s. Commercial harvesters could get nearly $19/pound for matsutakes and entered Karuk territory in 1991 (Richards and Creasy 1996). Both the influx of outsiders, whose norms and reasons for picking were very different, and the potential access locals suddenly had to easy money in an economically depressed region altered the relationships Karuk residents held with the mushrooms. When the price spiked to $300 per pound in 1993, the Karuk Tribe filed a complaint that their traditional gathering sites and family traditions were being threatened. A Forest Service study commissioned in response found that "as commercial pickers arrived in greater numbers, local tribal members expressed concern that their traditional gathering sites, most of which are on Forest Service land, were being overharvested by commercial pickers. Many of these commercial harvesters have been Southeast Asian refugee immigrants who live out of the area. At the same time, some local residents in the region, both Native American and non-Native American, had begun to gather matsutake commercially" (Richards and Creasy 1996, 362). The Forest Service issued permits for mushroom harvesting, and while at least some officers made accommodation for traditional Karuk noncommercial use (e.g., by issuing permits for the entire season rather than only one day or week and not collecting fees), even these accommodations reflect an imposition of the non-Native state regulation onto Karuk cultural norms and practices. In other cases, enforcement of mushroom regulations was a direct vector of state power. Karuk tribal member and cultural practitioner Achviivich relates how "there were two tribal members right up here and they had them sprawled on the ground with a gun on the back of their head because they didn't cut their mushrooms in half."[29] Just over 20% of those we asked in the 2005 survey indicated they had been harassed by game wardens while gathering mushrooms (see Figure 4). The importance of the natural world for racialization comes vividly to light in the context such stories and data.

In the 2005, Karuk Health and Fish Consumption Survey tribal members were asked whether members of their household had been questioned or harassed by agency law enforcement officials while gathering a variety of other cultural and subsistence items. Twelve percent reported such contacts while gathering basketry

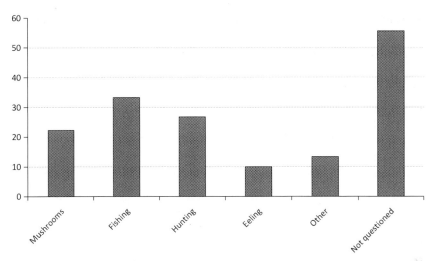

FIGURE 4. Percentage of households questioned/harassed by game wardens while gathering for subsistence or ceremonial purposes (Karuk Tribe 2005).

materials, and over 40% indicated harassment while gathering firewood (see Figure 4).

To be fined or have a family member imprisoned imposes a significant economic burden on families. This is a risk that many are unwilling or unable to take. Twenty percent of survey respondents reported that they had decreased their subsistence or ceremonial activities as a result of such contacts. As per the 2016 Klamath Basin Food System Assessment data collected from 286 Tribal households representing 843 Karuk people, 56.11% reported that state and federal regulatory *rules* related to hunting, gathering, and fishing were the greatest barriers to accessing Native foods. In addition, 68% reported that the *heavy degradation* of traditional gathering sites and *climate change* were the greatest barriers to accessing Native foods, followed by 60.91% reporting the *limited availability* of Native foods (KBRSA 2016, 26–27). Achviivich explains how, in the face of degraded forest conditions due to fire exclusion and possible harassment by law enforcement, many people give up hunting and buy store-bought foods instead:

> How much are society's laws preventing me from gathering? Well, 80%, 90%.... Do I want to go out there and be hassled about it? Why? I go down to the damn store and buy that stuff a lot, you know. It is going to cost you more to go hunt, to go out into the woods and get it. It is not like it is readily available any more. It is not like you have a gathering spot like we used to have a gathering spot. You know, you used to have a gathering spot to gather something and you would go there and gather. Now you don't. Now you can't burn there. You can't burn there every year

and every other year or however often you need to burn it in order to make your crop come up good. You can't do that. You can't burn. And you have to have a permit to get everything. Everything. You have to get a permit to get rocks off the goddamn river bar out here. Did you know that?

From altered ecological conditions due to fire exclusion to the outright illegality of Karuk practices, the reorganization of the Karuk subsistence economy has been both forcible and an extension of the process of forced assimilation. Assimilation is (en)forced when Karuk people are denied access to traditional land and food resources needed to sustain culture and livelihood. Assimilation is (en)forced even more overtly when, for example, game wardens arrest tribal members for fishing in accordance to tribal code rather than state regulation (see Table 2).

When Karuk people cease to harvest traditional foods for fear of retaliation, or when these resources are so depleted that they are forced to purchase store-bought foods or apply for government commodity food services, these are textbook examples of forced assimilation. The sheer absence of the species that Karuk people consider their relations and which have provided sustenance to the Karuk people since time immemorial is a particularly painful and sinister form of forced assimilation. Without salmon and tan oak acorns, Karuk people are currently denied access to foods that represented upward of 50% of their traditional diet. With the decline in access to once-abundant food sources such as deer, acorns, elk, salmon, and mushrooms, a significant percentage of tribal members rely on commodity or store-bought foods. As Karuk tribal member Jesse Coon Goodwin explains, "We can fish at the falls. Dipnet and that, you know, that's the only place we can fish really. But we're not able to go out and go hunting anymore, without getting in trouble for it or something, you know, so—now we have to go to the store to buy our food, and get different kind of foods that aren't suitable for our bodies, like food that was made here for our people, you know? So a lot of it has changed that way." David Pellow notes that we can understand these encounters with fish and game wardens as a form of racial profiling by the settler-colonial state on a continuum with the analyses by Black Lives Matter activists and scholars (Pellow, personal communication). This process of race making over the past 160 years both asserted whites and Native people as inherently real and quite different racial categories and set up the contours of each racial category in such a way as to support the logic of Manifest Destiny. At each step, the natural environment has been key to this process, providing ideological justification and notions of inevitability, on one hand, and supplying material wealth to support settler and state power, on the other. The importance of the natural world to racial formation is further underscored by the fact that natural resource policies continue to disregard preexisting Native land occupancy and title, as well as forced assimilation in the Klamath Basin today—a theme that will be the focus in chapter 2. State interventions in Karuk relationships with the natural environment occur in many ways: natural resource policies that set water quality standards so low that Karuk cultural uses of the river are unsafe (Kann 2008; Kann and Corum 2009), through direct prohibition of the practices of tending the landscape with fire (Hillman and Salter 1997) and through imposition of non-Native regulations for food acquisition such as hunting, fishing, and the harvesting of mushrooms. These policies reorganize the material flows of cells, calories, and organisms through the socioecological system of human and natural kinship relationships, thereby enacting racial violence onto people and the land. I next offer some reflections on the changing dynamics of each.

TRAJECTORIES OF ENVIRONMENTAL ALTERATION AND RACIAL FORMATION ACROSS TIME

I have described how racial formation on the mid-Klamath has taken place through a process that began with direct genocide of Karuk and other Native people in the region, the failure of Congress to ratify treaties. Racial formation continues into the present through the ongoing failure of the state to recognize Karuk land title and the imposition of land management policies that continue to sever ties across species and operate as vectors of forced assimilation. Through this process, the categories of white and American Indian came to be seen as "real." For each racial project, the ability of the state to manipulate the natural environment has been crucial. Capitalist primitive accumulation has meant Indigenous dispossession. And each racial project has simultaneously produced both racial categories and racialized landscapes. As Moore, Kosek, and Pandian, (2003) write, "Race and nature work as a terrain of power." Nature itself has been very much transformed in the process of race making. In the process of race making—itself deeply intertwined with global political and economic events as well as events across North America—white European presence in the region has become "naturalized," and Indigenous cultural systems have been disrupted. For white Euro-Americans, enormous wealth has been generated through the extraction of natural resources. For Native Americans, hunger and poverty have dramatically increased as the natural food system has become disrupted. "Nature" in the Klamath Basin has been transformed from a profoundly abundant ecosystem to its compromised condition today. The natural environment has been crucial on a material level for the consolidation of state power and the generation of capitalist wealth for whites, on one hand, and at the symbolic level in shaping the perception that these processes were both inevitable and good, on the other.

If it is through restructuring relationships with the natural environment that the process of racial formation has occurred historically, the ability of both Karuk and white people to enact their desired types of relationships with the natural environment continues to be a central part of both racialization and its resistance today. Neither the meaning of race nor the operation of racism are static. Both

TABLE 3 Dynamisms of Racial Formation and Natural Environment

	Changing relationship with natural environment	Changing natural environments (e.g., degradation, climate change)
Meaning of race (creation of the racial category)	Indian urban migration, as well as more violent changes to relationships with land (e.g., forced displacement) Develop new organization values and daily practices that become connected to pan-Indian racial category - Resulting in new identification of fry bread as Native food	What does it mean to be Karuk when there are no more fish? What does it mean to be Karuk when there are no more acorns? Potatoes, kishvuf, wild oats?
Operation of racism (lived-out experience)	Lack of understanding of self in absence of natural world for cultural context Diabetes and heart disease as diets altered, food insecurity	Contamination of Native foods in arctic and elsewhere Less access to environmental decision making in multiple policy contexts (racial state, colonial state) Environmental racism in urban neighborhoods: lead poisoning, toxic exposure or asthma in Black urban neighborhoods, exposure to uranium radiation for Navajo, pesticide positioning for Latinx - Race and stigma with environmental degradation - Racism via emotional dimensions of environmental degradation - Racism manifests as differential health problems (e.g., asthma and cancer rates)

changing environments and the changing ways that groups interact with natural environments remain very much a part of the ongoing, dynamic production of race and racism across the country and around the world.

Table 3 outlines the dynamism between race, racism, and the changing conditions in the natural environment. On one hand, human migration (either voluntary or forced displacement) changes the relationships between people and particular places, leading to new racial categories and identities, as well as new manifestations of racism. As Native people endure environmental decline and forced migrations, new contours of racial categories and meaning systems beget new structures of (environmental) racism. These have included artificially elevated rates of diabetes without traditional foods, asthma rates from proximity diesel fumes in urban areas, or suicide rates in the absence of access to ceremonies and traditional practices. For example, "new" panethnic constructions of what it means to be American Indian developed in urban settings in the 1960s and 1970s, while fry bread emerged as a new "traditional food" resulting from BIA commodity programs that in turn were a function of environmental degradation and the reorganization of Indigenous food systems. New manifestations of anti-Native racism emerged, too, in these urban contexts in the absence of relationships with land.

Today the degradation of the natural environment is an increasing vector for the changing relationship between people and the land. As these relationships change, the contents and contours of racial categories as well as the meanings of race shift. For example, chapter 5 will examine how constructions of masculinity are altered with the decrease of salmon in the Klamath river and how, in turn, environmental decline manifests as gendered and racialized forms of violence for Karuk families. Many of the examples in the second column of Table 3 relate to both climate change and the other aspects of the rapidly altered natural world denoted by the concept of the Anthropocene.

Goenpul Aborigine scholar Aileen Moreton-Robinson (2015) writes that the Indigenous perspective produces a particularly valuable contribution to the study of race and whiteness in particular. Moreton-Robinson notes that in contrast to and alongside emphases on the role of slavery, migration, and the development of capitalism on racial constructions, Indigenous scholars bring attention to how land and land ownership matter for the construction of whiteness. Early sociological theorizing did have this focus on land and the natural world (see, e.g., Du Bois, Marable), yet more recently, these dimensions have dropped out race theory. It is my hope that this detailed engagement with racial theory will highlight these dynamics. Perhaps nature has played less of a role in sociological notions of race because the modernist assumption that nature only matters in the daily life of so-called primitive people continues to play into the structure of existing theory. Perhaps part of why relationships with land and nature have been less emphasized by race scholars of the twenty-first century as compared to those of the twentieth century is because the overt struggles over land as wealth are no longer as widespread as they were during the "frontier" era when millions of acres at a time were visibly shifting in "ownership." The fact that the end of direct genocide was followed by other mechanisms of Native erasure and dispossession points to the changing ways that nature matters to the larger processes of racial formation in the advance of capitalism, colonialism, and associated environmental degradation. Presumably, the absence of attention to nature and the absence of scholarship on and from Native perspectives are linked. As David Pellow has pointed out, possibly the absence of this discussion of race and nature is due to a fear of conflating the two, of presuming linkages between racial difference and biology in light of the history of eugenics and ongoing narrative in white supremacist scholarship and politics (personal communication, 2018). Regardless of the reasons for its theoretical absence, the species we call nature are as important as ever in structuring the reality and meaning of all people's lived experience of race today.

Now that patterns and meanings of land "ownership" are no longer shifting as rapidly as during the "frontier era," the importance of nature for the material and symbolic aspects of race making manifests in new ways: through differences in bodily toxins and disparities in the quality of foods consumed (e.g., Guthman 2011).

The categories of race forged by Manifest Destiny and frontier genocide continue to structure the socialization of individuals, on one hand, and give meaning to a wide variety of state actions, on the other. These same categories of white and Native and their particular contours persist today, daily organizing land management decisions from forest plan revisions to timber policies. And the importance of nature for racial formation is becoming especially apparent with advancing environmental degradation, as is the importance of race and racial formation in the ways we imagine and exploit land and nature. Now in the face of climate change, the importance of the natural world for social outcomes is gaining some attention within the discipline of sociology.

Climate change may be the most serious ecological problem our world has faced. Climate change also evokes an urgent need to rethink many aspects of Western social, economic, and political systems from the organization of energy around fossil fuels to the sustainability of cultural values of excessive consumption and the relevance of epistemologies that presume a separation of the social and natural worlds. Climate change is fundamentally about race and racism. Racial inequalities and the racialized state have served as a mechanism to displace problems onto Indigenous communities and communities of color. Laura Pulido and coauthors (2016) underscore that "vulnerable communities, in this case communities of color, are essential to the functioning of racial capitalism" (26), which, drawing upon Robinson (2000) and Faber (2008), they define as "a distinct interpretation of capitalism that acknowledges race as a structuring logic.... Racism, as a material and ideological system that produces differential meaning and value, is harnessed by capital in order to exploit the differences that racism creates. In this case, devalued communities, places, and people serve as pollution "sinks," that enable firms to accumulate more surplus than would otherwise be possible" (Pulido et al. 2016). And as David Pellow (2009) notes with respect to racism and the environment,

> The very existence of the modern U.S. nation-state is made possible by the existence of toxins—chemical poisons—that permeate every social institution, human body and the non-human world. To be modern, then, is associated with a degree of manipulation of the human and non-human worlds that puts them both at great

risk. To be modern also appears to require the subjugation and control over certain populations designated as "others," those less than fully deserving of citizenship, as a way of ameliorating the worst impacts of such a system on the privileged. These two tendencies, the manipulation of the human and more-than-human worlds—are linked through the benefits that toxic systems of production produce for the privileged and the imposition of the costs of that process on people and non-human nature deemed less valuable and therefore expendable. (57)

Yet while contamination and resource exhaustion may be out of sight for the more privileged peoples, these problems have not gone away. Instead, the physical displacement and ideological divides that racial capitalism has produced have interfered with potential information feedback loops—Indigenous people have been warning the dominant society about the need for urgent action for human survival for some time, yet these warnings, if they are heard at all, have been dismissed.

The lack of visibility of the natural environment as a space for the enactment of power and the lack of visibility of Native Americans by sociologists are part of this problem. It is my hope that bringing more attention to the importance of the natural world into sociological theories of race will not only enrich this area of theory but also work against Indigenous erasure within sociology and expand interdisciplinary use of the powerful sociological framework of racial formation. For despite the limitations I have noted here, sociological theories of race and power are uniquely important, and there is much to be gained by their cross-disciplinary application.

If the scholars of today, Native and non-Native alike, can acknowledge that traditional ecological knowledge and management have made the ecology of the Klamath what it is today, then it follows that racism and cultural genocide produce further environmental decline. As they gain political and economic standing, Native American people, including the Karuk, have become increasingly visible in their natural resource management. Yet tribes are disadvantaged in these settings due to many reasons, including a lack of a broader social understanding of their unique cultural perspectives, the marginalization of traditional knowledge as opposed to "real" science, and a lack of acknowledgment of the violent history perpetuated against them—much less the continuing effects of this history today.

2 · ECOLOGICAL DYNAMICS OF SETTLER-COLONIALISM
Smokey Bear and Fire Suppression as Colonial Violence

Without fire the landscape changes dramatically. And in that process the traditional foods that we need for a sustainable lifestyle become unavailable after a certain point. So what that does to the tribal community? The reason we are going back to that landscape is no longer there. So the spiritual connection to the landscape is altered significantly. When there is no food, when there is no food for regalia species, that we depend upon for food and fiber, when they aren't around because there is no food for them, then there is no reason to go there. When we don't go back to places that we are used to, accustomed to, part of our lifestyle is curtailed dramatically. So you have health consequences. Your mental aspect of life is severed from the spiritual relationship with the earth, with the Great Creator. So we're not getting the nutrition that we need, we're not getting the exercise that we need, and we're not replenishing the spiritual balance that creates harmony and diversity throughout the landscape. —Ron Reed, Karuk cultural biologist

Fires cause damage to timber and other resources, and, as a corollary, that the object of fire protection is to prevent or reduce these potential losses.
—Show and Kotok (1923, 3)

What is colonialism but separation of people from the land?
—Taiaiake Alfred (2005, 9)

The alteration of the physical, "more than human" world we often call nature not only has facilitated the development of racial categories and the movement of wealth from one group to another as highlighted by racial formation theory in the last chapter, but also has led to a reorganization of the meaning of wealth, the alteration of relationships, human bodies, spiritual energies, and political possibilities. Physical changes in the land have supported and legitimized the emergence of the racial categories of white and Native as described in chapter 1. But the formulations of power taking place through these processes cannot be wholly conceptualized within the framework of race. And although capitalism has emerged in the context of colonialism, the dynamics experienced by Indigenous communities cannot be subsumed by either capitalism or even racial capitalism.[1] Indeed, as Fenelon and Trafzer (2014) emphasize, "Indigenous peoples represent the most complex social analytical issues in the world today, including invasion by foreign groups, outright genocide, culturicide and multiple forms of coercive assimilation, and ranging over half a millennium of modern colonization histories covering the Americas and globally" (3). In contrast to the racial formation framework that emphasizes the movement of resources and shifting power relations between groups within a given society, or Marxist frameworks that attend so adeptly to human exploitation via labor, here the social, economic, ecological, and political systems of one group of people are imposed upon—and to the extent they are successful—supersede those with another, with the goal of elimination and replacement of existing systems and whole societies. One goal of this chapter is to illustrate why we need the framework of settler-colonialism to understand such dynamics.

This chapter builds upon themes from chapter 1 regarding the importance of the natural environment to the operation of social, political, and economic power in North America. I will do this through a detailed examination of the alteration of human-ecological relationships with fire. Settler-colonial theory emphasizes that colonialism is an ongoing process rather than a past static event. Thus far, however, most attention has been paid to the human dimensions of that process. Yet if settler-colonialism reorganizes the material flows of organisms and bodies and alters the spiritual flow of energies that move in and out of human communities, these ecological dynamics are fundamental to social and political power in ongoing, nuanced, and complex ways. Indeed, many would explicitly say the actions of the other beings and entities in the so-called natural world themselves participate as animate actors in the dynamics of settler-Indigenous power (e.g., Todd 2014; Watts 2013). If settler-colonialism is about Indigenous erasure, that erasure involves not only Indigenous peoples but also the Indigenous ecologies within which people exist. I will trace this history of human-fire relationships in the Klamath Basin to highlight what Kyle Whyte calls "the ecological mechanics of settler-colonial domination."

In direct contrast to the notion that North America was a wilderness untouched by humans, Indigenous people in California and across the continent have systematically developed sophisticated methods of using fire to enhance the presence and production of specific plant species, optimize hunting conditions, maintain open travel routes, and generally support the flourishing of the species upon

FIGURE 5. Acorns, huckleberries, and fire. Photo credit: Jenny Stormy Staats.

which they depend (Blackburn and Anderson 1993; Anderson 2005; Hillman and Salter 1997; Lake et al. 2017; Lake 2007; Lewis 1973). Low-intensity, purposefully set fires are also used to provide protection from the fuel buildup that causes larger, hotter (potentially quite dangerous) fires that are currently burning across the West as this book goes to press. Fire is especially important in mountainous places like the mid-Klamath that are prone to significant lightning activity. The mid-Klamath forests and grasslands are considered to be "fire adapted" both due to the presence of frequent lighting strikes and also due to the systematic use of fire by Karuk people over thousands of years (Skinner, Taylor, and Agee 2006).

The practice of burning is also central to Karuk cultural, social, and spiritual practices—what my Karuk colleague Ron Reed calls "indigenous social management." Over three-quarters of the species essential for Karuk traditional food and cultural uses are enhanced by fire (personal communication; Tripp 2013; intergenerational traditional ecological knowledge; Norgaard 2014a, 2014b). Karuk Department of Natural Resources founder and Director Leaf Hillman describes this practice: "Right here in the valley, my grandfather, he burned this whole slope

FIGURE 6. Smokey Bear legitimating fire suppression in Orleans, CA. Photo credit: Jenny Stormy Staats.

that you can see on that side of Orleans, over on the redcap side. He burned that whole slope every three years. He'd burn it in early October and the rains always put it out. Some years the rain came sooner, some years it came later. So some years it went a little further.... But ultimately the rains always put it out." Whereas Indigenous land stewardship has been organized at the local level through tribal and family responsibilities to particular places, as well as guided by tribal knowledge informed by interactions with place, the capitalist-settler state created bureaucratic institutions to manage the land (Scott 1998). These natural resource institutions set comprehensive, often nationwide policies based on ecological principles that were believed to be universal. Fire suppression was mandated by the very first session of the California State Legislature in 1850 during the apex of genocide in the northern part of the state.[2] When the Klamath National Forest was established in 1905 together with the formation of the U.S. Forest Service on the national level, the new agency began a policy of fire suppression in an attempt to protect commercially valuable conifer species from being "wasted" in fires (Show and Kotok 1923).

This "exclusion"[3] of fire from the landscape has resulted in changes in species composition through stand conversions throughout Karuk Aboriginal Territory and homelands. The conversion of former oak woodlands into single-aged conifer-dominant forests has reduced their complexity and diversity, thereby increasing the risk of stand-replacing wildfires and pathogenic disease. Likewise, forest species composition has shifted from acorn-producing tanoaks to commercially valuable Douglas firs (Cocking, Varner, and Sheriff 2012). Former grasslands and meadows that were once prime harvesting areas for "Indian potatoes" have become encroached by conifers through the exclusion of fire and have all but disappeared. The exclusion of fire has created an overabundance of ladder-fuels and dense underbrush, which promote the hotter, larger fires that many people experience as "catastrophic." Today, these "fire excluded" forests are the equivalent of food deserts in the inner city. There has been a shift from what Leaf Hillman describes as "a forest that is productive for people, to a forest that is productive for timber." The exclusion of fire from the landscape has led to a dramatic reduction in the quality and quantity of traditional foods, negatively affected spiritual practices, threatened cultural identity, and infringed upon political sovereignty as Ron Reed describes in the passage that opens this chapter.

On an individual level, the altered forest conditions create social strain for those who hold responsibilities to tend to specific places and to provide food to the community for both subsistence and ceremonial purposes. While these dynamics matter profoundly at the local level, they also represent a new dominant order with major and widespread transformations of the natural world, cosmologies, economies, spiritualties, and human relationships with the land.[4] As described in chapter 1, this new order generated the wealth that has made what would become California one of the wealthiest economies in the world. Fire exclusion, then, has simultaneously produced Indigenous exclusion, erasure, and replacement. Land has been a key part of this process. Quantitatively, a given number of acres owned by the state translates into some equivalent degree of power in the hands of the colonizers. Qualitatively, the composition of species and the qualities of relationships that people are able to carry out with them inform political possibilities. The struggle over these different ecological, cultural, political, and economic systems is the central dynamic of colonialism. Whyte et al. (2018) describe how "'ecologies' are systematic arrangements of humans, nonhuman beings (animals, plants, etc.) and entities (spiritual, inanimate, etc.), and landscapes (climate regions, boreal zones, etc.) that are conceptualized and operate purposefully to facilitate a society's capacity to survive and flourish in a particular landscape and watershed. Waves of settlement seek to incise their own ecologies required for their societies to survive and flourish in the landscapes they seek to occupy permanently" (159). Centering these ecological power dynamics sheds light on a

FIGURE 7. Thick vegetation, lots of fuels. Photo credit: Will Harling.

number of theoretical conversations. My goal in this chapter is twofold. First, I aim to use this case study to illustrate the relevance of the settler-colonialism framework for my own discipline of sociology and the social sciences broadly. Settler-colonial theory emphasizes at least three key dynamics fundamental to Indigenous relationships with the state and the consolidation of state power that are highly relevant for sociological analyses but as yet underdeveloped in sociological theory: the notion that North American colonialism is an ongoing structure rather than a past event, the centrality of land to the operation of both Indigenous and state power, and the structuring of state relationships with Indigenous peoples in terms of elimination and replacement.[5] These dynamics are very different from those emphasized by the dominant theories of the discipline. Settler-colonial studies highlight both the role of the settler state as well as the settlers themselves in this process, who as Steinman puts it "are often the tip of the spear" (Steinman, personal communication), making the contribution of settler-colonial theory an especially relevant addition to the racial formation of both Native and non-Native peoples (see Klopotek 2011; Moreton-Robinson 2015; Glenn 2015; Fenelon 2014). Attending to these dynamics matter not only for better understanding of the experiences of Indigenous peoples today but also fundamentally informs basic tenets of the discipline as the chapters in this book will elaborate. Indeed, just as Roxanne Dunbar-Ortiz emphasizes how, in the course of writing *An Indigenous History of the United States*, she had "come to realize that a new periodization of US history is needed" and "to learn and know this history is both a necessity and a responsibility of the ancestors and descendants of all parties," the same can be said of sociology. And just as Aldon Morris (2015) describes how antiblack racism stifled the work of W.E.B. Du Bois, with the effect of both minimizing Du Bois's theoretical and methodological contributions and orienting the theoretical trajectory of U.S. sociology around white supremacy for decades, the negation of Indigenous experiences and knowledges has profoundly shaped sociological theorizing. Go (2017) writes, "Sociology embeds the metropolitan-imperial standpoint not necessarily because it was in the direct service of empire (though in some cases it was), but because it was formed in the heartland of empire, crafted in its milieu, and was thus embedded in its culture ... confronting inequality and marginalization in sociology cannot only be about achieving demographic diversity within the discipline. It must also involve an epistemic insurgency—an intellectual revolution—or least an epistemic shift. Without such an epistemic shift, sociological knowledge will continue to be limited" (196). Second, by detailing the ecological dynamics of settler-colonialism, I aim to expand attention to the importance of manipulation of the natural world for the development and maintenance of the settler-colonial state, as well as how entities in the natural world contribute to Indigenous resistance. Settler-colonial theory emphasizes that colonialism is a dynamic process ongoing in the present, and the nuanced dynamics of ecological change form a critical part of the complex and shifting terrain of Indigenous-settler power. *Just as colonialism is not a single event of the past, we must think beyond "land theft" and dispossession as single events of the past.* This chapter will illustrate these elements of colonialism as ongoing processes that take place through *the alteration of land, the alteration of species composition and ecological structure,* and *the alteration of relationships between people and the nonhuman entities collectively known as nature.* The interruption of relationships between Karuk people and fire is the means by which the state continues to "facilitate the dispossession of Indigenous peoples of their lands and self-determining authority" (Go 2017, 7) and expedite settler possession of wealth in the form of gold, fish, and timber. Whyte et al. (2018) write that "settler colonialism is a form of oppression in which settler permanently and ecologically inscribe homelands of their own onto Indigenous homelands" (158). This alteration of ecological relationships in turn becomes a mechanism for forced assimilation, the disruption of knowledge and cultural systems, the deterioration of physical and mental health (the central foci of chapters 3 and 5), forced assimilation, and economic and political dispossession. Within each realm, the dynamics of erasure and replacement, the importance of ecology, and the understanding that colonialism is an ongoing structure are all critical for the consolidation of state power and Indigenous resistance.

Following a discussion of the various colonial formations and introduction to settler-colonial theory, this chapter begins with descriptions of Karuk fire regimes and cosmologies, details the arrival of non-Indigenous or Western settlers in the region, describes how fire exclusion has become a vehicle for historical and ongoing colonial dispossession through altered ecology, and highlights the creative tribal resistance that takes place through political and cultural struggles over relationships with fire. While some settlers recognized the value of Indigenous fire management practices, federal forest managers began the policy of fire suppression as part of the alteration of the forest to commoditize timber. If fire is medicine in Karuk culture, fire suppression has been the embodiment of colonial violence on the Klamath today.

SITUATING THE NATURE OF POWER: RACE, CAPITALISM, INTERNAL COLONIALISM, POSTCOLONIALISM, EMPIRE, SETTLER-COLONIALISM

> Settler colonialism and its decolonization implicates and unsettles everyone. —Tuck and Yang (2012, 7)

Discussions about the relationships between race, colonialism, and empire have been exploding across the social sciences. This invigoration of perspective is a direct result of a blossoming of work in Native studies. Sociologists face somewhat of a theoretical void when it comes to conceptualizing North American colonialism. Sociologists developed the theory of internal colonialism (Carmichael and Hamilton 1967; Blauner 1972; Snipp 1986) and may use postcolonial frameworks to theorize dynamics in the global South (Go 2016, 2017; Magubane 2013; Steinmetz 2013a, 2013b, 2014). Compared to the application of racial theories or theories of social class, however, neither approach has received much scholarly activity by U.S. sociologists. Important sociologists from Duane Champagne, Wilma Dunaway, James Fenelon, Michelle Jacob, Joane Nagel, Matthew Snipp, Rima Wilkes, Thomas Hall, and Erich Steinman, to list a few, have made critical contributions to sociological understanding of Indigenous experience, yet there is no American Sociological Association (ASA) section on either Indigenous peoples or colonialism, and a word search for colonialism within sociological indices yields few entries for happenings in North America. Aside from a few lone voices, until recently, the limited work engaging Indigenous experiences in U.S. sociology has most frequently subsumed Native experience into the framework of race relations or discussed these dynamics in terms of internal colonialism.

Work on the concept of internal colonialism (Carmichael and Hamilton 1967; Blauner 1972) has been important for highlighting dimensions of coercion and wealth extraction from Indigenous communities, but as Omi and Winant (2014) and others note, theories of internal colonialism are nonetheless insufficient in that they blend the experiences of multiple racial groups, fail to theorize social and ecological relationships at large, and deploy the term *colonialism* as a metaphor rather than literal state structure (see Byrd 2011; Veracini 2011). Furthermore, while the internal colonialism framework effectively highlights the economic aspect of exploitation, it is less useful for understanding the political, social, cultural, or spiritual dimension of the operation of power. Nor does internal colonialism account for the dimension of relationality with the natural world that Native studies scholars such as Taiaiake Alfred, Richard Alteo, Eve Tuck, Kyle Whyte, Leanne Simpson, Glen Coulthard, and others put as central or provide a way to understand that colonialism is about subsuming and erasing existing epistemological and cosmological systems. Last, the internal colonialism frame emphasizes the economic dimension of exploitation within reservation communities as internal colonies, but how did Native people end up in only these physical spaces? The framework lacks a way to theorize intersocietal interactions in that it fails to account for the forced movement of Indigenous peoples from the rest of the continent or for the actions of the state and settlers in those spaces no longer acknowledged as Indigenous lands. Postcolonial studies has a broader spatial application of colonial processes, insists that critical analyses must start from the standpoint of the colonized, and more fully engages the cultural and political operation of power but is rarely applied to describe dynamics in North America, which, in the case of the United States, are instead analyzed through the frame of internal colonialism and empire. The very term *post* signals its orientation toward the ongoing challenges in societies where formal colonialism has ended, rather than dynamics in places like North America where the colonial occupying state remains physically present. Whether by happenstance or implicit bias, the failure of postcolonial studies to engage the ongoing dispossession described in these pages ironically colludes with colonial logics of erasure.[6]

Alongside this exciting resurgence of postcolonial theory within U.S. sociology, a number of sociologists have recently been advancing the notion of the United States as an empire (e.g., Steinmetz 2013; Adams and Steinmetz 2015; Go 2013, 2017; Bacon and Norton 2019). The framework of empire characterizes the malignant nature of state power, making visible relations between militarization at home and abroad. Recent works by Gurminder Bhambra, Julian Go, George Steinmez, Zine Magubane, and others provide critical disciplinary critiques of sociology, even advancing the notion that sociology is itself a product of empire. Zine Magubane (2013, 2014) traces the absence of attention to empire and colonialism by U.S. sociologists to the influential impact of Robert Park at the University of Chicago in the 1920s, detailing how various features of Robert

Park's personal biography, including his relationship with Booker T. Washington, structured his erasure of colonialism and imperialism from histories of the discipline. Magubane (2014) underscores the impact of these events into the present, noting that

> the inability of mainstream sociological approaches to adequately capture and account for the complexities of colonial history is not simply a function of inadequate methods. Nor is it due solely to sociologists' inability to use our existing methods well enough. Neither does it stem simply from our collective failure to ask the right questions—although all of these play a role. Rather, colonialism is so difficult for sociologists to account for because the idea of society itself—the concept that gave academic sociology both its focus and its identity—is predicated on suppressing, erasing, and evading the active role American sociology played in the colonial encounter. (579)

Go (2015, 2016) elaborates how a series of analytic bifurcations emerging from sociology's metropolitan-imperial standpoint constrain sociological theory and empirical approach. Drawing on work by Patricia Owens (2015) and George Steinmetz (2013b), he describes how "the very notion of the 'social'—as a space between nature and the spiritual realm—first emerged and resonated in the nineteenth century among European male elites to make sense of and to try to manage social upheaval and resistance from workers, women, and from so-called natives" (195). Go aptly traces the development and evolution of such ideas to their current manifestations, noting how "over the course of the twentieth century in the United States, sociology became increasingly tied to the worldviews and interests of white male elites in the new emerging American empire. Even seemingly innocuous concepts and theories like Park's race relations cycle and assimilation theory bear the imprint of this imperial standpoint" (195). While recent descriptions of the United States as an empire are congruent in many ways with the framework of settler-colonialism (e.g., analyses of economic and military force), Indigenous studies scholars, including Potawatomi philosopher Kyle Whyte, stress that theories of settler-colonialism include emphases on cultural dynamics of erasure that are not conceptualized in the empire frame. Australian political theorist Patrick Wolfe is often credited as a founder of this framework. In his formative articulation of the need for specifying "settler-colonialism" as a specific configuration of colonialism, Wolfe (2006) emphasizes that because of the permanent aspect of settlement in places such as North America, colonialism becomes a *structure of the new society* rather than a *single or series of past events* as it has more commonly been understood: "When invasion is recognized as a structure rather than an event, its history does not stop . . . when it moves on from the era of frontier homicide. Rather, narrating that history involves charting . . . [how] a logic that initially informed frontier killing transmutes into different modalities, discourses and institutional formations as it undergirds the historical development and complexification of settler society" (402). It is this notion that *colonialism continues to structure society at large* that makes it so important for sociologists to engage. Just as Morris (2015) brought attention to the racist context within which U.S. sociology developed, just as Steinmetz, Go, and others underscore the importance of imperialism and British colonialism for the founding categories and structures of sociological thought, it matters that U.S. sociology has been and continues to be imagined and developed in the wake of unacknowledged Indigenous genocide, from a standpoint of a nearly silent occupation. It matters that nearly all U.S. sociologists craft our theory within that space of colonial amnesia and willful ignorance. Essential concerns of sociological theory from assimilation to immigration or the nature of state power look very different if theorized through a lens of ongoing North American colonialism. Imperialism in the U.S. context, for example, is quite different from British traditions of empire: "U.S. imperialism is fueled by imaginaries of American dominance, American cultural hegemony, American freedom to trade or embargo any place in the world, American democracy, etc. There is an entire set of imperial desires and imaginaries. All of these are only possible if there is a uniquely American sense of entitlement to the land and a moral myth that land dispossession was not treacherous or was inevitable. The flow of goods out of the U.S. and into the U.S. will always require the capacity to continue to dispossess Indigenous peoples of their lands" (Whyte, personal communication). This is not to say that either postcolonial theories or theories of empire are intrinsically incompatible with settler-colonialism or Indigenous studies—quite the opposite, both frameworks are critical tools for conceptualizing colonial power. Rather, these areas of scholarship have yet to engage the ongoing presence and critical perspectives of Indigenous peoples. And because invisibility and erasure are such central mechanisms of settler-colonial power, this absence matters. In other words, just as empires and settlers themselves presumed the North American continent as *terra nullis*, so too have critiques of empire thus far failed to account for the presence of Indigenous peoples in North America. *Used on their own, existing fields of postcolonial studies and empire thus become complicit in settler-colonial erasure.* These differences in the conceptualization of the power at play become particularly apparent when we examine how each form of colonialism might be undone. Decolonization with respect to imperialism, postcolonial, and settler-colonial contexts looks quite different, as Tuck and Wang (2012) observe: "Decolonization in exploitative colonial situations could involve the seizing of imperial wealth by the postcolonial subject. In settler colonial situations, seizing imperial wealth is inextricably tied to settlement and re-invasion. Likewise, the promise of integration and civil rights is predicated on securing a share of a settler-appropriated wealth (as well as expropriated 'third-world' wealth). Decolonization in a settler context is fraught because empire, settlement, and internal colony have no spatial separation. Each of these features of settler colonialism in the US context—empire, settlement, and internal colony—make it a site of contradictory decolonial desires" (7).

Or, as Kyle Whyte puts it,

> Given how settler colonialism works, we know that solutions to exploitation or racism do not work for Indigenous peoples, and many other groups too. We know that an end to U.S. imperialism would not work. For an end to U.S. imperialism might even lead U.S. settlers to double down on the extraction of resources and the economic losses would cut Tribal programs . . . the desire to be permanent in the U.S. fuels the imaginaries, ideologies, and legal freedoms that depend on Indigenous land dispossession in the past and continuing and depend on a certain shape of the ecosystems to be able to make certain businesses and settlements more viable—they don't have to keep clearing and re-clearing the land as dramatically as they would have had during Indian occupancy. (Personal communication)

Thus, while there are many compatibilities between these frameworks, the standpoints of Indigenous peoples under settler-colonialism point to different theoretical analyses and emancipatory desires than for those of peoples under overseas external forms of colonialism or for whom explicit/formal colonization has ended.

As I detail in chapter 1, conquest and frontier genocide in North America took place in the context of global capitalist expansion. Naming capitalist social relations is therefore essential for any description of Indigenous-state power relations. A number of sociologists have usefully applied Marxist and world systems theory to theorize dynamics within Indigenous communities. Leanne Simpson (2017) emphasizes the importance of specifying capitalism: "I can't see or think of a system that is more counter to Nishnaabeg thought that capitalism, and over the past two decades I have heard elders and land users from many different Indigenous nations reiterate this, and it is part of our elder's analysis and thinking we ignore" (77). Like Fenelon, Dunbar-Ortiz, and other Indigenous theorists, Simpson underscores the importance of drawing from Indigenous perspectives for capitalism's critique:

> Indigenous peoples in my mind have more experience in anti-capitalism and how that systems works than any other people on the planet. We have thousands and thousands of years of experience building and living in societies outside of global capitalism. We have hundreds of years of direct experience with the absolute destruction of capitalism. We have seen its apocalyptic devastation on our lands and plant and animal relations. This in no way diminishes the contributions of other anticapitalism theories, thinkers and writers; rather I think it adds the beginnings of a critical reframing of the critique, one that is centered within grounded normativity. (73)

While essential, Marxist critiques of capitalism are also different from those emerging from Indigenous peoples. Sandy Grande (2004) observes, for example, "both Marxists and capitalists view land and natural resources as commodities to be exploited, in the first instance, by capitalists for personal gain, and in the second by Marxists for the good of all" (27). On this same point of the degree of objectification of the natural world present in non-Native anticapitalist theorizing, Leanne Simpson (2017) underscores,

> Capital in our reality isn't capital. We have no such thing as capital. We have relatives. We have clans. . . . My ancestors didn't accumulate capital, they accumulated networks of meaningful, deep, fluid, intimate collective and individual relationships of trust. In times of hardship we did not rely to any great degree on accumulated capital or individualism but on the strength of our relationships with others. . . . Resources and capital, in fact, are fundamental mistakes within Nishnaabeg thought, as Glenna Beaucage points out, and ones that come with serious consequences—not in the colonial superstitious way, but in the way we have already seen: the collapse of local ecosystems, the loss of prairies and wild rice, the loss of salmon, eels, caribou, the loss of our weather. (77)

Later she notes, "We didn't just control our means of production, we lived embedded in a network of humans and nonhumans that were made up of only producers" (80).

If we take a peek beyond the bounds of our discipline, sociologists can see how other disciplines have grappled with conceptualizing the dynamics of the U.S. nation-state and Indigenous experiences today. In the past ten years, the recent and rapidly developing theory of settler-colonialism in particular has sparked much scholarly activity and even a new journal devoted to the topic. Key Native studies scholars, including Taiaiake Alfred, Richard Alteo, Glenn Coulthard, Sarah Deer, Leanne Simpson, Jennifer Nez Denetdale, Eve Tuck, Mishuana Goeman, Audra Simpson, Andrea Smith, Rebecca Tsosie, and Kyle Powys Whyte, highlight the ways that colonial state power is generated via "genocidal practices of forced exclusion and assimilation" of Indigenous peoples. In their landmark piece, Tuck and Yang (2012) emphasize that "settler colonialism is different from other forms of colonialism in that settlers come with the intention of making a new home on the land, a homemaking that insists on settler sovereignty over all things in their new domain. Thus, relying solely on postcolonial literatures or theories of coloniality that ignore settler colonialism will not help to envision the shape that decolonization must take in settler colonial contexts" (5). Wolfe (2006) emphasizes that because in places like Australia, the United States, and Canada, settlers came to stay, legitimation of their claims to land required the *elimination* of Indigenous people (rather than "merely" their enslavement or other means of exploiting their labor). Whyte notes, "One thing settler colonial theory is trying to show is that one of the best ways to understand Indigenous oppression is as a structure of erasure and replacement. Indigenous peoples do not always need to be eliminated directly in the sense of death; settler states work to transform the ecology, laws, policies, mythology and education to make settlers feel as though they are 'Indigenous,' that is, legitimately, rightful to the land and without having any moral compunction against the obvious fact that their lives are only possible via Indigenous dispossession" (personal communication). The fact that settlement is permanent changes the character of this form of colonial relations, centering them on acquisition of land for settlement in rather than labor or extractable resources as with external colonialism. Indeed, the emphasis on the role of land is at the heart of the contribution of the settler-colonial framework, making it particularly well suited to theorizing the relevance of the so-called natural world on social action at large—a running theme throughout this book. Tuck and Yang (2012) describe how, "within settler colonialism, the most important concern is land/water/air/subterranean earth (land, for shorthand, in this article). Land is what is most valuable, contested, required. This is both because the settlers make Indigenous land their new home and source of capital, and also because the disruption of Indigenous relationships to land represents a profound epistemic, ontological, cosmological violence. This violence is not temporally contained in the arrival of the settler but is reasserted each day of occupation" (5). Coulthard (2014, 4) underscores how the "modus operandi of colonial power" involves "attempts to uproot and forcibly destroy the vitality and autonomy of Indigenous modes of life through institutions such as residential schools; through the imposition of settler state policies aimed at explicitly undercutting Indigenous political economies and relations to and with the land." Coulthard describes "a settler colonial relationship is one that is characterized by a particular form of domination; that is, it is a relationship where power—in this case, interrelated discursive and non-discursive facets of economic, gendered, racial and state power—has been structured into a relatively secure or sedimented set of hierarchical social relations that continue to *facilitate the dispossession of Indigenous peoples of their lands and self-determining authority*" (6–7, emphasis added). Yet we are in an exciting time. Sociological theorizing on colonialism and from Indigenous perspectives is on the rise. Longstanding contributions from James Fenelon are gaining more attention, and recent work by Michelle Jacob, Julia Cantzler, Erich Steinman, Rima Wilkes, J. M. Bacon, Matthew Norton, Vanessa Watts, Elizabeth Hoover, and others is contributing to the more frequent (although still minimal) use of the term *settler-colonialism* within U.S. sociology and beyond. As I emphasized with respect to Marxist critiques of capitalism, the framework of settler-colonialism does not replace postcolonial theory (after all, the framework of settler-colonialism too is subject to a host of important critiques). Indeed, many of the dynamics long critiqued by postcolonial theorists and other critics of modernity, including processes of knowledge production or the fallacy of binaries, play out vividly in settler-colonialism. Rather, postcolonial and settler-colonial theories emerge from unique political and historical standpoints. Indeed, capitalism, empire, and settler-colonialism operate together. Kyle Whyte notes,

> U.S. imperialism is supported by the sheer amount of control over such a vast array of resources, and the cheap capacity to access them.... Imperialistic economic and military systems feed the power of the settler society to continue in its dispossession of Indigenous lands and erasure of Indigenous histories. The wealth, access to expertise, immigration and military capacity bolstered by imperialism strengthen the grip of settler society against Indigenous resistance. Many extractive industries that support settlement are only profitable due to the U.S. capacity to secure markets and capital globally, for example. At the same time, settler colonialism shores up imperial hegemony, such as how the U.S. often offered reasons of national security as an excuse for dam projects that flooded Indigenous lands or energy independence as an excuse for continued extraction on Indigenous lands. (Personal communication)

Whyte notes how empire and settler-colonialism uphold one another discursively as well because "imaginaries of U.S. moral superiority and democracy that fuel U.S. imperialistic arrogance and paternalism and the erasure of U.S. treachery in the global imagination shroud the reality that the U.S. involvement with other countries is that of a genocidal state imposing itself using different means on another country" (personal communication). Nor does the lens of settler-colonialism replace the importance of race and racism. Rather, understanding settler-colonial relations can bring nuance to existing sociological approaches to race and racism. On this point, Tuck and Yang (2012) underscore how, "in this set of settler colonial relations, colonial subjects who are displaced by external colonialism, as well as racialized and minoritized by internal colonialism, still occupy and settle stolen Indigenous land. Settlers are diverse, not just of white European descent, and include people of color, even from other colonial contexts. This tightly wound set of conditions and racialized, globalized relations exponentially complicates what is meant by decolonization, and by solidarity, against settler colonial forces" (7). As Evelyn Nakano Glenn (2015) puts it, "The logic, tenets, and identities engendered by settler colonialism persist and continue to shape race, gender, class, and sexual formations into the present" (57). Thus, it is the context of colonialism with its dynamics of erasure and replacement that the *content* of the racial category "Indian" has been constructed as less than human, close to nature, static, and an artifact of the past as a means of justifying their elimination via the practice of genocide as discussed in chapter 1 (e.g., white supremacy is part of both racism and settler-colonialism). Similarly, constructions of the racial category "white" and the ideology of white supremacy emphasize qualities that legitimate entitlement to land, as detailed in chapter 1. Indeed, racism continues to matter, for as Fenelon notes, "The dominant society is still operating in terms of race" (2017, personal communication). Brian Klopotek (2011) describes how,

> Legally, politically and morally, indigeneity and nationhood set Native Americans apart from other racialized minorities. Tribal sovereignty has been paramount in Native American studies because it protects Native peoples from dissolving into the political mass of the United States—a central concern of indigenous groups living under Anglo-American racism and colonialism.... Native peoples have both a racial status and a political status in the United States; the many shared experiences of racial discrimination with other communities of color attest to the racial status of Indians, just as the many aspects of Indian experiences that are unique to indigenous groups justify a distinction. (7–8)

Yet as Tuck and Wang note, integration into the settler nation that draws its wealth from Indigenous communities and Indigenous ecologies cannot be seen as a positive path for resolution.

As I write, settler-colonial theory is dynamic, rapidly developing, contested, and the subject of much scholarly attention. James Fenelon, Hayden King, and others call for greater nuance in the use of concept. Nor is sociology the only social science discipline reticent to engage the implications of this framework. Alyosha Goldstein (2014) observes how "a critical analytic lens that takes into account the significance of colonialism for the various ways in which the geopolitical configuration of the United States has changed over time remains largely absent" (1). And in her recent progress report for the discipline, geographer Laura Pulido (2017) notes that geography has "engaged settler colonial theory primarily only in relation to whiteness, and Chicana/o studies has yet to effectively grapple with it" (2). Elizabeth Hoover (2017) challenges political ecology to draw upon Indigenous knowledges and frameworks to better theorize notions of health, food, political power, and justice.

My illustration of the value of colonialism (and settler-colonialism in particular) for sociological understandings of social relations and the consolidation of state power begins with descriptions of Karuk relationships with fire and then details how fire exclusion has become a vehicle for historical and ongoing colonial dispossession through altered ecology. Tuck and Yang emphasize, "In the process of settler colonialism, land is remade into property and human relationships to land are restricted to the relationship of the owner to his property. Epistemological, ontological, and cosmological relationships to land are interred, indeed made pre-modern and backward. Made savage" (5). Yet while much attention has been rightfully paid to more "social" aspects, details of the operation of settler-colonial violence vis-à-vis the continued alteration of human-ecological relationships are less well detailed. And while attention has been paid to erasure of Indigenous peoples, there has been less focus on the importance of, reasons for, or consequences of the erasure of Indigenous ecologies. These ecological dynamics of settler-colonialism are complex and nuanced, involving many different entities in particular settings in the natural world from the movement of wind and fire to the growth patterns of trees, the characteristics of Madrone bark, the population dynamics of acorn weevils, the length of time lamprey amniocytes burrow in river substrate, or the ability of Pacific giant salamanders to find shelter from wildfire.

Indeed, the notion that nature is separate from culture, and the nearly universal sociological premise that the experiences of "modern" peoples and societies can be understood without accounting for the beings and material processes known as "nature," is a direct product of the colonial worldview. Settler-colonial states have aimed to erase not only the presence of Indigenous peoples but also the ecologies with which they are embedded and, indeed, the relevance of ecology itself. And the near-hegemonic success this discourse has achieved has allowed settler-colonial states to erase their footsteps as they go, so to speak. Yet as Coulthard (2014) writes, "I believe that reestablishing the colonial relation of dispossession as a co-foundational feature of our understanding of and critical engagement with capitalism opens up the possibility of developing a more ecologically attentive critique of colonial-capitalist accumulation, especially if this engagement takes its cues from the grounded normativity of Indigenous modalities of pace-based resistance and criticism" (14). I hope this next section will be in the service of such an effort.

FIRE AS MEDICINE

> We are closely related to fire. Fire takes care of us and we take care of fire.
> —Leaf Hillman

Despite pervasive divides between nature and culture within European descended thought generally and the discipline of sociology in particular, human societies have coevolved with the other entitles we call "nature"—mutually influencing one another. On the Klamath, people have shaped the ecology and species composition of Karuk territorial landscapes since time immemorial through tribal traditional management and, since 1905, increasingly through the activities of the U.S. Forest Service (Crawford et al. 2015). Rather than a static or inert system as presumed in Western notions of "wilderness" (Anderson 2005; Blackburn and Anderson 1993; Cronon 1983), the natural world is dynamic and mutable, and humans have long been key dimensions of that dynamism. Around the globe, fire has been an especially important tool for the large-scale human alteration of species and landscape conditions (Gowlett 2016; Pyne 2016).[7]

The Karuk Draft Eco-Cultural Resource Management Plan (Karuk Tribe 2010) notes, "Fire caused by natural and human ignitions affects the distribution, abundance, composition, structure and morphology of trees, shrubs, forbs, and grasses" (4). Skinner et al. (2006) write that "Native people of the Klamath Mountains used fire in many ways: (1) to promote production of plants for food (e.g., acorns, berries, roots) and fiber (e.g., basket materials); (2) for ceremonial purposes; and (3) to improve hunting conditions" (176). U.S. Forest Service ecologist and Karuk descendant Dr. Frank Lake has oriented his career around bringing visibility to the sophistication and importance of traditional ecological knowledge with respect to the use of fire. Lake (2013) notes that while

> many fire and paleo-climate scientists are not accustomed to utilizing tribal traditional ecological knowledge (TEK) derived from consultation with contemporary tribal/indigenous people about climate and fire regimes or to using oral history or ethnographic information on these topics. A review of earlier ethnographic work and more recent oral history interviews of tribal elders conducted by the Karuk Tribe and myself suggest that tribal TEK encompasses a core area of knowledge about discrete fire events that contribute to landscape fire regimes. Tribal knowledge of fire ecology is closely coupled with subsistence economies, ceremonial practices, and individual or family adaptive strategies. Tribal TEK may also be able to describe how climate and weather influence fire behavior, from the yearly to decadal scale, with generalized understanding of century scale climate and fire regime changes. (4)

Ecologist Kat Anderson (2005) describes the ecological benefits of Indigenous burning practices across California:

> When Indian women or men set hillsides on fire, they not only spurred the growth of young sprouts from shrubs and trees but also opened up areas to increased sunlight, heightened the structural complexity of forest, woodland, and shrubland habitats, stimulated the seed germination rates of seral and serotinous species, recycled nutrients for the whole community, altered insect populations, and promoted increased biodiversity. Periodic burning encouraged native annuals, grasses, and herbaceous perennials to grow under shrubs and trees, creating a healthy understory that enhanced the permeability of the soil surface, checked surface erosion, increased rates of nutrient cycling, enhanced soil fertility, and provided food and habitat for animal species, thus increasing biodiversity and the possibility of mutualistic community interactions. (238)

Fire is especially important for maintaining open grasslands for elk; managing for key food sources, including tan and black oak acorns; maintaining quality basketry materials; and producing smoke that can shade and thereby cool the river for fish. Karuk fire regimes generate what ecologists call "pyrodiversity" on the landscape by extending the season of burn and shortening fire return intervals (Hankins 2005; Lake, personal communication). Karuk knowledge and practice to this end reflect a sophisticated non-Western ecological science. The multitude of foods, materials, and other products that come from Karuk environments are in turn evidence of the profound diversity of fire regimes that are required to maintain relationships with hundreds of animal, plant, aquatic, and mushroom species (Lake 2007, 2013; Anderson and Lake 2013). As Karuk Director of Natural Resources Leaf Hillman puts it, "Fire is a cultural resource."

Although the impacts of fire on the ecology of forest species are most immediately apparent, burning also affects the amount of water that moves into creeks and rivers, in turn providing profound benefits to key riverine species like salmonids. The Karuk Draft Eco-Cultural Management Plan outlines how "certain trees and shrubs utilize water more than others, fire affects this relationship" (Fites-Kaufman, Bradley, and Merrill 2006). Forest, shrub, and grassland distribution affects the process of infiltration from precipitation and resultant levels of evaporation with how those plants utilize water (DeBano et al. 1998). The balance of water in and water out, leading to the amount of moisture in the soil and the quantity and quality of springs is influenced by fire" (Biswell 1999, 157). Karuk fisheries technician and traditional practitioner Kenneth Brink describes this relationship: "We did our fire management, which enabled to put more water into the tribs [tributaries], say like on a drought year, you take all your understory out, like all these blackberries and stuff would never be here. These alders would not be all big. There might be one or two big ones making a shade instead on all these little suckers. I mean, you didn't see the alder, and didn't see willow trees, you saw willow brush. All this foliage takes up a lot of water." From an ecological standpoint, the use of fire has benefits on multiple scales ranging from landscape-level impacts to enhancing the conditions for specific species (Anderson 2005). Fire-adapted forests burned in smaller overall areas in mosaic patterns that contained patches of high-intensity fire (Mohr, Whitlock, and Skinner 2000; Skinner et al. 2006; Perry et al. 2011). Lake and Long (2014) note, "Traditional burning practices served as a disturbance that not only maintained desired growth forms of individual plants, but also promoted desired plant communities across broader scales" (179).

At the level of individual species, cultural burning enhances the growth and productivity of key food sources from huckleberries and mushrooms to the tanoaks that once provided nearly 50% of the protein and calories in Karuk diets. For example, Anderson (2005) describes the importance of burning to release phosphorus and encourage nitrogen-fixing plants. She notes the multidimensional benefits of burning, including:

> Burning opened up areas to increased sunlight, allowing shade-intolerant herbaceous plants—some of which fire ecologists dub "fire-followers"—to come to life from hidden seed banks and quiescent bulbs. Sun-loving plants such as lilies, brodiaeas, soaproot, and wild onions appeared and attracted numerous wildlife spe-

cies such as deer, bears, and gophers. The light fires characteristic of the indigenous style of burning increased the structural complexity of communities in two dimensions. Vertically, they increased the variety of plant physiognomies, helping to establish layers of herbs, shrubs, and trees at different, distinct heights. Horizontally, they increased the patchiness of the community, ensuring greater heterogeneity of leaf cover and species composition. For certain species, these fires also functioned to maintain a greater variety of age and size classes of individuals. (238–239)

Unlike widespread European conceptions of fire as "bad," fire is an essential component of a healthy Klamath ecosystem, as well as Karuk cultural and spiritual practices. Fire is medicine for the natural world, which *includes* people. Human social, cultural, physical, spiritual, and economic relationships are intimately interwoven with Karuk use of fire as Ron Reed's quote in the chapter opening illustrates. As people alter the landscape with fire, it not only enhances particular food and cultural use species but also connects people to their place on spiritual and psychological levels. The hard work of moving about in the forest is good for the body and mind (e.g., promotes mental and physical health). And the act of using fire promotes individual and group identity, as well as political sovereignty. As Frank Lake notes, "Burning is a spiritual obligation and becomes also an act of political defiance in the context of governmental oppression and regulation of such retained cultural rights."

Key to Karuk cosmology and values is the notion of human responsibility to an animate nature. Leaf Hillman references the Karuk Creation Story as he describes the intimate and serious social obligations Karuk people have to other species:

> At the beginning of time, only the spirit people roamed the earth. At the time of the great transformation, some of these spirit people were transformed into trees, birds, animals, fishes, rocks, fire and air—the sun, the moon, the stars. . . . And some of these Spirit People were transformed into human beings. From that day forward, Karuk People have continually recognized all of these spirit people as our relatives, our close relations. From this flows our responsibility to care for, cherish and honor this bond, and to always remember that this relationship is a reciprocal one: it is a sacred covenant. Our religion, our management practices, and our day-to-day subsistence activities are inseparable. They are interrelated and a part of us. We, Karuk, cannot be separated from this place, from the natural world or nature . . . we are a part of nature and nature is a part of us. We are closely related.

Use of fire is central to fulfilling these obligations. This worldview in which humans are related to species in the natural world has been referred to as "kincentricity" in the academic literature (Martinez 1995; Salmón 2000; Senos et al. 2006). At the same time, these activities of traditional stewardship such as burning the landscape form the centerpiece of living traditional ecological knowledge and cultural systems—what Frank Lake calls "pyro-kincentricity." Family interactions in the natural world are central for transferring traditional knowledge of how to care for the land through appropriate harvesting, fishing techniques, and cultural knowledge concerning the use of fire across generations. Kenneth Brink reflects on how these relationships are explicit sites of moral teachings and how this cycle was taught to him: "It might have been through picking mushrooms, it may have been through gathering acorns. It's certain, family values and family adventures like that you know. They get passed down from generation and family to family and on down the line."

Social ties between grandparents, aunts, uncles, and siblings are the avenue for the passage of not only specific forms of traditional Karuk knowledge but also values and identity. As Ron Reed put it, "That fish is a little bit more than a fish, that acorn is a little bit more than an acorn. It's all about our ceremonies and our worldview." Thus, human relationships are set in and sustained by an intimate ecological context.[8] As Whyte (2013a) writes, relationships across species "may be integral to the maintaining of multiple family, social and political relationships within the community; some species may even be the basis of clans and other important social groupings" (3). In the Karuk culture, villages—and the families who inhabit them—had ownership of certain resources in specific geographic locations (e.g., a particularly verdant sugar pine tree grove) over which they could not only exercise socialization-furthering management and harvesting practices but also exert their ownership status to build ties to other villages or families by sharing in these harvest gatherings or trading activities. "It's these types of events that are missing nowadays. People used to get together and work for days and nights at a time—women teaching girls to free the nuts from the cones through the use of fire, and men teaching boys to climb the trees and hook the cones. A good many inter- and intra-family relationships were built this way, promoting strong family bonds and inter-generational transfer of knowledge" (Lisa Hillman, personal communication).

Without direct access to Native food, fiber, and medicinal plant resources, strong social and cultural bonds are difficult to develop and/or maintain. Frank Lake describes how he misses the time with his family that he is losing as a direct result of the declining salmon run: "Fish is the essential thing that kind of bonds my relationship with my family. I know it's a big part of it. And, it's like, 'oh well fish aren't running so don't bother coming to the Klamath'. . . . I miss that contact with my family and my extended family around fish—I end up buying pork or meat at the store rather than spending the time it takes to collect the fish, be there with the elders." Whyte (2018b) describes how "spiritual relationships with nonhumans, the cultivation of places as sacred (or not), and social rules that commit people to help one another and repair fraught relationships motivate us to see ourselves as bound to a 'covenant of reciprocity'" (140). Ecological conditions thus underlie moral systems, and "moral qualities of responsibility facilitate resilience" (Whyte 2018b, 140). Lake (2007) explains how economic, political, spiritual, and health outcomes are interwoven with particular qualities in the landscape:

> Because many northwestern Californian tribes were dependent on fire-induced landscape level changes to adequately maintain cultural subsistence economies and religious functions, exclusion and suppression of burning shifted their dependence to other natural resources and income/capital sources, including government economies and (less nutritious) food support programs (Huntsinger and McCaffrey 1995, Stercho 2006). Effective fire suppression through both physical infrastructure and public campaigns, including fines, imprisonment, and threat of injury or death for arson convictions, facilitated the majority of changes in vegetation quality, which in turn has degraded formerly fire-induced conditions culturally valued by tribal people. (286)

Cultivating, harvesting, processing, preserving, and consuming traditional food and medicine are central for Karuk socialization, community ties, and religious practice. As Ron Reed explains, consuming traditional foods and participating in management activities are at the heart of Karuk culture and "being Indian": "You can give me all the acorns in the world, you can get me all the fish in the world, you can get me everything for me to be an Indian, but it will not be the same unless I'm going out and processing, going out and harvesting, gathering myself. I think that really needs to be put out in mainstream society, that it's not just a matter of what you eat. It's about the intricate values that are involved in harvesting these resources, how we manage for these resources and when." Activities in the environment from hunting, fishing, harvesting, gathering, burning, and more are the means through which families operate as sites of cultural reproduction (Willette et al. 2016). They are also a means of passing cultural role models or imparting what it means to be "male" or "female": "We've gone now from working together to dig potatoes—as females, you see, since that is really the domain of women, to driving alone to the super market on the coast. Indian potatoes used to be a large part of our diet, and we have several origin stories that feature them which also teach us how to carry out our role as women. It's just not the same" (Lisa Hillman, personal communication). As Rabbit, a traditional dipnet fisherman, explains, "You got to have fish to teach them how to fish. You got to have fish to teach them about fish. It'd be hard, you know to tell your son how to dip when there's nothing in there to dip." Native American genocide has been a simultaneous assault not only on people or the land but also on the spiritual order that nourished and governed an entire field of sentient beings and ecological relationships. It is for these reasons that some have used the term *ecocide*,[9] and Bacon (2018) coined the term *colonial ecological violence*.

SMOKEY BEAR AND FIRE SUPPRESSION AS COLONIAL VIOLENCE

In contrast to this integrated system of intimate relationships organized around fire, the European settlers who came to the Klamath region at the turn of the past century feared fire and set up land management policies to suppress it—a worldview epitomized today by the iconic character of Smokey Bear. Ecological changes and their scientific rationales became the means to perform Indigenous erasure and replacement, and they continue to serve as ongoing vectors of colonialism. Stephens and Sugihara (2006) describe these changes in their important work, *Fire in California's Ecosystems*:

> Since European explorers first touched the shores of California, their activities, shaped by their needs and values, have changed the state's fire regimes ... there was often a profound change on ecosystems in the areas where they [Native Americans] actively managed with fire, including many oak woodlands, montane meadows, coastal grasslands, and coniferous forests. These ecosystems now supported a different burning pattern, replacing the specific pattern of Native American ignitions and lightning with a new combination of settler burning and lightning. Coastal areas experience little lightning, and fire regimes in these areas were dominated by anthropogenic ignitions (Keeley 2005, Stephens and Fry 2005, Stephens and Libby 2006). *Removal of anthropogenic fire from these ecosystems has brought about wholesale changes in species composition, by encroachment of invasive species, conversion to other vegetation types, and increased fire hazards.* (431–432, emphasis added)

These changes in the landscape did not "just happen"; they took a major effort. Indeed, the exclusion of fire was so important to the settler-colonial project that it was established by the first meeting of the California State Legislature in 1850. Johnston-Dodds (2002) outlines how Section 10 of the "Act for the Government and Protection of Indians" specified that "any person was subject to fine or punishment if they set the prairie on fire, or refused 'to use proper exertions to extinguish the fire'" (29). A series of large fires across the country in the late 1880s and early 1900s that together burned substantial acreage and killed approximately 2,500 people reinforced the new settlers' position that fire was "bad" and should be controlled.

Notable features of Forest Service activity in these next decades include an increasing reliance on military science and military structure to justify and enforce non-Native visions of forest management. Lake (2007) writes, "Furthermore, in 1928 the US Forest Service's policy regarding incendiarism was still similar: 'The incendiary problem, however, is not a fire hunt but a man hunt; not fire, but the owner of the hand that lights it, is the public's enemy.... The hand of the incendiary is set against the public welfare and it is the duty of every citizen to help apprehend those who willfully set fires and to see that they are punished as they justly deserve' (Klamath National Forest 1928:14–15)" (273). One feature of the colonial framework is the explicit attention it brings to relationships between police activity at home and military activity abroad. Steinmetz, Schaefer, and Henderson (2017) describe the "colonial character of policing" in black ghettos marked by officers using techniques developed by the military; Fenelon (2017) compares the character and purpose of checkpoints in Palestine to those in North Dakota during the pipeline resistance at Standing Rock; Go (2017) places the 1960s transformation of the U.S. police force through "police science," surveillance techniques, organization, and fingerprinting as a result of militarization abroad, linking the police response in Ferguson and Standing Rock to military activity. Indeed, 2.6 million dollars' worth of military equipment, including MRAP vehicles and 1990 surplus military equipment, has been used against U.S. citizens at home. Just as Fenelon (2017) describes "fracking, pipelines and oil capitalism as the engines and infrastructure of empire," so too has the Forest Service increasingly developed its use of science and military command structure to change the forest type, marking fire policy as an important engine and infrastructure for colonialism in forested regions. And yet whereas a growing scholarship is addressing the process of enforcement of this colonial system onto people, less is said about the process of enforcement of white supremacy and colonialism onto land. Furthermore, the enactment of colonialism onto land is not a separate issue from what happens to people—targeting of the land has been about targeting and severing relationships between people and land. Whyte et al. (2018) detail how "the replacement of Indigenous ecologies with settler ecologies can inflict rapid changes. These changes, such as the destruction of rice ecology, undermine the plant and related species whose physical manifestations in ecosystems foster qualities of relationships (trust, redundancy, and others) that are important for our continuance" (163).

Frank Lake's (2007) extensive review of this local fire history describes how "initial reduction in the extent and frequency of Native American ignitions occurred from 1855 to 1870 during American settlement, more so under government rangers starting in the early 1900s followed by the establishment of fire lookouts in the 1920s" (275). Fire exclusion became formal policy at the national level by order of the nation's first forester, Gifford Pinchot, who declared that "one of the objectives of the National Forests was to make sure that 'timber was not burnt up'" (Pinchot, quoted in Stephens and Sugihara 2006, 433). In an 1899 essay on "the relations of forests and forest fires," Pinchot wrote "that fires do vast harm we know already, although just what the destruction of its forests will cost the nation is still unknown." Stephens and Sugihara (2006) describe how "the U.S. Forest Service was established as a separate agency in 1905 with Gifford Pinchot as its first chief. Under his direction, a national forest fire policy was initiated and the agency began systematic fire suppression including the development of an infrastructure of fire control facilities, equipment, fire stations, lookouts, and trails. The forest reserves were created partly because Congress believed the nation's forests were being destroyed by fire and reckless cutting (Pinchot 1907)" (433). Similarly, the second chief of the Forest Service, Harvey Graves, emphasized that "the first measure necessary for the successful practice of forestry is protection from fire" (Graves, cited in Stephens and Ruth 2005, 533). In 1910, the "Great Idaho Fire," in which seventy-eight firefighters were killed and two and a half million acres of national forest burned, further cemented the perception that wildfire was dangerous and prompted the Weeks Act of 1911 that expanded fire suppression efforts. Forest Service efforts to suppress fire extended to cooperations with other state and federal agencies to eradicate cultural burning practices, as seen in the following letter written in 1922:

> The forest people at Yreka, Thomas West, Supervisor, requested the aid of this office [Indian Service] to the end that these Indians be prevented from setting fires in the Klamath Forest next year. They say that these Indians cause many fires each year to great damage to the forest and because of the forest people's help to us in aiding these Indians to get their lands [public domain allotments], I promised Mr. West our whole-hearted support and cooperation in an endeavor, next year, to educate the Klamath River Indians against their practice of setting fires in the Forest. This matter will be taken up with Superintendent J. B. Mortsolf and the three offices and officials will get together on some plan whereby every Indian will be seen and cautioned.[10]

In 1935, the "10 A.M. policy" directed that "all fires should be controlled in the first burning period or by 10 A.M. the following morning" (Stephens and Ruth 2005, 34). Civilian Conservation Corps crews were deployed to rural communities across the country, including the town of Orleans in the heart of Karuk territory, and the "Smokey Bear" campaign was launched in 1942. Frank Lake's (2007) detailed review of the history of Indigenous burning practices and fire suppression in the mid-Klamath includes interview testimony detailing this unfortunate juxtaposition of fire suppression activities on sacred sites: "Beginning in the 1920s, fire observation stations, e.g., lookouts, were constructed and fire personnel were placed there to detect and report fires (Jackson pers. com. 2002). Some of these lookouts were constructed on tribal sacred sites used as prayer seats (Alfred pers. com. 1996). Occupation and use of these lookouts or field camps modified tribal land use practices, especially traditional setting of fires near these areas that were culturally significant habitats" (274). Interview notes from Karuk tribal member Harold Tripp reinforce Lake's statements: "Most of these mountains got their altars destroyed because that's right where the Forest Service put their lookouts" (Tripp, personal communication 2005). Ironically, many of the most culturally and spiritually important places throughout Karuk territory have been the site of particularly intense alteration as a result of fire suppression.[11]

Fire suppression efforts increased exponentially on the Klamath as well as nationally during the period following World War II, a time when timber prices were at an all-time high and wartime tools, from aerial retardant drops to "helitack crews," bulldozers, and smokejumpers, were employed. During this era and in addition to employing military hardware and tactics, the Forest Service also adopted the military command structure in its drive to suppress fire. With the deployment of tools developed from warfare, Stephens and Sugihara (2006) write that "the public was now well shielded from the history of human–wildland fire relationships and fighting fires had become 'The moral equivalent of war (Pyne 1997)'" (434). Fire suppression practices developed by the Forest Service were also applied to lands administered by the Bureau of Indian Affairs. Lake notes, "The BIA began mimicking the USFS and viewing all fires as bad, and extinguished them with a fury. Not only did they put out all fires, they made it a criminal offensive if Indians set fires. So now it became a crime for Indians to manage the forest the way that they were taught and were supposed to do (Colegrove 2005:43)" (cited in Lake 2007, 274). Such a wholesale shift in ecological practice, state structure, and lay public epistemology does not happen overnight. Indeed, as compared to discussions of Indigenous land dispossession as a past event, we can trace the 130-year legacy of fire suppression as a process that continues land dispossession into the present, thereby understanding what is at stake in the scale of Karuk resistance to Forest Service fire policy today. Fire suppression may be an engine of colonialism, but it is one that has been continuously and creatively resisted in a wide variety of ways. From the beginning, Karuk and other Indigenous peoples tried to educate the newcomers as to the importance of fire. This dialogue and the discourses employed by the state to justify their perspective on the matter make explicit the colonial logics of the time. In a letter written to the local paper in 1916, "Klamath River Jack" exhibits sophisticated traditional ecological knowledge as he explains the ecological function and importance of fire:

> Fire burn up old acorn that fall on ground. Old acorn on ground have lots worm; no burn old acorn, no burn old bark, old leaves, bugs and worms come more every year. Fire make new sprout for deer and elk to eat and kill lots brush so always have plenty open grass land for grass. No fire brush grow quick and after while choke out all grass and make too much shade, then grass get sour, no good for eat. No fire then too much leaf stay on ground. No grass can grow up. Too much dead leaf, ground get sour. Indian burn every year just same, so keep all ground clean, no bark, no dead leaf, no old wood on ground, no old wood on brush, so no bug can stay to eat leaf and no worm can stay to eat berry and acorn. Not much on ground to make hot fire so never hurt big trees where fire burn. Now White Man never burn; he pass law to stop all fire in forest and wild pasture.[12]

As Forest Ranger Jim Casey condescendingly responds to Klamath River Jack's questions regarding Forest Service practices in 1916, "Did you ever stop to think what made that brush field? Do you know that all through it there are old black stumps burned clear to the ground? Fire did it. When I asked Frank about it, he said that three big fires had run over that place in the last twenty years. Are there any berries in that strip; is there any grass in there; can you hunt in there? No, Jack. *Because your fire is bad medicine.*" Settlers did use fire to clear lands for grazing and to access mining sites, although these uses of fire were for much different purposes than Indigenous burning. Karuk people not only tried to educate settlers as to the ecological need for fire but also continued to use fire, despite prohibitions against burning. Lake (2007) draws on present-day and archival interviews with Karuk people together with Forest Service records to describe how, "following the aftermath of American settlement, genocide, and forced removal, Native people still tried to carry out traditional subsistence and burning practices" (272) and notes that "Native American fires were reduced after 1911 around the Orleans area because of the proximity to the USFS rangers and then in the 1930s with Civilian Conservation Corps who hired and/or enlisted rangers and fire crews to extinguish fires" (272).

Ongoing Indigenous burning practices were the source of much frustration by the USFS rangers. A 1918 letter from the Orleans District Ranger F. W. Harley to the forest supervisor in Yreka details the attitude of the new federal agency toward appropriate fire ecology, frustration, and disregard for locals' perspectives and illustrates the degree of overt hostility and willingness to use violence on behalf of the state to enforce Western conceptions of the appropriate place of fire. Harley writes, "The Forest Service in the administration of the forests have no more important duty to perform than keeping the fires down to a minimum." Harley then goes on to categorize the reasons people were burning and what he thought should be done about it: "Some set fires out of 'pure cussedness' . . . in the pure cussedness class the only sure way is to kill them off, every time you catch one sneaking around in the brush like a coyote shoot at them." Interestingly, this letter does recognize the problematic consequences of fire suppression: "They are succeeding to a certain extent and the consequence is that at the present time, there is more thick underbrush, windfalls and general humus in the forest cover than before the service was in effect." Lake (2007) reviews passages from a publication of the Klamath National Forest in 1928 that also characterizes Indigenous burning practices in a profoundly negative manner. Rather than recognizing the sophisticated ecological management represented by these actions, they were described as "selfish and malicious."

> Two primary sources of wildfires ignitions were recognized, lightning and human incendiarism (Jackson pers. com. 2002 recalls other ignition types). Lightning ignitions were an "act of nature" which could not be prevented or controlled like

incendiarism, described as "selfish or malicious motives." ... The incendiary problem, however, is not a fire hunt but a man hunt; not fire, but the owner of the hand that lights it, is the public's enemy.... The hand of the incendiary is set against the public welfare and it is the duty of every citizen to help apprehend those who willfully set fires and to see that they are punished as they justly deserve (Klamath National Forest 1928:14–15). (Cited in Lake 2007, 273)

A second passage from the 1928 report illustrates not only the need the agency felt to defend the policy of fire exclusion but also the use of scientific rhetoric and discourses of "practicality" rooted in a Taylorist managerial mind-set of efficiency to justify their actions: "The existing policy of the Forest Service in fire prevention and suppression has not been reached on the basis of guesswork. It represents continuous and critical study of forest fires. Fire exclusion is the only practical principle on which our forests can be handled, if we are to protect what we have and insure new and more fully stocked forests for the future (Klamath National Forest 1928:17–18)" (cited in Lake 2007, 302). Note that it was not only a military structure but also the rhetoric of Western "science" that has been used to enact and justify state actions. Seth Suman (2009) writes, "The idea that science and technology were among the gifts that Western imperial powers brought to their colonies was an integral part of the discourse of the 'civilizing mission,' one vaunted by both proponents and critics of the methods of colonialism" (373, see also Adas 1989). While many advocates of Indigenous burning practices now make use of Western scientific frameworks, the Western ideas of ecology that justified fire exclusion have themselves been instruments of colonialism. By 1938, the extent of burning had declined: "It is reported that in the past it was a general practice to burn timber and browse lands with the expectation that annual burnings would promote grass growth. Although this practice has been discouraged and is rarely followed now, there is still a degree of sentiment in its favor. It is believed that much of the browse cover has developed as the result of fires, and that most of the brush areas would eventually produce a fine stand of fir timber if fires were prevented and suppressed and grazing properly managed" (1938 report, quoted in Huntsinger and McCaffrey 1995, 62). Indigenous use of fire was nonetheless ongoing in the decades that followed, and this activity was a continued source of consternation for the Forest Service. The 1950 Six Rivers General Inspection Report included a focus on what they term the "Indian Incendiary Problem" and included this passage regarding the issue:

> One problem area exists; the "river strip." ... There is a fairly large Indian population here and the area is still "west of the Pecos." The State has apparently not yet decided to take fire control laws across the river. Previous attempts brought a threat of bloodshed.... It looks as if we will have to live with this problem a while longer—until the area becomes more civilized, lending the State any assistance

needed in developing an attitude toward protection among the local people. Perhaps the burning of basket grass areas and doe pastures would do the job. (USDA, Six Rivers National Forest 1950:27–28, cited in Busam 2006, 60)

Leading up to World War II, timber production increased in Karuk country, and along with this development, Indigenous burning practices were progressively restricted. Now in the face of half a century of increasingly intensive fire suppression, the forest structure began to change and fuels began to accumulate. As Lake (2007) notes, "The once common frequent low-intensity fires had been easy to extinguish compared to changes in vegetation resulting from densification and timber harvesting which decreased potential frequency for a given area and increased fire intensity due to the accumulation of fuels" (276). Within Karuk territory as well as nationally, timber harvest and fire suppression efforts amped up significantly following World War II. With rising technological capacities and an increasingly militaristic structure refined in the war effort and brought home to be used on people and the lands of the West, fire suppression continued to be effective even in the face of increasing fuel loads, and the structure of forests and grasslands continued to change substantially in the second half of the twentieth century as a result (Stephens and Sugihara 2006). Lake conducted interviews with local Karuk community members regarding fire history and reports that "after 1945, new roads or improvements to travel routes were constructed along former trails improving the government's ability to suppress fires and access timber (Gates 1995, Hillman pers. com. 2002, Allen pers. com. 2007, Ferris pers. com. 2007)" (276).

If Karuk use of fire created landscape abundance for Karuk culture and people, fire suppression efforts supported very different objectives and outcomes. Fire exclusion has altered species composition and diminished the production of hundreds of important food resources from the more commonly discussed examples like acorns, huckleberries, and elk to a wide variety of mushrooms and bulbs (Anderson and Lake 2013, 2016). By the 1950s, basketweavers faced a short supply of useable weaving materials. A 1954 article in the *Humboldt Times* reports, "They used to burn the wildgrass where the chosen brush grows and they gathered the bush the second year after burning. They burned the brush so 'good things will grow up.' It is different now. The U.S. forest service has strict laws regarding brush fires in the rich timber country. The Indians have to apply for fire permits and this is something the old women do not understand. They manage without the brush fires but they say the baskets are not as good" (*Humboldt Times Centennial Edition*, Section 9, page 8, February 1954). Lake (2007), who spoke with a number of people describing circumstances at this time, further observes that "the notion that Native Americans needed permits to burn was an additional cultural and institutional constraint. Tribal people having to ask the US Forest Service for permission to burn in their own traditional family-use areas was the equivalent of relinquishing tribal usufruct rights and tenure. Why should tribal people need to ask permission from the US Forest Service to do things they believed they retained the ancestral rights to do?" (276). Lake writes that "the threat or actual danger of federal enforcement actions leading to potential death, injury, imprisonment, or fines for arson (incendiarism) was a considerable factor (M. McCovey pers. com. 2002, Peters pers. com. 2005). Risks for Indians to continue burning were often too great for tribal families to attempt (Aubrey pers. com. 2005)" (276). By contrast, the new fire regime in turn supported the installment of white settlers by enabling the establishment of an economic system based in capitalistic commodity production. While I have thus far emphasized state legal actions, the actions of settlers as both individuals and profit-making commercial ventures reinforce state actions hand in hand—although not necessarily in a totally top-down or tightly coordinated manner. The dynamic between the state and settlers plays out in varying sequences. The state sets regulations, and then the settlers pursue (often economic) actions within a new legal and social terrain; the state then further regulates to protect this new settler presence and their economic interests and so forth. In 1860, mills in Humboldt County produced 9.6 million feet of lumber; "by 1875 the output had ballooned eightfold to 75 million feet" (Raphael and House 2007, 196). The legal structures and state actions of genocide and forced removal detailed in chapter 1 made way for the arrival of settlers who in turn provided essential labor for the growing economy of California. A 1900 census of Humboldt County reported that 22% of the population was foreign born (Cornford 1983).[13] By 1905, the county's lumber mills were considered the most technologically advanced in the world (Cornford 1983). By the 1950s, lumber in Humboldt County[14] was the largest source of tax revenue in the state of California, and Humboldt County ranked second in logging output for the entire United States (Wilson 2001, 4). At this time, Wilson (2001) also reports that loggers would make about one million dollars from the lumber produced in one square mile of Douglas fir land (6).

Land dispossession too is a process and not an event. New cultural logics of relating to the natural world via extraction, on one hand, and the trivialization of the natural world, on the other, came to supplant longstanding Karuk understandings of animals and other species as respected teachers and kin.[15] New forms of knowledge (Western forestry techniques) upheld these practices and began to replace traditional ecological knowledges (at least as the dominant "legitimate" discourse). Today, the narratives of loggers, pioneers, and gold miners as "first peoples" enact settler replacement on the land as the real "first peoples." Fire suppression became a central mechanism to the performance of this erasure by disrupting human-fire relationships that were central to sustenance as well as spiritual and social practice. Fire regime alteration matters because it shifts economic, cultural, and political power relations.

Describing conditions slightly downstream in Yurok Country, Huntsinger and McCaffrey's (1995) observations that "control of reservation and allotment natural resources has been withheld from them [Yurok people] under the auspices of scientific forest management" are equally fitting for Karuk country (155). They note that "managing the reservation for a 'fine stand of timber' precluded most Indigenous modes of subsistence, as well as crop production and grazing. Environmental shifts resulting from fire suppression and the forest professionals' focus on maximizing tree growth meant that allotments along the Klamath were becoming an increasingly poor source of direct support for their owners" (174). Thus, Huntsinger and McCaffrey emphasize both the overt and more subtle processes at play in colonialism with respect to land:

> Two major mechanisms can be identified by which the Yurok were divested of their forest resources: (1) by straightforward expropriation of their lands, as Yurok property rights were ignored and access to gathering sites was cut off; and (2) through ecological change brought about by a shift in management regimes. American management changed the forest, even on lands still owned by the tribe or its members. In these cases, the simple title to a piece of land was preserved, but the land itself was changed. In United States forestry programs, the land tenure rights remaining to Indian owners included the right to alienate the land but not to manage the vegetation. Vegetation management and Yurok culture and economy were closely linked. The increasing unsuitability of the changed forest for Yurok subsistence helped push the Yurok to sell their land. (166–167)

Fire exclusion has led to the exclusion of Karuk people and continues to do so today when, as Ron Reed describes with the changed forest structure, "the reason we are going back to the landscape is no longer there." In his introductory essay to the new journal *Settler Colonial Studies,* Veracini (2011) writes that in contrast to other forms of colonialism, settler-colonialism "is characterised by a persistent drive to ultimately supersede the conditions of its operation. The successful settler colonies 'tame' a variety of wildernesses, end up establishing independent nations, effectively repress, co-opt, and extinguish Indigenous alterities, and productively manage ethnic diversity. By the end of this trajectory, they claim to be no longer settler colonial (they are putatively 'settled' and 'postcolonial'—except that unsettling anxieties remain, and references to a postcolonial condition appear hollow as soon as indigenous disadvantage is taken into account)" (3). Because the dominant Western views of "nature" are universalized and tend to objectify a generic notion of "nature" and "land," the details of dispassion *as an ongoing process manifested through land transformation* are easily overlooked.

COLONIALISM AS ONGOING STRUCTURE: FIRE CONFLICTS TODAY

In contrast to the notion of postcolonialism, the settler-colonial framework highlights the *ongoing* operation of Indigenous erasure and settler replacement through legal, political, and cultural structures. Native and Indigenous studies scholars emphasize that colonialism is not limited to the overt genocide or forced assimilation of the past but *is very much enacted in the present* through both state institutional structures that continue to impoverish communities and erode sovereignty and cultures and practices of interaction of the dominant "settler" society (Alfred 2005; Goldstein 2014; Steinman 2012, 2015; Smith 2012; Steinmetz 2014; Veracini 2011; Wolfe 2006). Over the past decade of working with the Karuk Department of Natural Resources, I have observed continual examples of the ongoing operation of colonialism through land management process related to herbicide use, water quality, dam relicensing, and forestry practices.[16]

In some rural communities, the local conflicts are about resource extraction such as mining or timber. But on the Klamath, the debate that raged in the local paper one hundred years ago at the time of Klamath River Jack remains the most contentious and critical land management issue. Environmental decline from fire exclusion forms an extension of colonial violence to the Karuk community by simultaneously disrupting ecological and social reproduction. As such, fire exclusion continues to be a major force of land dispossession.

Dr. Frank Lake shared the following story by way of illustrating the multilayered impacts that come from non-Native agency actions. Lake notes that Karuk territory formerly had many fire-resistant ridge systems characterized by open grasses, manzanita, and sugar pines. These species are present as a result of highly tuned traditional fire management carried out over decades. In 2017, while working as a USFS Heritage natural resource advisor, Lake was visiting one particular ridge and found evidence of "prehistoric" family camping areas and manzanita processing sites. From the large open characteristic of the manzanita, he could see that it had grown up under more open conditions. This would have been a site characterized by large open canopy and understory of bunchgrasses. However, over the past decades, this ridge, like many others, had been a target of past fire suppression activities. With fire suppression, Douglas fir and other brush species had invaded the site,[17] but there were a few remnant sugar pines as evidence of the prior vegetation characteristics and legacy of prior tribal stewardship. Frank explains, "So now in 2017 the incident management team wanted to put in a fire line on this ridge system, but because there weren't enough Type I 'Hot Shot' firefighter crews available this was done with a bulldozer. So the food processing tools and other artifacts were broken by dozer tracks. The area gets trampled as it is run over by the bulldozer, and firing operation burn out causes heat fracturing of the artifacts. In this way, the legacies of Karuk food and fire stewardship are erased from the land, and replaced by the 'safety' needs of the dominant society." Here not only were the physical artifacts damaged, but the distinct cultural legacy of vegetation mosaic of the ridge system, which stands as a cultural knowledge archive of past land stewardship, also disappeared. Lake (2007) describes these as "third order fire effects" (343) because there is an erasure of cultural landscape, of particular artifacts, and of the future ability to learn from the ancestors and the land. "Everything gets re-set. You lose not only the biological legacy of the forest structure, or the damage to the stone artifacts, but you have lost the ability to learn and teach in that site. You can no longer take the next generation to that site and say 'look, see this is what we did.' Nature is our teacher and we just lost an important instructor" (Frank Lake, personal communication, 2018). From the agency's point of view, the loss of the artifacts is, however, the only harm that can even be recognized. Even those impacts are considered "mitigatable" from the agency perspective—a harm that can be traded out for some other tribal benefit. Yet neither the decades of time it took Karuk people long since passed on to create that ridge system nor the knowledge developed over tens of thousands of years that was evident in that landscape can be traded.

This poignant example illustrates how, like settler-colonialism itself, land dispossession is a process and not an event. Even the term *land* objectifies, generalizes, and glosses over the complexity of animate beings and their relationships to one another, while the term *dispossession* is much too rooted in a capitalist logic of ownership in contrast to Indigenous sensibilities of responsibility and kinship. Whyte et al. (2018) emphasize, "Indigenous ecologies physically manifest Indigenous governance systems through origin, religious, and cultural narratives, ways of life, political structures, and economies" (159).

Today, the politics of wildland fire play out in terms of the desire of Karuk people and other local non-Native actors to return fire to the land and, ironically, through community and ecological impacts from firefighting techniques such as "backburning,"[18] the creation of fire lines, and the use of chemicals and fire retardants. Finally, the absence of fire coupled with climate change generates concern over so-called catastrophic fires, which—at the extreme end—have the potential to convert forests into brush fields, resulting in a buildup of hazardous fuels that make the next fire uncontrollable when burning under adverse summer hot and dry conditions.

This last dimension takes the Karuk Department of Natural Resources into the realm of climate change politics, as will be detailed further in the Conclusion. Throughout the "politics of fire," the arc of colonial dispossession continues to sever the relationships across species that have sustained the Karuk community since time immemorial. To the extent that it limits tribal political sovereignty, cultural practices and spiritual energy, and the alteration of the forest, rivers, and grasslands of the mid-Klamath, the exclusion of fire constitutes a very literal means of Indigenous erasure. Frank Lake notes that "literally the ecological vegetation characteristics in both species' structure and composition are altered. This minimizes the legacy of Indigenous management, of the 'tending' of such vegetation for production of the many basketry, food, medicinal resources that can be visibly recognized in the vegetation characteristics." Karuk cultural biologist Ron Reed eloquently explains this process in this quote I repeat from the chapter opening:

> Without fire the landscape changes dramatically. And in that process the traditional foods that we need for a sustainable lifestyle become unavailable after a certain point. So what that does to the tribal community? The reason we are going back to that landscape is no longer there. So the spiritual connection to the landscape is altered significantly. When there is no food, when there is no food for regalia species, that we depend upon for food and fiber, when they aren't around because there is no food for them, then there is no reason to go there. When we don't go back to places that we are used to, accustomed to, part of our lifestyle is curtailed dramatically. So you have health consequences. Your mental aspect of life is severed from the spiritual relationship with the earth, with the Great Creator. So we're not getting the nutrition that we need, we're not getting the exercise that we need, and we're not replenishing the spiritual balance that creates harmony and diversity throughout the landscape.

Indigenous erasure manifests as well through erasures within people's collective imaginations. Erased from the dominant sensibility is the possibility of an animate world, the possibility that humans and the other species we often call "nature" might work together to create abundance (Fenelon 2015a, 2015b; Watts 2013). Erased are notions of belonging, responsibility, and reciprocity. Now in the highest esteemed institutions of the land, it has become difficult to acknowledge or imagine that the natural world matters for the social, that there are spirits in all living things, or that there are viable forms of social organization beyond capitalism.

Whereas existing sociological theory engages power dynamics with respect to economic exploitation, racism, and sexism, Coulthard (2014) notes that "there is much more at play in the contemporary reproduction of settler-colonial social relations than capitalist economies; most notably the host of interrelated yet semi-autonomous facets of discursive and non-discursive power identified earlier" (14). Furthermore, Coulthard states, "The colonial relations should not be understood as a primary locus or 'base' from which these other forms of oppression flow, but rather as the inherited background field within which market, racial, patriarchal, and state relations converge to facilitate a certain power effect—in our case, the reproduction of hierarchal social relations that facilitate the dispossession of our lands and self-determining capacities" (14–15). In this next portion of the chapter, we can see the threads of ongoing structural dispossession and its focus in uprooting and forcibly destroying the bases of Native power through impacts to altered ecology and its consequences for the interruption of knowledge systems and cultural reproduction, physical and mental health, economic conditions, and political sovereignty.

It is important to note that none of the politics of fire as I have described them would be so visible were it not for the proactive vision of the people with whom I have been working. The offices of the Karuk Department of Natural Resources are a small set of buildings densely packed with people holding clear and practical visions of how to move forward. Despite a significant lack of resources as compared to other federally recognized tribes, the Karuk have been at the forefront of achieving innovative policy change at the local, state, and national levels. While the emphasis in this chapter is on the forces of colonialism, I am indebted to Beth Rose Middleton for pointing out that decolonization requires attention to and recognition of the visions in Indigenous communities of how to move forward. In her call for a specifically Indigenous political ecology, Middleton notes that existing mainstream political ecology frameworks ascertain "how indigenous people are politically, economically, culturally and ecologically marginalized, but rarely provides a way forward from this plight. Indeed, indigenous people and struggles are only visible within Western framings—resulting in a situation whereby they are seen but not heard" (2015, 573). Karuk traditional knowledge and vision with respect to the use of fire is ongoing. As I write, my Karuk colleagues are featured in the present issue of *Yes Magazine* for their innovative and proactive work bringing fire back to the landscape. Bill Tripp and others within the Karuk Department of Natural Resources have very specific visions of how to move forward on these issues, expertly navigating federal and state policy in multiple realms. These visions are being put into action as Karuk practitioners lead an annual two-week fire training program and co-coordinate an ambitious effort through the Western Klamath Restoration Partnership that is bringing fire back to the land and through their climate adaptation planning efforts.

ECOLOGICAL ALTERATION, INDIGENOUS DISPOSSESSION, AND THE CONSOLIDATION OF STATE POWER

Western ecologists and Native practitioners alike now describe how fire exclusion and suppression practices, combined with forest management activities (e.g., harvesting older forest, establishing plantations), have increased the density of trees, shrubs, and fuel loading (Lake 2013; Odion et al. 2004). Studies of the Klamath mountain region note "two periods with distinctly different fire regimes: (1) the Native American period, which usually includes both the pre-historic and European settlement period, and (2) the fire suppression period" (Skinner et al. 2006, 176). The authors also note that

> over the 400 years prior to effective fire suppression, there are no comparable fire-free periods when large landscapes experienced decades without fires simultaneously across the bioregion (Agee 1991; Wills and Stuart 1994; Taylor and Skinner 1998, 2003; Stuart and Salazar 2000; Skinner 2003a, 2003b). Along with these changes in the fire regimes are changes in landscape vegetation patterns. Before fire suppression, fires of higher spatial complexity created openings of variable size within a matrix of forest that was generally more open than today (Taylor and Skinner 1998). This heterogeneous pattern has been replaced by a more homogenous pattern of smaller openings in a matrix of denser forests (Skinner 1995a). Thus, spatial complexity has been reduced. (178–179)

Not only were the impacts of fire exclusion significant alterations of the land, but following World War II, fire prevention programs utilized herbicides to control undesired shrubs and hardwoods such as oaks in forests and roadsides. The use of forestry herbicides is another pertinent example of the colonial state developing military technology abroad (during the Vietnam War) and then applying it against people and the natural world at home. In this practice, herbicides are used on logged areas to prevent the growth of broadleaf hardwood trees and brush, species that compete with newly planted conifer seedlings after clear cutting. A combination of 2,4-D and 2,4,5-T—chemicals similar to Agent Orange—was aerially sprayed onto forests via helicopters for what is known as a "conifer-release" because when the broadleaf hardwood trees are out of the way, the conifers will grow faster (LeBeau 1998; Bowcutt 2011). This program brought military technology, as well as surplus chemicals from the Vietnam War, back home for use in rural communities. Dioxin, the active ingredient of 2,4,5-T and Agent Orange, has been linked to hormonal and endocrine disruptions in Vietnam veterans and their wives and children (Nham Tuyet, Thi, and Johansson 2001; Ngo et al. 2006, 2013). Incidents of water supply contamination, late-term miscarriages, and unusual cancers and birth defects have been documented in the Karuk community (McCovey and Salter 2009; Norgaard 2007). There are numerous stories of deer killed for meat whose livers and internal organs were deformed and abnormal. The fact that a common herbicide for use in forestry in the region is not registered for use on food shows the implicit cultural assumption that people do not get their food from the forest. This was explained to me by a staff member of the California Indian Basketweavers Association:

> Garlon, this is the most frequently used chemical in the county. It has pretty long persistence.... What's interesting and shocking about this chemical is that it's not registered for use on food crops at all ... and there is no drinking water safety limit either. They're spraying about 100,000 pounds of this in the County every year, and there is no drinking water safety limit. It is just totally under the radar for the Safe Drinking Water Act. The issue about it not being registered for use on food

crops, when there are people getting food plants out of the forest is pretty disturbing, too. That is just one way that the traditional lifestyles aren't being taken into account when they register these chemicals.

The assumption that food and water supplies do not come from the forest, and thus that forestry herbicides need not be tested for use on food or water, puts traditional Native American people at greater risk (Wofford et al. 2003; LeBeau 1998). Indeed, Karuk people with whom I spoke experienced past herbicide spraying and high rates of miscarriages as one more event in a series of acts of genocide. Karuk elder Mavis McCovey was a practicing nurse at the height of the aerial spraying. She shared with me a letter she wrote to the Regional Forester in 1981: "I guess it's easy for such a large organization to ignore such a small group of people. But it is not as easy for us to ignore the Forest Service when its actions cause such terrible damage to us. There are only 800 of us left. When we lose one baby, it is the same proportion as if you lost 275,000 babies. The herbicide spraying is clearly threatening our very survival as a people. Our cultural group is already endangered enough as it is." While forestry herbicides are still used today on private timberlands in the neighboring Yurok reservation (Wofford et al. 2003; Matthewson 2007; Bill 2006),[19] airborne application in Karuk territory was forcibly halted by a 1983 injunction on aerial spraying on public lands.

These ecological changes produced and continue to drive a reorganization of the region's economies, meaning and knowledge systems, and political possibilities. One key feature of settler-colonial theory missing from existing sociological understandings of race and gender is the central attention to land and ecology as bases of power. Colonization involves the extraction of wealth through the material separation of Indigenous communities from their lands as emphasized by Alfred (2005). A focus on human-fire relationships highlights how the disruption of relationships is not a one-time event but an ongoing process. Coulthard (2014) writes, "Colonial domination continues to be structurally committed to maintain—through force, fraud and more recently so-called 'negotiations'—ongoing state access to the land and resources that contradictorily provide the material and spiritual sustenance of Indigenous societies on the one hand, and the foundation of colonial state-formation, settlement, and capitalist development on the other" (7). Yet even existing theories in Native studies may benefit from a further detailed explication of the way that this process is related to changing characteristics of the land. A number of scholars have emphasized that access to land is a central goal of settler-colonialism; less has been detailed about how the *alteration of the land* and the *alteration of human relationships with land* are key dimensions of the consolidation of state power. Wolfe's characterization of settler-colonialism puts the importance of land centrally on the table, yet this conceptualization takes land in a generic, quantified, and objectified way. By contrast, Karuk and other Indigenous peoples understand the land as animate and hold very specific relationships to individual species as kin and particular sections of forest and river as family gathering places. The alteration of these places and the kinds of beings who flourish there become key parts of the operation of power as detailed through nuanced impacts to the gender structure of the community in chapter 4 and emotional harm in chapter 5. Land matters not only in a static quantitative sense of acres, but there is also an essential qualitative dimension. And it is by paying attention to the qualitative dimension that we focus on nature as animate and bring into view the possibility that the alteration of the land is also a process that diminishes the spiritual integrity of the land and the people. Tracing the arc of human relationships with fire over the past century and a half provides a vivid means to understand such dynamics. Through an examination of fire policy as a mechanism for structural dispossession, we can see how colonialism is not only about territorial access but also the reorganization of species and relationships within that territory. Kyle Whyte (2017a) coins the term *containment* to describe this dynamic of colonialism whereby "cultural and political institutions designed to inhibit or 'box in' Indigenous capacities to adapt to environmental change" (5).

Colonialism as alteration of the land persists by denying people access to the practices needed to care for the land. North American genocide and colonialism have been assaults on a spiritual order that nourished and governed an entire field of ecological relationships. The destruction of these relationships has a circular effect with both cultural and ecological impacts reinforcing one another. In neighboring Yurok country, Huntsinger and McCaffrey (1995) describe how "landscape change resulting from the displacement of indigenous management regimes has been a major factor in divesting the Yurok people of natural resources, land, and indigenous lifeways. The direct effect of federal Indian land tenure policy on Indian lifeways has long been recognized, but the role of ecological change resulting from suppression of tribal control of natural resources has received less attention" (155). The alteration of the land leads to the alteration of human relationships with the land. Denied access to stewardship practices as traditional management makes impossible the social and cultural practices described in the first part of this chapter. Denying people the use of fire for clearing around acorn trees, for example, becomes quite literally the present face of forced assimilation and cultural genocide. Not only do important cultural practices not occur, but food sources also become diminished (without burning, acorns, for example, become infested with bugs and are difficult to gather), and the next generation of youth fails to learn burning techniques so traditional knowledge is lost. Frank Lake adds, "And you have to watch your wildlife relations: bear, deer, elk, squirrels, birds all suffer from lack of quality food resources, and this affects your psychological health, when you know they—animals and your own human family—would be better if acorn orchards were properly burned and tended."

Colonial violence manifests as a wresting of people from the natural world. As Leaf Hillman describes, "Every project plan, every regulation, rule or policy that the United States Forest Service adopts and implements is an overt act of hostility against the Karuk People and represents a continuation of the genocidal practices and policies of the US government directed at the Karuk for the past 150 years. This is because every one of their acts—either by design or otherwise—has the effect of creating barriers between Karuks and their land." Alteration of human-Karuk fire relationships is an ongoing manifestation of what Bacon terms "colonial ecological violence."

It is for these reasons that resistance to land-altering policies is a cornerstone of Indigenous colonial resistance. "The theory and practice of Indigenous anticolonialism, including Indigenous anti-capitalism, is best understood as a struggle primarily inspired by and oriented around *the question of land*—a struggle not only *for* land in the material sense, but also deeply informed by what the land *as system of reciprocal relations and obligations* can teach us about living our lives in relation to one another and the natural world in non-dominating and non-exploitative terms—and less around our emergent stats as 'rightless proletarians'" (Coulthard 2014, 13).

FIRE PRACTICES, KNOWLEDGE SYSTEMS, AND CULTURAL REPRODUCTION

> Confronting inequality and marginalization in sociology cannot only be about achieving demographic diversity within the discipline. It must also involve an epistemic insurgency—an intellectual revolution—or least an epistemic shift. Without such an epistemic shift, sociological knowledge will continue to be limited. —Go (2017, 196)

As Go (2016, 2017) notes, epistemic exclusion matters for sociology as a whole, yet such exclusions are taking place in more ways than we have acknowledged. Today, the colonial state continues to exclude Indigenous forms of knowledge in part through alteration of the natural world that is a source of information and forms the basis of knowledge regeneration. Karuk and other traditional knowledge is embedded in and emerges from the practices of traditional management and stewardship. Thus, ecological alteration in the form of changing fire regimes is a fundamental contributor to epistemic exclusion. This is so because ecological alteration reworks the way people interact with knowledge and the land. Maintaining Karuk traditional knowledge requires the practice of cultural management—in order for each generation to know how to use fire, for example, they have to get out and burn and then gather and hunt in the postfire area to learn the beneficial effects. Through this active engagement, the next generation not only learns how to use fire in the current conditions of the ever-changing landscape but also learns cultural values and responsibilities, as well as reaffirms their identities as Karuk people. And by contrast, when the Tribe is unable to carry out traditional management, both species and culture decline. Thus, the process of cultural reproduction is contingent upon, and embedded within, material practices in the landscape. Coulthard (2014) coins the term *grounded normativity* to describe these "modalities of Indigenous land-connected practices and longstanding experiential knowledge that inform and structure our ethical engagements with the world and our relationships with human and nonhuman others over time" (14). Whyte et al. (2018) eloquently describe how, "for many Indigenous peoples, collectives are not anthropocentric. That is, they do not exclude animals, plants, and ecosystems as members with the responsibilities of active agents in the world. In many cases, plants, animals, and ecosystems are agents bound up in moral relationships of reciprocal responsibilities with humans and other nonhumans" (155). Different qualities of relationship promote responsibility and moral accountability:

> The qualities of a relationship are the actual properties of that relationship that motivate humans to care for rice and to gain and protect knowledge of rice ... the more humans take responsibility, the more the other parties or relatives reciprocate (e.g., flourishing rice harvests) if the appropriate causal relationships are also in place (such as causal relationships known via Indigenous knowledge systems about the impact of certain human practices on the growth of rice and the impact of certain ceremonies and educational practices on motivating and training humans to engage in stewardship practices skillfully). This reciprocity further secures and strengthens human motivation as the benefits of taking responsibility are physically manifest. (Whyte et al. 2018, 160)

In contrast to epistemological systems that are static enough to be recorded in books, Karuk traditional knowledge is a living system that must be enacted and carried on through active ongoing stewardship in order for it to continue (Norgaard 2014a, 2014b). In the face of continued fire exclusion, however, Native American cultural identity and traditional ecological knowledge are both at risk (U.S. Department of Agriculture 2012, 30). And ironically, the longer landscapes have gone without fire, the more dangerous fire can be, leading to more difficult barriers to its use.

PHYSICAL AND MENTAL HEALTH DIMENSIONS OF FIRE

Fire suppression also affects human mental and physical health in terms of smoke exposure. All fires create smoke, which has widespread and pervasive respiratory impacts on human health (Mott et al. 2002). Mental health impacts of smoke exposure include irritability and fatigue. High-severity large-scale fires burn for

FIGURE 8. Smoke over Orleans Valley. Photo credit: Will Harling.

much longer than traditional cultural burning of the past, leading to particularly significant health impacts (Long, Tarnay, and North 2017). In their work on pyrohealth, Johnston, Melody, and Bowman (2016) note this contrast: "While no studies on smoke exposure from traditional indigenous landscape burning exist, the smaller mosaic of patch burning promotes small low intensity fires, which overall produce relatively lower emissions, due to the smaller spatial size and lower fuel loads under such fire regimes" (3).

As noted by staff in the Integrated Wildland Fire Management Program, "With fire exclusion we have a wider pendulum between fires and no fires, between smoke and no smoke, such that when fires occur there may be very large with heavy smoke for periods of weeks at a time. These circumstances tend to be particularly difficult for respiratory problems." Johnston et al. (2016) further write, "The cessation of indigenous burning, active fire suppression, introduced species, and a warming climate are all contributing to increasingly frequent, large-scale, intense fires in many flammable landscapes. Emissions from large landscape fires can be transported for long distances affecting large and small population centres far from the fires themselves. Smoke episodes from severe landscape fires result in measureable increases in individual symptoms and in population indices of ambulance call outs, admissions to hospital and mortality" (3). Thus, while all smoke is harmful to human health, smoke from cultural burning is less impacting to human health than the high-severity fires that result in its absence.

As larger fires have become more frequent in recent decades, so too has awareness of the negative health effects of smoke increased. But given that nearly all Karuk traditional foods are enhanced by the use of fire, the physical and mental health dimensions of fire are much more complex than smoke inhalation per se.

Food matters for human health—as these foods become less abundant, the inability to access traditional food not only impacts the Karuk tribal members due to decreased nutritional content of specific foods but also results in an overall absence of food, leaving Karuk people with basic issues of food insecurity. The 2016 Klamath Basin Food Assessment conducted by the Karuk Tribe indicates that 45% of respondents said they got a portion of their food from hunting, gathering, or fishing. Traditional foods and medicines support physical and mental health in multiple ways (Alves and Rosa 2007). Recent U.S. Department of Agriculture studies show that while roughly 85% of the U.S. population is food secure, only 77% of Native Americans in the United States are food secure (Gordon and Oddo 2012). Self-report data from the Karuk Health and Fish Consumption Survey indicate that 20% of Karuk people consume commodity foods, and another 18% of those responding indicated that they would like to receive food assistance but do not qualify.

On another level, damage to the ecosystem, as well as to particular plants, fungi, and animals, can be particularly disturbing to Karuk people. Brave Heart and DeBruyn (1998) note that "for American Indians, land, plants, and animal are considered sacred relatives, far beyond a concept of property. Their loss becomes a source of grief" (62). During the 2008 fires, Karuk elder Marge Houston was visibly upset due to the impacts on both people and animals of the larger fires:

> You know, it's [fire suppression] doing detrimental damage to each and every person who is breathing all this smoke. Whereas instead of a short time in the fall, now it is damn near all summer. And that's got to have some effect on people, it's got to have the shortage of game, some effect on that. Because you, these people, the fire is over here, they are back-burning all around. What about those animals that are in the middle. Where do they go? No place but into the fire. They either die of the smoke inhalation or they eventually burn.

The relationships between colonialism, land alteration, and physical and mental health are the central focus of the next chapter.

FIRE POLICY AS A DRIVER OF ECONOMIC REORGANIZATION

A central motivation for the alteration of the ecology has been for the purposes of imposing a new economic regime. Through fire suppression, Native stewardship practices that have been and continue to be oriented around species complexity, subsistence activity, and long-term sustainability have been forcibly replaced by extractive management activities undertaken with the intent of withdrawing commodities that in turn support the escalation of capitalism and monetary wealth for non-Native people. Stephens and Sugihara (2006) describe how "industrialization and urbanization of America from the late 1800s to the mid-1900s generated an increasing need to manage wildlands for commodity production to supply an increasingly urban population. Protection of forests and rangelands from wildland fire was a central part of this evolution in management philosophy. The application of a fire protection philosophy that was developed to protect European forests was applied to every plant community in North America during this period" (Wright, Wright, and Bailey 1982, 441). Thus, while we know that Karuk people were wealthy prior to European invasion, that poverty in the Karuk community is now very high, and that the land management policies enforced upon Karuk Aboriginal Territory were the mechanism for this transfer of wealth, we cannot describe specific dollar impacts to Karuk people from the exclusion of fire. Karuk poverty has been the result of a series of events through which traditional Karuk management practices have been interrupted, wealth from the land has been transferred to non-Indian hands, and the environment has been degraded. Exacerbating the effects of poverty are the symptomatic embodiments of the intergenerational trauma resulting from these events as well as those resulting from genocidal acts experienced across several generations: substance abuse, domestic violence, and breaks in the natural transfer of intergenerational knowledge. The absence of food and cultural use species in this overgrown forest undermines the subsistence economy. Subsistence economic activities, including managing for and gathering acorns, mushrooms, berries, and basketry materials, continue to be outright illegal, impacted by the exclusion of fire, or regulated in other ways that limit or prohibit Karuk access. Laws regarding how the land would be understood (animate vs. inanimate, treasured relative or objectified property), who could hold what kinds of relationships on the land (owning vs. tending), and what activities could be carried out there (tending, burning, hunting vs. agriculture and commercial forestry) were implemented by the state of California and the federal government specifically to advance this new economic order and achieve a transfer of wealth to non-Native settlers in the region. The exclusion of fire continues to be instrumental to this economic reorganization—a theme that will be evident as well in the next section on forced assimilation.

LAND MANAGEMENT AS FORCED ASSIMILATION

State actions ranging from changed ecological conditions in the face of fire exclusion to the outright illegality of Karuk practices make clear the range of ways the reorganization of the Karuk subsistence economy has been officially sanctioned—indeed, an extension of the process of *forced assimilation* as described in chapter 1 (Hoxie 1984; Fixico 1986). Forced assimilation is a particularly important theme specifically conceptualized via settler-colonialism as opposed to theories of race. Whereas sociologists have engaged assimilation as a mostly beneficial or at least benign force, settler-colonial approaches illuminate the very destructive nature of these state actions, as well as the reasoning behind them. Beyond direct genocide, the next tactic for Indigenous erasure was cultural erasure. On the Klamath, forced assimilation happens overtly when game wardens arrest tribal members for fishing according to tribal custom rather than state regulation. Forced assimilation is overt when schools punish students for attending ceremonies, for while there is an official statement that allows for tribal students to participate in traditional ceremonies (lasting ten days), many teachers have a strict attendance policy that is directly tied to the grades students receive. As a mother of six Karuk tribal youth and wife of a world-renewal ceremonial leader puts it,

> We try very hard to raise our kids to learn and practice their Karuk cultural heritage, but when they get poor grades because of it, we have to make concessions. Right when it was time for the older kids to really take over some of those leading roles in ceremonies, they are in their Junior and Senior years. I guess no one expects Indian kids to go to a decent college, anyway, so maybe that's the point. Meeting your responsibilities as a tribal leader makes it impossible to keep a high grade point average. I'm not even talking about harvest season—there's no way for the kids to really help us when the main season for harvesting comes around. Naturally, that's the time school is in full session. (Lisa Hillman, personal communication)

In 2016, the Klamath Basin Food System Assessment surveyed Karuk households, again finding that over half of households listed "limited by rules" as a barrier to their consumption of traditional Karuk foods, and many people reported decreasing subsistence or ceremonial activities as a result of contacts with game warden law enforcement. Being fined or imprisoned is a risk that many are understandably unwilling or unable to take. Over 60% of those responding to the 2016 survey reported that removal of legal barriers to gathering would increase their access to traditional foods. As Janet Morehead, Karuk tribal elder, puts it,

> I know we're supposed to be able to gather berries and harvest mushrooms, but me and my sister have been harassed so many times that sometimes we wonder how long we're going to be able to keep this up. We both wear our tribal IDs around our neck—and heck, we're too old ladies that are obviously Indian—but I don't think people know that we are supposed to be able to do it on public lands. And then there are the marijuana growers. I'm afraid to get shot and I'm sick of being treated like I'm doing something wrong. (Personal communication)

While overt criminalization of Karuk methods of caring for the land and procuring foods is more readily identifiable, forced assimilation happens more insidiously as the land management actions of the state such as fire suppression deny Karuk people access to the land and food resources needed to sustain culture and livelihood. Fire-excluded forests are brushy, making travel through the area very difficult. Brush has overgrown traditional trails. In other cases, species of importance may be physically present, but not growing in the appropriate form for use or producing foods. Ron Reed describes this situation:

> You have deer meat, elk and a lot of times associated with those acorn groves are riparian plants such as hazel, mock orange or other foods and fibers, materials in there that prefer fire. The use of those materials is dependent upon those prescribed burns. So when you don't have those prescribed burns it affects all that in a reciprocal manner. It's a holistic process where one impact has a rippling effect throughout the landscape. We can only have that for a certain amount of time before the place becomes a desert without cultural burns, because the plants are no longer soft and the shoots are no longer food, instead they become these intermediate stages where they are just taking up light and water and tinder for catastrophic fire. So it has an impact not only on the species we are talking about, but how you harvest and manage and hunt those species as well.

Forced assimilation reaches its perhaps most insidious form when the food species that Karuk people would like to gather, hunt, or fish for are simply not there. Sadly, in 2017, Karuk people were not even allowed to catch fish for ceremonial purposes—because the species return was so low that the Karuk Tribal Council had to issue a moratorium—for the first time in history. The council felt they had no choice but to let the few that have survived ongoing decimation swim upstream to spawn.

> I would love to have been a fisherman yesterday. To be going down there, and catching as many fish as you want before noon. Just shut down the fishery at noon, because everybody in the area had enough fish. To be able to go give your fish to the medicine man. On an annual basis. Every year, every year, but now we fish like crazy to do that. To extend out the giving arm to people wherever they need fish. It's no longer there. The acorns are no longer there; the traditional food isn't there. So you can't extend out to your elders like you're supposed to. You don't connect your babies to your elders during the harvest seasons.

The forces of altered ecological conditions, responsibilities to tribal law, economic necessities, cultural practice, and contested political status all come together in a vicious convergence. Steinman draws upon Veracini's (2011) mapping of denial of access to habitat as an "indirect material" mechanism of transfer. "Denial of access" is more than putting up a fence but is a dynamic that emerges from the severing of relationships. This is the form of settler-colonial power that is resisted through tribal land reclamations, the revival of traditional knowledge, and more.

Thus, colonialism as the separation of people from the land is not a one-time event of the past but an ongoing present that takes place in part through the alteration of the quality of the land.

"OUR POWER COMES FROM THE EARTH": FIRE EXCLUSION, INDIGENOUS RESISTANCE, AND TRIBAL SOVEREIGNTY

Winona LaDuke's business card reads, "Our power comes from the earth." The alteration of species and ecological functions in the forest shape political dynamics and possibilities for Karuk and non-Indian people alike. While few sociologists attend to the influence of the natural environment on political power, geographers developed the subfield of political ecology to specifically emphasize such relationships (see, e.g., Robbins 2004; Escobar 1999). And, despite the fact that too few sociologists have engaged with the importance of tribal sovereignty as a political or socially relevant fact, the Indigenous conceptions of power entailed in these debates are illuminating (Champagne 2008; Middleton 2015). As Steinman (2012) notes, these concepts can inform social movement scholars and theories in political sociology and social movements alike: "While the Indian Sovereignty Movement has been described by some social scientists (Wilkins 2007; Champagne 2008), it has not received analytical attention from social movement scholars and political sociologists. This inattention reflects limitations of the existing conceptualizations of both the nature of power and domination in the United States and of political power and contestation more generally" (1074). It is my hope that these examples of fire policy will contribute to not only an illustration of the ongoing process of colonialism but also the relevance of the natural environment for sociological understandings of the nature of power.

For example, the legitimacy of Karuk sovereignty and management jurisdiction is grounded in the fact that Karuk people have continued to carry out cultural and spiritual practices of caring for the land and species they consider their relations since time immemorial. Karuk sovereignty and land management authority rest as well upon the fact that Karuk people have never ceded title of their ancestral lands. Although treaties were signed in 1851 and 1852, these were never ratified by the U.S. Congress (Norton 2014; Baldy 2013; Salter 2003). As a federally recognized tribe, Karuk political sovereignty and their ability to develop laws governing their lands, resources, and members are recognized by the United States. Since having their federal recognition restored in 1979, the Karuk Tribe has developed a constitution and numerous tribal departments, including a Department of Natural Resources, with thriving fisheries and fire and food security programs. Tribal environmental management authority is linked to political sovereignty. Legal political sovereignty, however, is limited by tribes' status as "domestic dependent nations" (see Tsosie 2013).[20] This, together with the fact that only a fraction of the Indigenous people of North America are federally recognized, leads many Native activists and political scholars to call for a rejection of what Coulthard (2014) calls the "colonial politics of recognition."

Indeed, Indigenous people themselves repeatedly emphasized deeper and more generative sources of power through their relationships with the earth. Tom Goldtooth (1995) writes, "Before colonization, Indigenous nations possessed complete sovereignty. Many Indigenous people today argue that Indigenous sovereignty remains in force and regard all federal laws limiting tribal sovereignty as illegal" (142). Yaqui-descended legal theorist Rebecca Tsosie (2013) writes, "The political sovereignty of indigenous peoples under U.S. federal Indian law is grounded in a more ancient sovereignty, which is an 'internal, culture-and-community-based model of sovereignty' that reflects the identity of Native peoples as the first Nations of this land. The concept of cultural sovereignty is a valuable basis or the construction of an indigenous right to self-determination because it is constructed from within Native societies, rather than from the outside by the federal courts of Congress, who struggle to determine the limits on inherent sovereignty" (243–244). The importance of the natural world for this richer source of power is underscored by the term *biocultural sovereignty* (Varese and Chirif 2006; Baldy 2013). Whether one considers traditional political or "biocultural" conceptions of sovereignty, relationships with the land are the basis of political power. And on either count, fire suppression diminishes Karuk political agency. In the immediate legal sense, fire suppression and firefighting activities such as backburning and fire-retardant drops hold the potential to erode sovereignty over tribal lands and resources because they interfere with the ability of the Karuk Tribe to exert management authority and for tribal members to access the land and carry out cultural practices. Ecological alteration may diminish the legal basis for political sovereignty if places in the landscape are no longer accessed and managed.[21] Yet perhaps more important, fire suppression alters the deeper bases of Indigenous political power in that it alters ecological functions, species composition, and diminishes the quality of relationships that may be sustained by that place.

In her articulation of the concept of biocultural sovereignty, Baldy (2013) draws from Varese (2006), who asks, "How could the indigenous outlast the European military invasion, the massive biological warfare, the systematic ecological imperialism and the meticulous destructuring of their institutions, and still initiate almost immediately a process of cultural and sociopolitical recuperation that allowed for their continuous and increasing presence in the social and biological history of the continent?" (Varese, quoted in Baldy 2013, 5). The answer for Varese and Baldy is that the power to survive both overall acts of hostility, as well as day-to-day sustenance and culture, comes from the earth.

CONFLICTS DURING FIRES: FIREFIGHTING IMPACTS TO TRIBAL SOVEREIGNTY

Not surprisingly, the activities that take place when fires happen represent one of the most obvious dimensions of how colonialism extends into the present via land management actions. As a mountainous fire-adapted region prone to lightning, the forests of the Klamath mountains experience frequent fire. Lightning strikes start multiple fires every year in a fire season that has expanded in recent decades to a period from June through October in Karuk territory. Just as the general forest fire suppression *policy* emerged from a European sense of how the forest is valued, so do firefighting *activities* reflect the economic, political, and cultural values of the dominant non-Native world. As I write, these fire policies and firefighting activities are at the center of direct struggles over tribal sovereignty and management authority. Karuk management authority is grounded in tribal occupancy and presence, including the longstanding practice of traditional management. As a sovereign government, the Karuk Tribe claims jurisdiction over membership, lands, and territory, including the right to manage air, lands, waters, and other resources as specified in the Karuk Constitution. This jurisdiction is recognized in Article II, Sections 4 and 5 of the Karuk Constitution, which states: "The laws of the Karuk Tribe shall extend to: 4. All activities throughout and within Karuk Tribal Lands, or outside of Karuk Tribal Lands if the activities have caused an adverse impact to the political integrity, economic security, resources or health and welfare of the Tribe and its members; and 5. All lands, waters, natural resources, cultural resources, air space, minerals, fish, forests and other flora, wildlife, and other resources, and any interest therein, now or in the future, throughout and within the Tribe's territory."[22] The Tribe operates within a complicated cross-jurisdictional terrain in which Karuk management authority is often unacknowledged and misunderstood, contested, and/or ignored by a complex slew of agencies, including the Environmental Protection Agency (EPA), U.S. Fish and Wildlife Service (USFWS), BIA, National Resources Conservation Service (NRCS), USFS, California Department of Forestry and Fire Protection (CALFIRE), the State Water Board, and California Department of Fish and Wildlife. And as cross-species relationships are disrupted by the actions of non-Native land management agencies, so too are Karuk political powers and cultural capacities diminished. In particular, Karuk Aboriginal Territory is located within the National Forest System. The Karuk Tribe has never relinquished possession of these lands, yet a lack of recognized ownership or jurisdiction limits of the Tribe's ability to care for traditional foods and cultural use species, as well as establish and maintain effective tribal programs. "These so-called Land or resource management agencies," states Karuk Department of Natural Resources Director Leaf Hillman, be they federal or state, refuse to acknowledge or respect the legitimate territorial boundaries of the Karuk Tribe. Agencies are simply an extension or arm of the governments they serve—in this instance, the United States of America and the State of California. As such, to acknowledge or recognize the territorial boundaries of the Karuk Tribe would be tantamount to an explicit admission that the continued occupation and rule by the United States is illegal and illegitimate. . . . Make no mistake, the ugly reality that we, the Karuk People, face every day is that we live in the Occupied Territory of the Karuk Tribe—and that occupation is brutally repressive by nature. Every expression or assertion of Karuk Sovereignty over our Territory necessarily represents a diminishment of both federal and state sovereignty—and vice versa.

When wildfires occur, large numbers of nonlocal firefighters with little or no knowledge of Karuk presence—much less the federal tribal trust responsibility—descend upon the area. If the Forest Service has adopted the view that fires are to be "fought" and have developed the military command structure to do so, it is no wonder that residents on the ground describe the experience as like being in a war zone. Simply the terms *firefighting* and *firefighter*, as well as maxims such as "fight fire with fire," belie the relationship the Western world has with this natural element. Further, there are no working synonyms for these words in the English language, and they are used without further reflection. In contrast, the Karuk Tribe emphasizes the "use" of fire as a "management tool" and its import for "fire-adapted" and "fire-dependent" species.

Given that Karuk sovereignty is linked to the ongoing ability to manage the lands, many actions taken by the state threaten tribal sovereignty on both direct and indirect levels. For example, during wildfire events, firefighting activities such as ridgeline fire-line construction and aerial retardant drops may directly damage cultural resources and interfere with the ability of Karuk people to hold ceremonies and take part in cultural practices. Decisions about what is to be protected and how to protect it, including backburning, constructing fire lines, and using chemical fire retardants, all too rarely support tribal values or perspectives. In fact, the outcomes of these decisions often profoundly damage the ecology of the region and culturally important food species and gathering areas. These impacts are blatant, are often immediate, and have lasting consequences.

During the large wildfires that now occur almost annually, agencies adopt a crisis management mode that often overrides existing mechanisms for tribal input. Fear of fire, coupled with agency and individual ignorance regarding tribal trust responsibilities, supports this crisis mentality in which emergency actions are taken without considering the long-term implications these actions may have for the local tribal community or the ecosystem. For example, major actions in the forest known to cause environmental damage such as road building and cutting trees near riparian areas normally require formal review and evaluation through the National Environmental Policy Act (NEPA). During wildfires, however, the decision-making process moves to a short timeframe with "emergency status" and hierarchical structure in which the NEPA is not required for significant management actions, including cutting trees in riparian areas or building roads. As Bill Tripp explains, "Individuals within the U.S. Forest Service and California Department of Forestry and Fire Protection (CALFIRE) make decisions with long-term consequences for the Tribe and ecosystem very quickly with little information about Karuk knowledge, values, or presence. These individuals do not have long-term connections to the watershed, and base their decisions according to non-Tribal criteria. Actions such as falling trees or use of heavy equipment or road building that are not normally allowed in riparian or wilderness areas will be given emergency exemptions during a fire." Unilateral decisions and actions taken during wildfires by the state agency CALFIRE or the USFS without tribal consultation also impact Karuk sovereignty and management authority. These decisions often originate from nonlocal incident commanders who are unfamiliar with local terrain, the local communities, or Karuk jurisdictional authority. Leaf Hillman describes how "noise and intrusion from the use of helicopters when ceremonies are under way and the use of 'federal closures' that denies people access to public lands are a de facto form of martial law, especially when armed officers enforce closures by arresting people for trespassing on their own lands."

Bill Tripp describes how what are known as "cultural vegetation characteristics" such as mature stands of tan oak trees that are remnants of generations of past traditional Karuk fire management have been bulldozed or intentionally burned during wildfire suppression efforts to create "fuel breaks" (Bill Tripp, personal communication). Ridge systems with significant beargrass components (important for weaving) have been bulldozed to create fire lines. In contrast, Tripp notes, "In a cultural management scenario they would be frequently burned in a manner that the vegetation type itself would serve as a natural fuel break." USFS fire ecologist, Karuk descendent, and cultural practitioner Frank Lake has also observed how fire lines and roads may be placed through culturally important areas, vital tree species such as tanoak or black oak may be intentionally cut down or severely burned as fire breaks, and snags that are important habitat for other cultural use species, including Pacific fisher, are likely to be cut as a preemptive tactic justified for "human safety" (Lake, personal communication). Not only is the sophistication of Indigenous fire knowledge ignored in these decisions, but Indigenous cultural resources are also directly impacted. Moreover, the longer that fire continues to be suppressed, the greater the fuel buildup (deferred risk), and thus the greater the difficulty in returning to Karuk fire regimes. Returning fire to the land is riskier under the new "artificial" situations of extreme fuel buildup and climate change.

The impacts and scale of any particular fire take place in a complex terrain in which both humans and other species exert agency. Within this terrain, Native and non-Native ideas of how the forest should be used have played out over time and are manifested in a given forest structure and species composition in the present. As a result, any present-day threat to tribal management authority during a particular fire event is also shaped by the history of past agency management actions on that site. Dr. Frank Lake has described how culturally significant historic trails along ridges are places where Karuk people have used fire frequently to keep open travel routes and access gathering sites (Lake 2013, Lake personal communication). Ironically, in the early days of fire suppression, these ridges and trails were the most easily accessed and became the focus of fire suppression activities, in turn causing particularly large fuel buildup over time (Lake 2007, 2013). Now during fire events, these same ridges and trails are often used for fire lines, meaning that sections will be "blackened"—a practice that causes direct mortality even to fire-adapted resistant cultural use species and may sterilize the soil. As a result of this combination of past fire suppression and the creation of fire lines, culturally significant trails and ridges have some of the highest degree of imposed alteration of their historic cultural fire regimes. Lake notes that "most of these structures that are being protected are white people's homes that are located on top of old tribal village sites. And it is incredibly ironic that many of these ridge sites where fire lookout towers have been placed are amongst the most spiritually important sites."

The use of fire retardants is a particularly important issue that directly impacts tribal people and species of importance—especially given that a high percentage of the local community gets at least some portion of their food and drinking water directly from the forest. Presumably because of the assumption that people do not live in the forest, criteria for retardant use are set without regard for human occupancy, and there is no long-term monitoring or consideration of cumulative effects on water supplies.

Use of chemical retardants occurs without awareness or regard for this circumstance by people from outside the area, and the Karuk Tribe has little to no ability to assert tribal authority regarding the use of retardants. One Karuk elder, Marge Houston, described how the Forest Service dropped fire retardant onto her prime acorn and mushroom gathering area, a site that was just 100 to 200 feet from her home:

> This summer that there was a less than an acre fire here on the Indian allotment and what happened was the Forest Service came in after it was under control they come in and did a borate drop, or some fire retardant. They wanted to come down and do another one 'till we had to get out there and start screaming at them. And basically they said, "Well, we'll only do two drops." But you know there's some big issues there. Because first of all that fire was under control. Second, it was less than an acre of fire. And now we have a contaminated subsistence harvest area along

FIGURE 9. Fire retardant spraying over Karuk lands. Photo credit: Will Harling.

with other culturally sensitive areas. . . . They cut down my acorn trees. And they missed the fire to begin with. . . . Sprayed it everywhere but on the fire. I couldn't even breathe for three days. All the oyster mushrooms that I got up here on this [fire] I cannot eat. 'Cause they come up and they're pink. Just like the chemical that they sprayed. I can't eat that. I'm not going to be able to eat a mushroom off that tree again. Damn.

Karuk staff report that recordkeeping for locations of retardant drops is haphazard at best (e.g., helicopter pilots estimate locations of retardant drops rather than using GPS). The Karuk Tribe's water quality program manager has described how even when retardant drops have occurred into the river or other water bodies, tribal staff have been prevented from accessing the site to gather follow-up water samples (S. Fricke, personal communication).[23]

The command and control organization of the Forest Service fire response, together with frequent personnel turnover, also acts to limit the ability of Karuk tribal staff to communicate and influence management decisions. Karuk ecocultural restoration specialist and traditional practitioner Bill Tripp describes the devastating emotional impacts of trying to communicate Karuk perspectives on fire and to protect cultural resources in the face of Forest Service presence fighting the large fires of 2008.

In my situation I find myself quite a few times just to the point of asking why am I even here trying to do this? I should just go and be happy somewhere. On these fires, every two weeks you are dealing with new people, and you're going over the same things, and you are trying to re-justify every decision that was made where you were barely able to hold onto protection of one little piece of something. And then you're losing a piece of that because new people came 14 days later. And then you're losing another piece of that and another. And you spend your whole time going over everything that you just went over again, and again, and again. And losing a little bit every time. And it causes some serious mental anguish.

Fire modifies the landscape in a variety of ways. In recognition of this fact, Indigenous peoples across the continent developed sophisticated methods and practices for using fire as a key management tool. However, the sheer amount of fuel buildup and single-age stands—a direct effect of over a century of fire suppression—poses a directly visceral threat to culture resources, reduces tribal management authority, and threatens tribal sovereignty long term. In addition to these direct threats to tribal sovereignty and management authority, there are significant indirect, longer-term impacts when emergency mode overrides numerous mechanisms for tribal input or causes immediate impact damage to key species. These longer-term impacts to tribal management authority continue via both direct and indirect changes in the landscape. For example, nontribal management actions such as reseeding and replanting, sediment control, road building, and salvage logging may occur while a state of "emergency" continues to take precedence over adequate government-to-government consultation. A state of emergency also allows management officials to forego sufficient environmental review. The effects of these postwildfire reseeding and replanting activities often result in large-scale forest conversions to conifer-dominant, mono-aged stands. These radical modifications in forest type reduce biodiversity and create conditions that compromise culturally critical trees and herbs. Fire breaks are often later converted into new logging/access roads, which affects Karuk management authority by further enabling commercial resource extraction. Salvage logging undermines Karuk management authority by removing woody forest material that is vital to various species of cultural importance. In the past ten years, the two national forests occupying Karuk Aboriginal Territory have proposed timber sales in order to "salvage" trees burned during high-severity fire. This trend is consistent throughout the Pacific Northwest; the majority of timber sales now take the form of salvage logging operations. In 2016, the Westside salvage timber sale (following the 2014 July Complex fire) proceeded with emergency water quality exemptions from the EPA despite opposition from the Tribe.

Over the long run, the continued crisis orientation to fire may shape management decisions, precipitate exemptions in existing regulations on logging and other post–fire management activities, and inhibit the ability of the Tribe to conduct cultural burning. And, as more funding is earmarked for fire suppression, less is available for proactive fire management such as prescribed burns that can be carried out safely at appropriate times of the year. Furthermore, the fact that high-severity fires are often replanted with commercial conifer species creates a negative feedback loop leading toward more high-severity fires in the future (Odion et al. 2004) since plantations burn at higher severity than do "natural" forests (Key 2000; Weatherspoon and Skinner 1995).

On the more dramatic level of ecological alteration, repeated high-severity fires hold the potential to alter the ecosystem so substantially that much of Karuk Aboriginal Territory could be reset to an early seral condition that has a tendency to burn at high severity over and over again. The process for such conversions can occur if very severe fires occur in rapid succession before successful seedling recruitment, germination, and establishment for canopy cover can develop. At that point, the forest is unable to develop a fire-resistant stand structure and will remain as a fire-prone brush field. Along with this potential is the loss of species that may cause a domino effect through the entire ecosystem.

Although the politics of fire and fire suppression come vividly to light especially when coupled with climate change, state interventions in relationships with the natural environment remain a leading edge of colonial violence in other ways as well. This occurs through natural resource policies that set water quality standards so low that Karuk cultural uses of the river are unsafe, through direct prohibition of the practices of tending the landscape through fire, and through imposition of non-Native regulations for food acquisition such as hunting, fishing, and gathering.

FIRE, CLIMATE CHANGE, AND COLONIALISM

> Though the climate destabilization described in Anthropocene futures may be a distinct ecological challenge for indigenous peoples, we experience it nonetheless as associated with an iteration of patterns of industrial settler strategies and tactics that is very familiar to us from our experiences with and memories of the other kinds of anthropogenic environmental change.
> —Kyle Whyte (2017a)

Many have now described the particularly poignant cultural, political, and spiritual impacts for Native people resulting from climate change—including impacts to tribal hunting, fishing, and gathering rights as species move and the loss of physical land due to sea level rise and from increasing storm surges in the face of reduced ice and coastal erosion. In addition, tribes face loss of political standing through shifting jurisdiction in light of the changing climate. On the Klamath, the politics of fire exclusion has gained momentum in the context of climate change more generally. Climate change both increases the prevalence of catastrophic fires (and all their associated social and ecological impacts) and has brought these issues to the table with a new sense of immediacy. In the context of both fire exclusion and the changing patterns of temperature and precipitation, fires have significantly increased in severity and size. The average number of fires over 1,000 acres has doubled in California since the 1970s at the same time as high-severity fires have become more frequent. These changing patterns of fire behavior affect specific Karuk tribal traditional foods and cultural use species, create infrastructure vulnerabilities for Karuk tribal programs, and pose broader implications for the Tribe's long-term management authority and political status as I have just described.

Yet as real as a threat as climate change may be, Karuk and other Indigenous peoples tend to have a different perspective on it than either the dominant public or the environmental community. Not only is climate change often experienced as one more problem on a continuum of settler-colonial actions threatening the survival of Indigenous communities and ecologies, but it is also often understood as a manifestation of the untenable nature of the settler-colonial worldview and political economic system. That is, both global climatic drivers in the form of carbon emissions and the local management actions that shape changes in fire behavior are experienced as vectors of colonialism.

Climate change is anthropogenic, or human caused. But humans have existed on earth for a long time. In the big picture, the organization of economic activity around fossil fuel extraction and use results from specific and very recent management decisions regarding, for example, the extraction of coal as a fuel source, the organization of an elaborate global and national transportation infrastructure, the globalization of economies, and militarization—each of which is in turn undertaken within the logics of capitalism and colonialism. Predicted climactic changes described by the scientific community are not "inevitable" acts of nature, but equally, climate change is not an inevitable outgrowth of human activity. Not only can climate change be understood as an outgrowth of colonialism, but ongoing colonial dynamics also shape the ability of Indigenous people to respond: "As in the past, industrial settler campaigns today also obstruct the efforts of indigenous peoples to respond [to climate change]—from legal and diplomatic failures to mitigate dangerous climate change by lowering emissions to the enactment of laws, policies and bureaucratic institutions that stymie indigenous efforts to adapt within current confines such as reservations" (Whyte 2017a, 4). I will elaborate on this discussion of climate change in the final chapter.

CONCLUSION

My aim here has been to use a detailed history of human relationships with fire in the Klamath Basin to illustrate the importance of three dimensions of settler-colonial theory that are highly relevant for sociological analyses and the social sciences generally but as yet underdeveloped in our theories: (1) the notion that North American colonialism is an ongoing structure rather than a past event, (2) the structuring of state relationships with Indigenous peoples in terms of elimination and replacement, and (3) the centrality of land and other beings in the natural world to the operation of both Indigenous and state power. Taken together, the alteration of Indigenous ecologies through fire exclusion has become a key mechanism through which the state has attempted to "uproot and destroy the vitality and autonomy of indigenous modes of life" (Coulthard 2014, 4). Indigenous resistance has been organized around and draws strength from deep ties with species in the so-called natural world. Indeed, on the Klamath, it is through human relationships with fire that colonial dispossession and its resistance are most clearly ongoing today. In particular, I have emphasized ecological alteration as a dynamic and ongoing process that continues to produce what Bacon (2018) calls "colonial ecological violence" through its diminishment of tribal political power today. Furthermore, *just as colonialism is not a single event of the past, we must think beyond the notion that "land theft" and land dispossession are single events of the past.* Instead, colonialism is an ongoing process that takes place through the alteration of land, the alteration of species composition and ecological structures, and the alteration of relationships between people and the nonhuman entities known as nature.

These dynamics of colonialism and Indigenous resistance clearly matter for specific sociological subfields—for example, understanding Indigenous experiences, or the nature of state power as I have emphasized here. And the incorporation of settler colonial theory into sociology as a whole is fundamentally critical. Just as gender scholars have articulated the broader social importance of sexism for a wide range of social dynamics, just as race scholars note that race and racism structure institutions, culture, and sociological understanding at large, colonialism is best understood as "the inherited background field within which market, racial, patriarchal, and state relations converge" (Coulthard 2014, 14–15). If, as Go (2017) writes, sociology "was formed in the heartland of empire, crafted in its milieu, and was thus embedded in its culture" (195), it is now due for an epistemic shift. In order to do this, Go argues that we must "recognize that sociology's dominant theoretical concerns, categories, analytic operations, and assumptions are not devoid of history and social location but are socially situated (like all knowledge is)" and "acknowledge that, because of this, sociology sometimes bears the imprint of the worldview, interests, and concerns of the socially dominant, hence of a very tiny if not miniscule group of people" (196). As I write, many such insurgencies are under way. Twenty-five years ago, Snipp (1992) observed that "American Indians have remained outside ordinary sociological inquiry" (352), a trend that has unfortunately continued. Over a decade ago, Wilkes and Jacob (2006) noted that Indigenous people are underrepresented in the academy and furthermore, "with the exception of anthropologists and to some extent historians, it is clear that for the most part, social scientists have not been particularly interested in Indigenous peoples" (423). And the two issues are linked. The exclusion of Indigenous voices and perspectives from the discipline is linked to larger epistemological limitations for the field as a whole. As Go (2017) observes, "After all, the exclusion of certain social groups in sociology has both reflected and been a part of this marginalization. It has contributed to the suppression of certain social standpoints, epistemic frames, social concerns, and categories, and therefore certain social knowledges" (194). These trends are especially notable in sociology, which lacks an ASA section on Indigenous peoples or colonialism, and few scholars identify as Native American. Content analyses of sociological scholarship on Native people indicate a marked tendency toward pathologizing their experiences through emphasis on drug and alcohol use and domestic violence (Bacon 2017). Scholarship by and about Indigenous peoples continues to be dramatically underrepresented in the discipline. Indeed, Indigenous sociologists and non-Native researchers whose work focuses on Indigenous experience describe receiving an interesting reception within the discipline. Papers submitted to generalist journals receive desk rejects with notes indicating the work was outside the scope of material published in the journal. One reviewer of a recent grant to the ASA on the topic of settler-colonialism and Indigenous peoples indicated that they had "difficulty seeing the sociology in this proposal" even though the reviewer "recognized several sociologists on the invited list."

Why does this all matter? If one aim of my project is to illustrate the need for sociologists to more fully conceptualize the circumstances of both Indigenous peoples and the natural environment, what messages do we take home from this investigation of fire policy and colonialism? Certainly, it may be important to take steps to incorporate Indigenous peoples for better sociological understanding of Native American experience and topics like colonialism and race, but how do these topics apply to the field broadly? Indeed, taking seriously the settler-colonial nature of the United States reorients a wide range of sociological topics by extending the terrain of inquiry and fundamentally challenging default paradigms. The broader application of colonialism for sociological theory will continue to play out with respect to topics of sociology of health in chapter 3, the alteration of gender structures and understanding of masculinity in chapter 4, sociology of emotions in chapter 5, and climate change in the Conclusion.

3 · RESEARCH AS RESISTANCE
Food, Relationships, and the Links between Environmental and Human Health

> Indigenous research is vital for challenging deeply embedded colonial assumptions and practices that have long constituted Native Peoples as objects of research rather than as authorities on their own lived experiences.
> —Dwanna Robertson (2016)

> Our understandings of the world are often viewed as mythic by "modern" society, while our stories are considered to be an alternative mode of understanding and interpretation rather than "real" events. Colonization is not solely an attack on peoples and lands; rather, this attack is accomplished in part through purposeful and ignorant misrepresentations of Indigenous cosmologies.
> —Vanessa Watts (2013)

Most academic knowledge generated today advances settler-colonial power, even when its authors might desire otherwise. This is so because deeply embedded cultural conceptions concerning what counts as knowledge, who can and cannot count as knower (and in turn who is research subject), and the entities and forms of knowledge shape both academic theories and methods. Questions of the nature of agency and the animacy of the natural world in turn underlie differences between Indigenous and so-called Western science conceptual frames.

My own relationship with the Karuk Tribe is a direct outgrowth of the politics of the struggle between settler and Indigenous knowledges. Power companies are granted licenses to operate dams in the public interest by the Federal Energy Regulatory Commission (FERC). In 2006, the license application for the Klamath River dams expired. In advance of this expiration, the power company that owned and operated the Klamath Hydroelectric Project applied for a new license. Their application initiated a relicensing process that was to involve years of scientific review, technical input, and participation from impacted communities—

all to evaluate the public benefits and impacts of the dams. Beginning in 2001, my colleague Ron Reed served as the Karuk tribal representative in the hydroelectric relicensing process, attending meetings for one week of every month for three years. Due to the remoteness of the area, these meetings were held in locations two, three, and even four hours' drive from his home—thus requiring extended time away from his family. During the years that Ron was attending meetings and articulating the tribal impacts of the declines in salmonid and other species, his mother and several of his aunts passed on—all of them in their seventies or younger. Ron became convinced that the lack of healthy food, specifically the loss of salmon, was directly impacting the health of his people. He began making links between the lack of healthy food and the high rates of diabetes, heart disease, and decreased life expectancy in his community and spoke passionately about this problem in the meetings.

The Klamath River Hydroelectric Project consists of a series of four run-of-the-river hydroelectric dams built between 1908 and 1964 that together produce a maximum potential of 150 megawatts of power. In contrast to dams designed for storage, irrigation, or flood control, these dams have had electricity generation as their primary function. Unfortunately, these dams have no fish passage, and little if any of the minimal electricity they produce has remained in the Klamath Basin. Instead, for some 45 miles of Karuk Aboriginal Territory along the main highway and river corridor, and another 38 miles of the river corridor through the Yurok reservation, the inhabitants of the basin generate their own power or do without. In contrast, however, the dams have major environmental and cultural impacts. Among the most important of Karuk foods have been the multiple runs of salmon that provided nearly half the total calories and protein of the pre-contact Karuk diet. Karuk fish consumption prior to the genocide of the 1850s is estimated at the enormous figure of 450 pounds per person per year (Hewes 1973). Over the past century, populations of salmon and other riverine foods, including lamprey, sturgeon, and steelhead, have been dramatically reduced by the diversion of flows, degradation of water quality conditions, disease, overfishing, and especially the series of dams that block access to over 150 miles of spawning and rearing habitat. Yet remarkably, despite 100 years of environmentally damaging practices, salmon continued to provide a significant food supply to Karuk people until about a decade after the construction of Iron Gate dam in 1964, which blocked access to 90% of the spawning habitat of this critically important salmon run. In fact, at least during fishing season, many Karuk people continued to consume salmon up to three times per day into the 1980s. Now, given that per person salmon consumption has recently averaged less than 5 pounds per person per year, Karuk people appear to have experienced one of the most dramatic and recent diet shifts of any tribe in North America (Reed and Norgaard 2010; Norgaard 2005).

FIGURE 10. Ron Reed speaking on dam removal at Sacramento rally. Photo credit: Kari Marie Norgaard.

In February 2004, PacifiCorp filed its final license application. Despite years of input from the Karuk and other tribes, commercial fishermen, scientists, and environmental groups who gave extensive testimony as to the ecological and cultural impacts of the dams, the power company claimed that there were "no downstream impacts from their operation below the dams." In the words of Ron Reed, "The document was five feet tall and contained no mention of our needs."

It was at this time that Ron and I began working together. I had a postdoctoral position at UC Davis and was living part-time in the Klamath area conducting a project on community mobilization regarding herbicide use. Karuk basketweavers were key in the organizing against herbicides, and I had interviewed a number of weavers in the process. Ron and I were introduced and began to work together. Over the next several months, we carried out a study and produced a report that made visible the relationships Ron had identified between ecological and human health.

In October 2004, the Karuk Tribe filed our preliminary report entitled "The Effects of Altered Diet on the Health of the Karuk People" with the Federal Energy Regulatory Commission (FERC). This report, the initial draft of which was written in a few short months, used a design and approach that combined community perspectives and understanding with traditional ecological knowledge and Western science. Our central finding was that Karuk people face significant and costly health consequences as a result of *denied access* to many of their traditional foods, especially salmon. Diabetes rates in particular had skyrocketed as salmon consumption dropped over the past several decades. Whereas diabetes was unheard of prior to 1950, it became more common in Karuk country in the 1970s. By the time of our survey, the total diabetes rate was estimated at 21%—four times the national average. Everyone we spoke to reported having a family member with diabetes. Furthermore, our report described how the lack of traditional food affects tribal members through a marked decrease in the nutritional content of specific foods and the physical activity that would have been exerted to procure them. Food security has become a major issue. We described how the Klamath River dams had been generating energy for users outside the basin and wealth for the power company but yielding no local benefits. Not only have the Karuk and Yurok Tribes remained without the benefits of the power generated, but the dams have deprived the tribal communities of the salmon that provided food, health, identity, cultural continuity, and the basis for their subsistence economies.

Our work was innovative in the policy world for the links that it made between environmental and human health—no other Tribe had claimed that dams were giving them artificially high rates of diabetes in a federal process! This report and the media attention it received generated significant public attention and became a key piece in the ongoing move toward dam removal (Gosnell and Kelly 2010; Leimbach 2009). In contrast to the widespread neoliberal emphases on education and personal responsibility within the health field, the Karuk Department of Natural Resources had successfully articulated how natural resource policies structured trial health. Since the release of the report and the political pressure it brought to bear, the Karuk Tribe went on to be part of the successful negotiation of a settlement agreement for dam removal.[1] From there, they effectively influenced the California state water quality standards for the Klamath River to include Native American Cultural and Subsistence Beneficial Uses (Stoll 2016) and have begun to make visible the role of human health in forest policies regarding fire suppression and climate change (Karuk Tribe, 2016; Norgaard 2014a, 2014b). A decade later, the report was again an important piece in the environmental impact statement for dam removal.

Since introducing the link between environmental health and environmental justice into the public discourse, the Karuk Tribe has carried on its work to reveal the profound connections between humans and their environment. Utilizing a combination of traditional knowledge and Western science, the Tribe continues to articulate the environmental contexts of human health in policy struggles related to water quality, fire suppression, and climate change (Diver 2016; Karuk Tribe 2016; Kann, Corum, and Fetcho 2010; Norgaard et al. 2013; Stoll 2016).

Decades of frustration and thousands of disrespectful encounters have led to a virtual cannon of literature on the pitfalls of academic work with Indigenous communities (Smith 2013; Cochran et al. 2008; Wilson 2008). My aim in this chapter is to twofold: first, to show how Western knowledge systems justify, maintain, and further settler-colonial preeminence and, second, to reveal the untapped potential value embedded in Indigenous knowledge systems. There remains more to be said about how scientific knowledge systems themselves serve colonial power and how different methodologies and disciplinary trajectories contribute more or less to the power of the settler state. There is also more to be said about the innovative ways that tribal communities have circumvented such knowledge politics to successfully transform policy to their advantage. Our work serves as an illustrative example of these dynamics and a positive illustration of how tribal communities are seeking to change this trajectory. Furthermore, stories of collaborative Indigenous environmental health research hold valuable lessons for the fields of environmental health, environmental justice, and food sovereignty—each of which could do much more to center Indigenous perspectives in their approach, framing, and tactics (Grey and Patel 2015; Middleton 2010; Vickery and Hunter 2016; Whyte 2015, 2018d).[2] Initially, Indigenous perspectives played vital roles in the formations of both environmental justice and food sovereignty movements, but these perspectives have since become subsumed by non-Indigenous outlooks and experiences. This development is particularly troubling because not only does it further settler-colonial dominance, but it also weakens the potential of these important movements.

In their key essay, "Decolonization Is Not a Metaphor," Tuck and Yang (2012) describe how such erasure of Indigenous contributions and perspectives is an ongoing aspect of settler-colonialism itself: "Settler colonialism is built upon an entangled triad structure of settler-native-slave, the decolonial desires of white, nonwhite, immigrant, postcolonial, and oppressed people, can similarly be entangled in resettlement, reoccupation, and re-inhabitation that actually further settler colonialism. The metaphorization of decolonization makes possible a set of evasions, or 'settler moves to innocence,' that problematically attempt to reconcile settler guilt and complicity, and rescue settler futurity" (1).

ACADEMIC RESEARCH AND SETTLER-INDIGENOUS KNOWLEDGE POLITICS

If settler-colonialism is legitimated by the erasure of Indigenous peoples, much past academic research has contributed to this project by erasing Indigenous perspectives and knowledges. In the twenty-first-century world of today, scientific notions of land management (still) mediate raw state power in the organization of species and entities in the natural world. Water quality monitoring can detect sediment loads or forms of contamination that may prevent further developments, and the need for endangered species habitat affects proposed timber sales. Science has been an important vehicle for affected communities to document impacts to land management projects. But if, as Azibuike Akaba (2004) writes, science is "a double edged sword," then this is even more so for Indigenous peoples. In the politics of fire and landscape restoration, divergent ways of knowing (i.e., knowledge systems) are at odds. This is revealed not only through different worldviews but also through different social, economic, and spiritual systems and values. If settler-colonialism is legitimated by the erasure of Indigenous peoples, much past academic research has contributed to this effort to this project by erasing, ignoring, and/or disrespecting Indigenous perspectives and knowledges and now by appropriating, commoditizing, and subsuming such perspectives and knowledges. Even beyond the particular knowledge frameworks, the pathways to knowledge acquisition serve different players in the balance of Indigenous and settler-colonial power. While serving as the Karuk Tribe's cultural biologist, Ron Reed had given years of oral testimony regarding relationships between the declining salmon runs, changing diet, and rise in diet-related diseases. He had described in detail the relationships between state policy and community and environmental health. Ron's observations and knowledge did not, however, count until his words were published in a report written by a non-Native academic with a PhD (myself).

For a community with contested land status pursuing survival in the twenty-first century, Western science ways of establishing truth are the dominant force legitimating various possible ecological relationships within Karuk Aboriginal Territory from timber sales to dam removal or the tactics of responding to forest fires. Faced with this reality, the Karuk Tribe has become an adept player in effectively utilizing Western science to support tribal perspectives. "It's the only way they'll listen. We gotta keep proving our traditional knowledge is right—has always been right—by having some 'ologist' do some study."[3] The Tribe has committed substantial resources to their fisheries, forestry, and water quality programs. External academic researchers like myself have been a key in not only translating across forms of knowledge but also legitimating Indigenous knowledge within Western frameworks.

Collaborative work by Elizabeth Hoover et al. (2012, 2015) and Hoover (2018), Katsi Cook, Nina Wallterstein, Bonnie Duran, Vanessa Watts Simmonds, and

Kelly Gonzales makes key contributions to understanding Indigenous perspectives on health, the influence of settler-colonialism on health, and the benefits of collaboration across very different epistemological and ethical frameworks. Elizabeth Hoover (2017) describes how "such partnerships do not merely change how we understand—they also alter *what* and *for whom* we learn" (5). Yet while collaboration has been at the core of environmental justice scholarship from its origins, too few such collaborations have been with Indigenous communities. And while there is an established literature on the benefits of collaborative community-based science for environmental health research, less has been written about the benefits in drawing upon Indigenous cosmologies and knowledge frameworks for reshaping either research frameworks or policy debates. Extremely few sociologists have been a part of this conversation. As Kimberly Huyser put it at the American Sociological Association (ASA) meeting in Montreal, "American Indians are over-researched and under-published" (2017, personal communication).

What might happen if academic researchers listened to what Indigenous communities have to say about health, justice, and ecology? Health studies rarely start with Indigenous perspectives on health, theorize the impacts of colonialism on Indigenous physical or mental health, or account for how health practitioners and researchers themselves enact colonial violence on communities—although this is beginning to change thanks to the hard work of a handful of key scholars.[4] In particular, the pathologizing of Native people is pervasive across most academic disciplines. At the more subtle unconscious end of the spectrum, it is so ubiquitous as to be nearly universal. Given that stigmatization has long been a key mechanism in the legitimation of oppression, and given that pathologizing of the body has been used against many groups (women, anyone not conforming to sexual norms, people of color), it should not be surprising that health research has been particularly fraught in this regard. With its connection to the natural sciences and its focus on the body, health research has long wielded a unique ability to reify deterministic notions of biological inevitability and social inferiority. Statistical documentation of suicides, disease rates, and interpersonal violence powerfully naturalizes a picture of a "weak" or "inferior" people in need of the control and intervention of the paternalistic state. Highlighting the challenging circumstances in Indigenous communities (high rates of diabetes or substance addiction) contributes to the pathologizing of Indigenous peoples. Cochran et al. (2008) aptly note, "No community wants to have the reputation of having the most alcoholics or the most people with mental disorders" (23). This is not to deny that real problems exist in Indian Country, but context and framing fundamentally matter. Researchers can be as well meaning as they please, but documenting conditions like alcohol abuse, interpersonal violence, or mental anguish without linking them to the longstanding structural forces of state-sanctioned genocide and colonialism launched against Indigenous peoples reifies the very notions of biological or cultural inferiority that have long justified the racist and colonial actions of the settler state. Here sociology has been particularly remiss. Jules Bacon (2017) describes how

> a keyword co-occurrence analysis of data from the Web of Science (WoS) Social Science Citation Index demonstrates the relative paucity of research on Indigenous peoples in the US, as well as an abundance of medical and risk associated discourses within the social science literature regarding Native peoples. The high centrality of the keyword "risk" along with the under-utilization of keywords that take colonialism or genocide into account reveals a system of presenting and organizing research about Native peoples in ways that accentuate medical, individualistic, and ahistorical approaches. These ways of thinking, by virtue of their erasure of colonial genocide, are complicit with the very problems they seek to address through the intellectual reification of the Indian as problem.

When sociologists promote particular theories, they have been powerful agents in the reproduction of white supremacy and the class system across a wide sweep of time and space. These frames are formidable means of legitimating state actions, as Aldon Morris's (2015) work on W.E.B. Du Bois powerfully illustrates. The negation of Du Bois's work as the first American sociologist was far more than a struggle between the egos or career aspirations of men at the time. The differences between the Atlanta and Chicago schools of sociology included the extent to which they theorized class and capitalism or biological deterministic arguments regarding race. Despite the arguably greater methodological rigor of work coming out of the Atlanta school, power, privilege, and white supremacy prevailed. Morris describes how Chicago became recognized as the leading school for sociology, setting up an early trajectory for U.S. sociology embedded in white supremacy. And just as Morris describes the battle between Du Bois and the well-heeled white scholars at University of Chicago at the origins of American sociology, there is a divide today between scholars at the center of the ASA and the few Indigenous sociologists and those working with Indigenous-centered theories such as settler-colonialism. Just as in Du Bois's times, this divide is significant far beyond the individual careers of the players but shapes the disciplines' ongoing relationship to the settler-colonial state. Ideas about race, about the other entities and beings known to many settlers as simply "nature," have been (and continue to be) promulgated by sociologists. What difference might it have made for Indigenous sociology today had Du Bois's approach and perspectives prevailed? And just as Morris changed the conversation with respect to black sociological perspectives, showing how disciplinary biases rooted in racism affected not only anti-black racism but also undermined sociological theorizing, it is my aim in this book to illustrate how Indigenous perspectives and knowledge offer radical, fresh insights for sociology, the social sciences, and academic knowing writ large.

In the past twenty or so years, there has been an explosion of literature on the benefits of Indigenous traditional ecological knowledge (TEK) for numerous aspects of natural resource management. Non-Native scientists and resource managers in the areas of fire ecology, forest management, fisheries, and climate change now draw upon these non-Western frameworks in their work illustrating how Indigenous knowledge can add details at spatial scales, bringing temporal depth and highlighting elements and relationships that Western-trained scientists overlook (Armitage, Berkes, and Doubleday 2010; Jacob and Blackthorn 2018; Leonetti 2010; McGregor 2008; Middleton 2011; Verma et al. 2016; Whyte 2013b; Williams and Hardison 2013). While both the fields of environmental health and environmental justice have a strong tradition of community collaboration and citizen science, a parallel recognition of the potential value of traditional Indigenous knowledge for research design or process has been slower and less transformative in the social and health sciences. Scholars, including Katsi Cook, Elizabeth Hoover, Bonnie Duran, Vanessa Watts Simonds, and Kelly Gonzales, are doing important work illustrating how powerful and transformative such collaborations can be, but the incorporation of Indigenous perspectives within environmental health is in its infancy. Not only do Indigenous communities have specific knowledge and unique perspectives to offer the research process, but incorporation of Indigenous environmental knowledge into social science research can also bring the added benefits of additional epistemological frameworks, as well as Indigenous values and ethics (Hoover 2017, Simonds et al. 2013).[5]

ENVIRONMENTAL JUSTICE AND FOOD SOVEREIGNTY

While early self-identified environmental justice efforts included important Indigenous activists, it has taken longer for the centuries-long fact of Indigenous resistance to colonialism to be understood as environmental justice struggles (Gilo-Whitacker 2017; Jacob 2013, 2016; Ranco and Suagee 2007; Manning 2018; Weaver 1996, 1997; Voyles 2015). Indigenous values, worldviews, and goals are yet to be reflected in broader conceptions of environmental justice. The erasure of Indigenous knowledge, presence, and leadership within the food sovereignty movement is among the most glaring erasures given that the origins of the term and movement tactics come from the Indigenous movement Via Campesina and the fact that in the United States, American Indian sovereignty is such a potent political force. Indigenous notions of sovereignty draw upon rich, morally grounded ecological relationships across species that have co-created the abundance needed for human survival and flourishing. What is called "food" is but the tip of the iceberg of these cross-species relationships, responsibilities, and moral systems (Reo and Whyte 2012; Whyte et al. 2018). As my colleague James Fenelon (2015) puts it, "What some analysts are calling 'food sovereignty' is really and interactive holistic set of relationships of food, land and more not fully understood. This is the base of production, consumption, and of life itself" (24). Nonetheless, the term *food sovereignty* is widely used with minimal understanding of its larger political meaning or of what Indigenous food sovereignty movements are fighting for or against. Instead, much of the food justice movement stands as evidence of the ongoing dynamics behind Tuck and Yang's (2012) essay, "Decolonization Is Not a Metaphor," in which the authors note the widespread erasure of Indigenous movements and theory that takes the form of appropriating terms with very specific political meanings like "sovereignty" and "decolonization":

> One trend we have noticed, with growing apprehension, is the ease with which the language of decolonization has been superficially adopted into education and other social sciences, supplanting prior ways of talking about social justice, critical methodologies, or approaches which decenter settler perspectives. Decolonization, which we assert is a distinct project from other civil and human rights–based social justice projects, is far too often subsumed into the directives of these projects, with no regard for how decolonization wants something different than those forms of justice. Settler scholars swap out prior civil and human rights based terms, seemingly to signal both an awareness of the significance of Indigenous and decolonizing theorizations of schooling and educational research, and to include Indigenous peoples on the list of considerations—as an additional special (ethnic) group or class. . . . Yet, we have observed a startling number of these discussions make no mention of Indigenous peoples, our/their struggles for the recognition of our/their sovereignty, or the contributions of Indigenous intellectuals and activists to theories and frameworks of decolonization (2–3).

There are many benefits of "decolonizing" environmental justice, environmental health, and food studies. Given that environmental justice, environmental health, and food sovereignty movements are about fixing relationships between people and the natural world, any understanding of what is wrong with those relationships would do well to look to the Indigenous ethics of care and responsibility that have shaped the flora and fauna of this continent for at least the past 10,000 years.

Furthermore, as environmental justice scholars, including Laura Pulido, David Pellow, Raoul Lievanos, Jill Harrison, Jonathan London, and Hilda Kurtz, have all noted, the reliance of environmental justice activists on the state for protection has not been effective. If we are to understand what is inhibiting environmental justice, environmental health, or food sovereignty movements today, we need to have stronger critiques—stronger explanations for the disproportionate exposure to hazards faced by communities of color and even stronger system critiques than existing analyses on the failings of commercial agriculture, capitalism, or racial capitalism per se.

The very existence of both the fields of environmental health and environmental justice reflects important developments regarding acknowledgment of interconnections between human and the environment, and the fields have continued to evolve toward more expanded notions of what counts as nature in recent decades (e.g., bringing notions of the environment into urban spaces and human bodies). Nonetheless, both fields remain caught in neoliberal frameworks in which individuals are the key unit of social analysis (Alkon and Mares 2012; Pulido et al. 2016; Swyngedouw and Heynen 2003). Whereas dominant health frameworks tend to decontextualize the social and political context of illness or disease (Krieger 2001, 2011), from Indigenous perspectives, social, political, cultural, and especially environmental dimensions of human health are key (Joe and Young 1994; Ferreria and Lang 2005; Kuhnlein, Erasmus, and Spigelski 2009). Whyte (2018b) notes from Anishinaabe perspectives, "It is common to look at the world as interrelated in ways that some people outside the Anishinaabe world do not always grasp, such as the complex interrelation of human health, storytelling, gendered and intergenerational relationships, cultural and ceremonial life, the intimacy of human relations with plants, animals and entities (e.g. water), and the moral responsibilities that come with family, clan, and band memberships." Part of the theoretical and broader public significance of this story of collaboration concerns the important connections between food and environmental health. In the absence of such context, work on tribal health in particular has negatively stigmatized and implicitly blamed Native communities for the health challenges they experience (Bacon 2017; Huyser 2017). While scholars have identified a neoliberal trend in health research, less has been said about the role of colonialism in shaping health outcomes or the environmental degradation that so often underlies them. In contrast, the Karuk Department of Natural Resources has built on academic collaborations to situate health outcomes from diabetes and obesity to mental anguish and food insecurity as outcomes of failed federal forest and riverine management policies.

To underscore the many benefits of incorporating Indigenous TEK within environmental justice and environmental health research, I will describe not only our findings but also our research process. The first part of this chapter engages knowledge politics. I will detail the methodology and findings of the altered diet study to illustrate how Karuk (and many other Indigenous peoples) notions of health, traditional ecological knowledge, and analyses of colonialism and health converged as an illustration of the powerful potential benefits of collaboration with Indigenous communities. The second part of this chapter discusses how Karuk and other Indigenous perspectives on health, as well as the framework of settler-colonialism, powerfully reframe conversations about food, health, and state power within environmental justice and food sovereignty movements.

METHODS: WEAVING KARUK TRADITIONAL ECOLOGICAL KNOWLEDGE AND WESTERN SCIENCE

In the summer of 2004, Ron Reed and I set out to assess impacts of the dams on the health, culture, and economy of the Karuk Tribe. Our work built on the environmental justice tradition in which affected groups collaborate with academics to conduct community-based science (Brown and Mikkelsen 1997; Brown 2013; Cordner et al. 2012; Lynn 2000; Pellow and Brulle 2005; Lockie 2018). Both Western science and Karuk traditional knowledge were fundamental to the framing of our research questions, as well as all aspects of our study design and interpretation. For example, one dimension of Karuk and many other Indigenous people's TEK is an emphasis on interconnection and attention to context. In this case, Ron's identification of the connections between salmon population declines, diet change, and community health shaped first our research questions and, from there, our methods. Our survey process, analysis, and writeup too were centrally influenced by Karuk TEK and local community knowledge in general. That this work was community based was fundamental. We knew we would need a historical approach to capture the change over time that Ron and others in the community described. As an "insider," Ron's local knowledge allowed him to identify data sources and provided a level of trust from the community we needed to gather information. Direct observations were also key. The fact that Ron and other community members could recall when salmon and other traditional foods had declined, the rise in diet-related diseases, and the temporal patterns when diabetes had become prevalent was essential for knowing what kinds of data we needed and what kinds of time frames would be meaningful to use in our survey design. Western science was also critically useful for documenting both health conditions and salmon populations—to that end, we examined tribal health records, used archival material, and conducted in-depth interviews and a survey.

Along the way, we encountered a number of methodological challenges that we were able to successfully navigate by using a combination of Western science, Karuk TEK, and community expertise. These challenges included constraints of time and money, minimal availability of historical data on either salmon populations or disease rates, the challenge of talking about changing diet without essentializing Karuk people into a static past, and the challenge of parsing out what Western scientific terminology calls "intervening factors" in diabetes occurrence such as alcohol use and decreased exercise. Sociological methods books extoll the benefits of triangulating across methods (essentially using more than one method to gain information on a question). Here we describe the benefits of triangulating across cosmological systems.

As the person responsible for academic context, the first challenge I encountered was an absence of comparable studies. When Ron Reed first approached me about documenting the heightened rates of diabetes and heart disease that he was seeing in the Karuk community, the relationship he hypothesized between the loss of salmon and increased disease made perfect sense. We knew that salmon are a healthy food, high in protein, iron, zinc, and omega 3 fatty acids and lower in saturated fats and sugar. I was also familiar with a number of tribal environmental justice cases involving the health and cultural importance of traditional foods. I assumed that other tribes were making links between environmental decline, the loss of salmon, and increase in diet-related diseases and figured that with a little investigation into the work of others, I would be able to find a template to draw upon in designing our project. Much to my surprise, however, this was not the case. While there were many examples of tribes grappling with the health effects of the loss of their traditional foods, their lack of access was due to environmental contaminants such as mercury and persistent organic pollutants rather than declining food availability (see, e.g., Kuhnlein and Chan 2000; Hoover et al. 2012). This literature, which had a justice framework, was not well linked with another body of medical and epidemiological literature describing high rates of diabetes and other diet-related diseases in tribal communities. In this second set of literature, tribes such as the Navajo had well-developed diabetes research programs that were carrying out extensive biological studies linking traditional diets with insulin levels and other aspects of blood chemistry. However, these studies had no mention of how diets or disease rates might be linked to environmental degradation or failed natural resource policies. Instead, this literature was based in the biomedical model, tending toward an individualistic focus, and sometimes even pathologizing participants. A third literature from environmental justice studies addressed the failure of the state to protect communities of color from environmental harm. This literature dealt with both state policy and disproportionate access to resources, but like the work on contaminants in traditional foods, the attention was on communities facing toxic exposure rather than species decline. Furthermore, work from non-Native communities lacked the depth of description of how identity, culture, and spirituality were connected to health or as dimensions of injustice voiced by Karuk people.

The focus of our work, while building upon all these sources of material, was different. Karuk people described their situation as a case of "denied access to traditional foods." Diet change—rather than being inevitable—was tied to ecosystem decline that was in turn a function of the improper management of the rivers and forests by federal and state agencies. As I illustrated in chapter 2, such policies are the leading edge of settler-colonialism today. The inability of Karuk people to continue to eat their traditional foods—and hence the corresponding rise in diet-related diseases—was not coincidental or the fault of Karuk people, who had somehow mysteriously become poorly educated with respect to healthy foods but had instead been produced through the failure of the state to protect tribal trust resources, despite their mandate to do so. Consequently, instead of having an existing template to follow, our study design and data analyses built upon and brought together well-established frameworks and findings within these three sets of existing literature.

We began with an analysis of medical records, archival material, and a small set of face-to-face interviews. This initial work clearly supported the premise of a link between altered diet and diabetes. In 2004, we received additional funding and expanded the project by conducting eighteen open-ended, in-depth interviews and a survey of all adult tribal members living in Karuk territory. The interviews provided detailed information from tribal members regarding health, diet, food access and consumption, and economic conditions over time. We purposely included people who had more access to fishing due to their location near the ceremonial fishery and others who had less access either because they lived farther away or came from families who were more assimilated. We spoke with people in their early twenties to their late sixties and from a range of backgrounds throughout the 100-mile length of the watershed within Karuk territory. Interviews were recorded, transcribed, and coded.

Initial themes of identity, community, responsibility, tradition, and resistance were immediately apparent from our interviews. Furthermore, Ron Reed and I spent many hours in conversation during the period we conducted interviews (often in the course of driving the long distances up and down the Klamath River to reach people). These conversations were important opportunities to discuss relationships between themes and to bridge gaps between my understandings as a non-Native academic and Ron's understandings as a Karuk insider.

While interviews allowed for in-depth descriptions of impacts, Western scientific methodologies hold out that there is always the possibility that the information received in this forum is not representative of the experiences of tribal members on the whole. Furthermore, there were limitations in the official medical records, including inaccuracies in the medical database and the fact that historical data in particular were not considered valid (not all persons with conditions are reflected in the medical records since an unknown number of tribal members are undiagnosed, use private insurance, or do not use the mainstream medical system). Statewide tribal diabetes analysts at the California Rural Indian Health Board indicated the Karuk as one of about five Tribes in California who were believed to have underreported diabetes rates. To augment both the interviews and medical data, we conducted the Karuk Health and Fish Consumption Survey in the spring of 2005. Here too, Karuk TEK was vital to our ability to develop questions that could lead to meaningful results. The survey contained sixty-one questions designed to evaluate the range of economic, health, and cultural impacts for Tribal members resulting from the declining quality of the Klamath River system, including questions designed for comparison with fish consumption data and self-reported prevalence of health conditions, including diabetes, heart disease, high blood pressure, and obesity. The survey questionnaire contained both questions drawn from previously conducted surveys by other tribes (e.g., by the Columbia River Intertribal Fish and Water Commission) and questions constructed specifically in the Karuk context as a result of information gathered during the in-depth interviews. The Karuk Health and Fish Consumption Survey allowed for the collection of quantitative data regarding economic patterns, health conditions, and fish consumption that have been long absent in the broader discussion of tribal impacts of riverine health. Family history information on these conditions was included, as well as information on age of death of family members. Medical and self-report data from the survey generally corresponded, suggesting accuracy in findings.

WHAT WE FOUND: THE EFFECTS OF ALTERED DIET ON THE HEALTH OF KARUK PEOPLE

The central finding of our research was the link between the dams, declining salmon populations, and changing diet. Overall, our report discussed the impacts of the severe decline in salmon runs using a combination of Western monetary economic values and Indigenous perspectives on the relationships between physical health, spirituality, culture, and identity. In later work, we also touched upon how changing salmon populations affected gender constructions and emotional health. These two themes are detailed in chapters 4 and 5 of this book. Our report was organized into five topical chapters that together laid out descriptions of people's current and pre-contact diets, the emergence of diet-related diseases, and the reasons for this shift. We wove different kinds of data sources together throughout our report and placed interview passages alongside figures and statistics on disease rates to better illustrate the context for medical and other quantitative data. This approach highlighted Karuk interpretations and explanations for the relationships between what Western science calls "variables." This presentation format itself thus presented information from traditional ecological knowledge and Western science side by side.

PRESENT AND HISTORICAL DIETS

Prior to the gold rush and influx of settlers to the Klamath region in the 1850s, salmon are estimated to have made up close to 50% of the energy and total protein in Karuk diets (Hewes 1973). Deer (*púufich*), elk (*íshyuux*), and tanoak acorns (*xunyê'ep*) are also of primary importance. Using anthropological reports and oral histories, we described how earlier generations ate from the land and the more recent limited access to traditional food that has forced the present Karuk population to buy most of their food in stores and rely on government commodities. These changes represent a major dietary shift. We did not have the data that might have better illustrated these relationships (e.g., there were inadequate predam fisheries population data for salmon returns). But through our construction of survey questions rooted in community expertise, we were able to use self-report data on household fish consumption by decade and thereby gain a rough idea of the pattern of change.

This chapter touches upon many issues, and at the core lie two facts: spring Chinook have been a particularly important food source for Karuk people, as well as the salmon run whose decline is most visibly linked to the construction of the dams given that nearly all the spawning habitat for this species had been blocked. But many species of traditional foods are affected by the dams and other forms of environmental degradation. Until quite recently, there was an abundance of riverine and upslope forest food resources to which Karuk people had access. This provided a safety net of foods should one species fail to produce a significant harvest in a given year. Thus, while salmon were centrally important, other food resources, including sturgeon, trout, eel, freshwater mussels, deer, elk, acorns, mushrooms, brodiaea ("Indian potatoes"), wild oats, huckleberries, and many more, were consumed fresh and also preserved to provide food throughout the year. Yet as of 2017, every riverine food species consumed by Karuk tribal members is in a state of decline. Now so few fish exist that even ceremonial salmon consumption is limited. Using a combination of data from interviews, the survey, and ecological studies, we found that at least twenty-five species of plants, animals, and fungi form part of the traditional Karuk diet to which Karuk people are currently denied or have only limited access. Furthermore, we stressed that the foods that were most central in the Karuk diet, providing the bulk of energy and protein—salmon and tan oak acorns—were among the missing elements.

SITUATING STRUCTURAL CONTEXT: WHY HAVE KARUK DIETS BEEN ALTERED?

Our report next laid out reasons for the diet shift. While our first goal had been to provide descriptions of the past and present Karuk diets, addressing the issue of why diets had changed led us to another set of methodological challenges. Existing literature on tribal diets, disease rates, and the health benefits of traditional foods was primarily from epidemiological traditions that excluded structural context, focusing instead on individual choice. Part of why structural factors have been so invisible in this framework is a function of the methodological framework itself.

Wing (1994) describes in his essay on the limits of epidemiology, "The dominant mode of epidemiological explanation takes place fully within the limits of a scientific practice that has been termed Cartesian reductionism, an analytical approach characterized by a focus on factors considered in isolation from their context" (74). Wing notes that "according to this logic, it is only by excluding the context and focusing on particular factors considered independently of historical conditions that science can produce objective knowledge that has a greater claim to authority than other forms of knowledge." Yet as detailed in this and preceding chapters, Karuk diet changes occur in the context of massive environmental degradation in which the habitat for key food species has become so damaged that their populations are now in decline. Ignoring this context in the name of "good science" would seem laughable were it not such a pervasive and serious problem. For it is when such context is excluded that Indigenous peoples are pathologized. Thus, in Indigenous communities across the continent, Native people experience stigma and internalize shame in the face of labels such as overweight and obese even as they continue to fight for their relations, their communities, and to carry on cultural practices. Such experiences of stigma wrought by academic researchers are yet another layer of colonialism beyond the original/ongoing genocide.

On another level, had we not paid attention to historical context, the key community explanation for the diet change would have be rendered invisible. Yet while we knew from Ron's experience and the testimonials in our interviews that the loss of traditional foods in the community was a major factor for diet change, how could we describe the complex of reasons for that change without essentializing Karuk people into some static past? How could we account for the many so-called intervening factors in diet change such as people moving out of the area or changing desires for traditional foods? What if a primary reason for diet change was that people had grown tired of eating salmon? Even had we wanted to use it, the traditional biomedical approach of controlling for external variables as explanations for behavior change was not an option due to funds and time. This represented a second area where we overcame a methodological challenge using a combination of Western science, traditional ecological knowledge, and community insight. Diet change and cultural changes are, after all, to be expected. Here, based on community perspectives from both Ron Reed's personal experience and our initial interviews, we knew that many people wanted to continue eating salmon and other foods but were unable to do so. This desire and frustration were impressed upon me again and again with heartfelt testimony. Thus, because we were aware of the local context, we were able to design survey questions addressing alternate reasons for diet change specifically, including questions about both changing patterns of consumption of various traditional foods and the reasons for these changes. We asked in detail about the timing and reasons for diet shift specifically in both our interviews and on our survey.

Through our survey, we learned that despite the reduced availability of salmon and other fish, a high percentage of Karuk families report that someone in their household still fishes or hunts for food, as indicated in Figure 11.

Unfortunately, whereas traditional foods from acorn to salmon, deer, elk, and mushrooms comprised major food sources in the living memory of adults, by the time of our survey, most families were unable to hunt, gather, or harvest adequate supplies of these species. For example, the 2004–2005 fishing season produced record-low harvest yields for a variety of riverine species, including

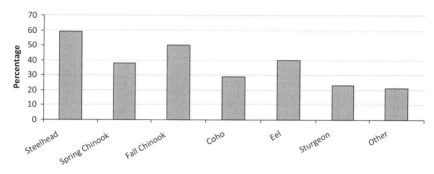

FIGURE 11. Percentage of Karuk households who fish for various species, 2005.

eels (Pacific lamprey and other lamprey species), spring and fall Chinook salmon, coho salmon, and green sturgeon. Through our survey, we also learned that over 80% of households were unable to gather adequate amounts of eel, salmon, or sturgeon to fulfill their family needs. For example, most households that caught salmon, steelhead, eels, and sturgeon report catching ten or fewer, and no households report catching more than fifty eels or fall or spring Chinook salmon total for the season. Furthermore, 40% of tribal members report that there are species of fish that their family gathered that they no longer harvest at all. For most of these species, the decline was quite recent; over half of respondents report that spring Chinook became an insignificant source of food for their families during the 1960s and 1970s, although other families continued to gather significant amounts of these foods into the 1980s and 1990s.

We also drew upon archival material, in-depth interviews, and historical context. According to both Karuk observations and scientific literature, a number of factors either deny or limit the access of people to their traditional foods. We detailed how genocide and forced assimilation over the past 150 years have led to a loss of traditional knowledge of relationships with the land (including the preparation and acquisition of traditional foods) and, to a lesser extent, changes in the tastes and desires of people. Yet despite these dramatic earlier events, the testimony of elders about foods they ate until recently indicates that considerable changes have also occurred within the past generation. We were able to situate these most recent changes as largely due to denied access to traditional foods as salmon populations have plummeted, mushrooms have been overharvested by competing commercial interests, and acorn yields have been reduced in the absence of proper fire management. Just as Wing (1994) describes how Latin American scholars identified global inequality and scholars from communities of color have called for attention to institutional racism, Indigenous health researchers have called for attention to the settler-colonial structures that shape research methods and theoretical frameworks (see Hoover 2017; Hoover et al. 2012; Voyles 2015; Simmonds and Christopher 2013). Powerfully holistic Indigenous health perspectives point to the need to expand not only attention to the context of colonialism as a negative health driver but also understandings of health as embedded in cultural and community relationships, including community members in the natural world (Whyte 2015, 2016d). Neither this research on denied access nor the policy impact it generated would have been possible without this context because none of the relationships we were discussing would have been visible. And only through attending to community-based knowledge and historical perspective were we able to understand or communicate the meaning of the impacts of the dams—far beyond mere health impacts.

LINKING HUMAN AND ECOLOGICAL HEALTH: THE EMERGENCE OF DIET-RELATED DISEASES

The next challenge in our report was to describe the relationships between changing diets, changing ecological conditions (the loss of salmon in particular), and the occurrence of diabetes and other health conditions. Here again, the combination of methodological approaches and the Indigenous understanding of whole systems was essential. Medical records indicated a 21% diabetes rate for the Karuk Tribe, nearly four times the U.S. average. The estimated rate of heart disease for the Karuk Tribe was 39.6%, three times the national average. But we knew from larger context and interviews that such conditions were relatively new and wanted to show the trajectory of this situation. Here we encountered another key methodological problem, lack of historical data on health conditions over time. Tribal medical records only went back several years. There was nothing as far back as the 1970s and 1980s that would correspond with the time period in which we were aware that the diet change had been so rapid. Here too we relied on collective understanding and community knowledge to situate the information we had available and to direct our efforts toward alternative data sources (i.e., the self-reported information via the survey). Self-report data on specific years when diseases first appeared in families seemed unlikely to be reliable, and therefore we felt that asking people to recall the decade in which those conditions first showed up in their families would be a reasonably accurate time scale. Using this approach, we were able to trace the fact that even though diabetes rates were epidemic at the time, diabetes has only recently appeared in the Karuk community. Self-report data from the Karuk Health and Fish Consumption Survey indicated that diabetes first appeared in most families (over 60%) beginning in the 1970s, as shown in Figure 12.

Survey and medical data together thus indicated as the presence of healthy traditional foods declined in the Karuk diet, a series of new health problems emerged. The loss of spring Chinook salmon appears to correspond most closely with the rise in diet-related diseases. During the 1960s and 1970s, spring Chinook dropped out of the diets of most Karuk tribal members, and shortly after, diabetes was reported in high numbers. Diabetes begins to appear in about 30% of Karuk

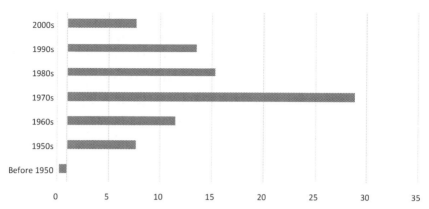

FIGURE 12. When did diabetes first appear in your family?

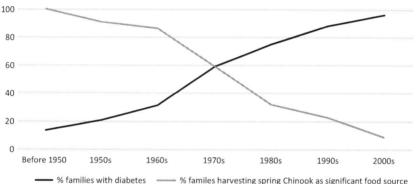

FIGURE 13. Spring Chinook decline and diabetes rates.

families roughly ten years following the loss of spring Chinook salmon as a significant food source. Through this combination of medical data, interviews, and self-reported medical information, we were able to detail how disease rates tracked with diet change over time, as shown in Figure 13.

But the relationship between the loss of large amounts of salmon in the Karuk diet and the emergence of diabetes is a correlation that may or may not be causal. The drop in spring Chinook harvesting and the rise in the diabetes rate could be happening by chance or due to some other outside force (such as people moving away from the river or some dramatic event). Causality cannot be determined from the statistical analysis itself but can be inferred by induction. From the literature, we knew that the loss of traditional food sources is recognized as being directly responsible for a host of diet-related illnesses among Native Americans, including diabetes, obesity, heart disease, tuberculosis, hypertension, kidney troubles, and strokes (Boyce and Swinburn 1993; Kuhnlein and Humphries 2017; Joe and Young 1994). Indeed, diabetes is one of the most significant health problems facing Native peoples today. Diabetes is described as a new disease among this population and is the consequence of drastic lifestyle and cultural changes that have occurred since World War II (Joe and Young 1994; Joe and Gachupin 2012; Garro and Lang 1994). Native peoples around the world experience skyrocketing rates of diabetes with the shift from a traditional to a Western diet. Ironically, the salmon lost by the Karuk people is an ideal food—both a preventive and curative for diabetes. Potential causality is indicated through this global pattern, together with the physiological specifics from medical literature and the temporal association between the change in diet and rise in diabetes. Based on information from interviews and other context, as well as the temporal relationships between diet change and diabetes emergence, we surmised that these health consequences stem from changes in the specific nutrient content of traditional foods such as salmon and acorns, as well as a decrease in the physical benefits of exercise associated with their gathering.

WHAT ABOUT ALCOHOL AND EXERCISE AS "INTERVENING FACTORS"?

A fourth conceptual problem concerned the potential for alcohol use and decreasing exercise as reasons for the rise in diabetes. Certainly, alcohol use is prevalent, and many people report exercising less than a few generations back. We had neither time nor funding to carry out the kind of study that might be used to tease out the extent to which disease rates were a function of diet change, alcohol use, or reduced exercise. Whereas the Western medical model emphasizes disease, Native American cultures traditionally define sickness as an imbalance in the physical, spiritual, emotional, and social realms. For many Native peoples, health is also felt to be influenced by the interactions between people and natural elements, since humans originated from and with the assistance of beings of the natural world (Simpson 2003; Trafzer and Weiner 2001). Within this framework, stress, grief, or anxiety could weaken one's well-being and make one vulnerable to disease. For example, in Betty Cantrell's (2001) study of the Plains people, many participants cited examples of themselves or others being diagnosed with diabetes during or after a stressful life event. Rather than seeing alcohol consumption, depression, and decreased exercise as separate factors that had to be distinguished from one another and diet change as within a biomedical framework, we felt they were interconnected and mutually reinforcing. Cantrell describes how even the preparation of traditional foods is healthy for people both physically and mentally: "A great deal of human energy must be expended to dry foods: the fruits, vegetables and berries must be gathered in the wild; the game must be hunted or trapped; the foods must be prepared for drying. All of these activities provide healthy exercise. In addition, it was believed that the emotional state and attitude of the person preparing the food was passed along to those who ate the food. Therefore, the cook tried to maintain a positive attitude before and during food preparation and songs of celebration were sung during food preparation" (Cantrell 2001, 71). Similarly, there are strong parallel traditions in the Karuk culture: "If someone isn't in a good space, they aren't supposed to be at ceremony—let alone cook. This is also the case for basketweaving or gathering, but especially when we cook. We have to have good thoughts—not talk smack about someone, you know?" (Sopie Neuner, 2018, personal communication). Both ceremonies and daily activities surrounding food provide meaning and identity that are fundamental to emotional well-being and cultural continuity (Kuhnlein and Chan 2000). This framework fits with the observations of Western science, too. There are impacts of loss and severe reduction in access to traditional food sources on other indicators of life stress, including, for example, rates of physical conditions such as tuberculosis, ulcers, and evidence for emotional stress, including suicide, depression, and high school dropout rates.

By drawing upon Indigenous cosmologies of connection, we were able to contextualize both alcohol use and the decline in traditional food availability within the framework of genocide and colonialism. Interviewees reported that loss of salmon was clearly a tremendous emotional burden that people described motivating negative behaviors such as alcohol use. And without salmon in the rivers, people spent less time engaging in activities like fishing for them. In very pointed comments shared with us concerning the emotional impacts of loss of salmon, people attributed this loss to genocide. Likewise, newer reports regarding the imminently approaching sudden oak death in Karuk Aboriginal Territory has sparked similarly expressed reactions. Leaf Hillman explains, "While many non-Indians are aware of the U.S. government's destruction of the North American bison as a tool to subjugate the Indian, to Native people, it all looks the same. It's understood to be the same phenomenon—destruction of the salmon, destruction of the acorns—they all lead to the same thing. Genocide." A Karuk mother in her early fifties drew parallels to the warnings against eating "white man's food":

> When the white people came here, the elders warned about eating the white men's "kêemish'ávaha"—poison food. I don't know if that was before or after all those Shastas were poisoned at Fort Jones, but I think about that a lot. Seems like that was a good way to kill us off—put small pox onto blankets they give out or invite a whole village to come for a great feast, poisoning them all at the same time. I heard they stood there and made sure no one crawled away—there were a couple of Indians late for dinner, and that's how they knew what happened. Anyway, seems like here we go again! Now they're killing us by killing off the tan oaks, killing off the salmon, choking out the huckleberries. It can't be a coincidence. They're killing our relations because in a white man's world, that isn't called murder.

These sentiments, combined with the fact that hunting, fishing, and gathering are physically intensive activities, led us to an alternative explanation that *combined* these multiple components into one factor rather than trying to separate them out. Instead of trying to think about alcohol use, exercise, and diet change as separate "variables" whose independent effects needed to be measured and accounted for, we listened to the voices of people who emphasized their common origins. We therefore described how the current health predicament stemmed from changes in the specific nutrient content of traditional foods such as salmon and acorns, as well as the decrease in the physical benefits of exercise associated with their gathering and the emotional toll of the loss of individual identity and collective culture. Rather than seeing the emotional despair that led to alcohol use and diet change as two independent forces, we understood them both as a function of environmental decline. Mental, emotional, cultural, and spiritual health benefits of eating and harvesting traditional Karuk foods were discussed as people experienced them: as interrelated and mutually reinforcing. These sentiments hold true today. A Karuk elder and ceremonial leader expresses it this way: "We're now in this period of time where the serving of traditional foods at ceremonies serves as the surrogate for the individual family, as a source of pride and reinforcing our identity as Karuk people. When we're unable to provide Native foods at ceremonies, the impact on the collective psyche—our identity, our pride—isn't an individual wound. It represents this collective shame." Community perspectives and traditional knowledge led us more generally to a focus on the interconnection between health and the social, cultural, and spiritual impacts of the loss of traditional foods. We therefore included a chapter highlighting the cultural and spiritual significance of the decline in salmon. This chapter explained the importance of traditional foods for cultural and spiritual practices, as well as emphasized the importance of culture and spirituality for human health. We underscored the mental, emotional, cultural, and spiritual health benefits of eating and harvesting traditional Karuk foods. Indeed, traditional food and food-gathering activities are at the very heart of Karuk culture. There are also creation stories that establish the foods that Karuk people are supposed to eat, including stories about salmon, acorns, eel, and deer. And there is a warning from Phoebe Maddux transcribed and published in 1932 that white food is "poison food" and "world come to an end food."[6] Furthermore, the activities of managing, gathering, preparing, and consuming traditional foods serve the functions of passing on traditional ecological knowledge, stories within the community and from one generation to the next. Food-related activities serve as social glue that binds the community together; they outline social roles that provide a sense of identity and serve as the vehicle for the transmission of values. Indeed, traditional food and food-gathering activities are at the very heart of Karuk culture. Creation stories establish the foods that Karuk people are supposed to eat, including stories about salmon, acorns, eel, madrone berries, "Indian" potatoes, and deer. Furthermore, the activities associated with managing, gathering, harvesting, preparing, preserving, storing, sharing, and consuming traditional foods serve critical roles in transmitting intergenerational knowledge and practices. These food-related activities also serve as social glue that binds the community together, outlining social roles that provide a sense of identity and serve as the vehicle for the transmission of values.

> We don't tell stories when there's no snow in the High Country, but when we used to go gather Indian potatoes, for example, we often joked about things that someone did—like a kid trying to take the whole thing instead of leaving the root crown—as something that the Orleans Maiden[7] did. Now we hardly ever do that [harvest brodiaea species]—the old patches are just about as grown over as you can imagine. So none of that gets passed on—we don't talk about what the kids are doing, how they're growing, how someone stores their potatoes, when the best time to gather is—you see, we've lost so much.

Diet change thus leads to a loss of culture and identity, a form of what Elizabeth Hoover (2017) calls "environmental reproductive injustice." Just as ceremonies surrounding fish create and maintain community ties and provide identity, so too does their absence and decline lead to further cultural disruption. Originally prescribed through Karuk creation stories, Karuk ecological knowledge is developed and passed along through practices of food management, harvesting, and preparation. As families work together, information and stories are shared. In addition to the transmission of knowledge, these stories provide bonds and cohesion that hold the family and community together. In addition to the loss of both culture and ecological knowledge with diet change, major social disruptions occur from the loss of traditional foods and all of their associated practices—from management to consumption. Traditional food has an important place in the fabric of community life (Kuhnlein and Chan 2000; Kuhnlein and Receveur 1996). When elders die young, they cannot pass love and knowledge on to the younger generations. When multigenerational family and social food-gathering activities disappear or are diminished, an entire field of social relations and their combined network are lost.

TRADITIONAL FOODS AND FOOD SECURITY

Communities are defined as food secure when all members have access to nutritionally good, safe, and culturally acceptable food through local nonemergency sources at all times. The traditional Karuk foods people desired may be particularly healthy, but it is also the case that in their absence, Karuk people face basic issues of food security. Before the impacts of dams, mining, and overfishing, Karuk people drew a large portion of their caloric and nutritional intake year-round from salmon for tens of thousands of years. Likewise, the impacts of ever-increasing wildfires, timber exploitation, introduction of plant pathogens, ongoing settler occupation, and loss of access to traditionally communal groves have dramatically reduced the consumption of acorns—a resource that matches the historic value of salmon subsistence. Hundreds of other food resources have been affected by these same events, and all are now threatened by a number of more recent resource-extracting activities and climate change. At this time, it is no longer possible for people to subsist on these foods. Ron Reed, traditional Karuk fisherman, notes the impossibility of feeding the current tribal population from the one remaining fishing site at Ishi Pishi Falls: "We had over 100 villages up and down the Klamath River, with fishing sites associated with each village. Now we are trying to feed our people off one fishery. It's not possible." Instead, poverty and hunger rates for the Karuk Tribe are among the highest in the state and nation. The Karuk Tribe's poverty rate is between 80% and 85%. In the absence of salmon and other foods from the land, people must purchase foods in grocery stores or go hungry. Coupled with today's indigent conditions, covering basic needs is a real issue for many community members. Self-report data from the Karuk Health and Fish Consumption Survey indicated that 20% of Karuk people consumed commodity foods, and another 18% of those responding indicated that they would like to receive food assistance but do not qualify.

Food insecurity is not only a major problem for current nutritional needs but is also a risk factor for long-term health. Difficulty in meeting basic needs results in overwhelming physical and psychological stress. The health challenges associated with food insecurity exist across the life span. Food plays an important role in cultural continuity and identity—the loss of which in turn fosters substance abuse, suicide, violence, and other so-called antisocial behaviors. The severe socioeconomic stress experienced by many Karuk families, together with emotional trauma, has been documented to cause serious social problems, including higher rates of substance abuse, school dropouts, psychological problems, and violence (Beauvais 2000, 110).

Our last chapter turned to more traditional economic analyses and emphasized the fact that when Karuk tribal members are denied access to the healthy foods that supported them since time immemorial, this also costs society. When an entire Tribe faces epidemic rates of expensive conditions such as diabetes, sizable state, county, and tribal medical resources will be used to address this problem. The American Diabetes Association reports that diabetes patients have an average annual per capita cost of health care at $13,243 per person per year in the United States. Given the 148 diabetic Tribal members within the aboriginal territory in 2004, our report included a calculation of the annual cost for Karuk Tribal members at over $1.9 million.[8] The report also noted that these increased medical care costs, paid by society as a whole, were not reflected in PacifiCorp's dam operation expenditures, nor were they withdrawn from the profits PacifiCorp receives from the production of electricity in a manner that damages the health of the Klamath riverine system. Instead, the higher health care costs of increased diabetes in the Karuk population are borne by society as a whole. PacifiCorp does not reimburse the Karuk Tribe or Siskiyou or Humboldt counties for the increased cost of health care that comes from the destruction of an abundant source of healthy food in the Klamath River. We contended that any cost-benefit analysis of the dams should include the $1.9 million annually to provide medical services for the artificially high incidence of diabetes in the Karuk Tribe.

RESHAPING POLICY: NATURAL RESOURCE MANAGEMENT AND THE ALTERATION OF DIET

When this information was compiled, it was submitted to the Federal Energy Regulatory Commission docket. We also released the report publicly. Coverage of our work in the front page of the *Washington Post*, a story on National Public Radio, and another widely reprinted Associated Press story, as well as dozens of additional news stories and coverage by local and regional radio stations over time, all helped keep the pressure on the power company.

COLLABORATION WITH INDIGENOUS COMMUNITIES

One take home of our work is the power of incorporating Indigenous and Western science for policy work. While collaboration has been a central aspect of the environmental justice and environmental health movements, collaboration with Native communities remains relatively uncommon. Muscogee (Creek) Nation scholar Dwanna Robertson (2016) writes that "centering Indigenous research in mainstream sociology is difficult at best" and notes that "Native studies, in general, meet with certain resistance from Western academia and its politics of knowledge and inquiry. Western thought often deems Indigenous research as irrelevant within the construction of knowledge. Indigenous methodologies are dismissed as nonconforming, invalid, or inapplicable" (248). Working with my Karuk colleagues, I have come to understand much about the interconnection of multiple realms that academic researchers rarely if ever discuss side by side—from the relationships between riverine and human health to the interconnection between fire policy and spiritual practices or the mutual influences between subsistence economies, family integrity, and political sovereignty. Not coincidentally, these themes of interconnection and relationship are key to Karuk and other Indigenous cosmologies. The knowledge structures of Western societies are by contrast based in atomism, or "the premise that systems consist of parts that do not change and that systems can be thought of as the sum of their parts" (R. Norgaard 1994, 54–55). According to this atomistic view, knowledge comes in discrete packages that can be added together like building blocks to advance understanding. Karuk TEK, by contrast, is highly dynamic, contextual, and relational. The very concept of "traditional ecological knowledge" as a stand-alone phrase implies that Karuk "knowledge" is a discrete *entity*. However, while the Western scientific cosmology presumes that the world can be categorized into "facts," that knowledge exists in the abstract outside particular contexts, and that observers are interchangeable, the kind of knowledge that has co-created the mid-Klamath ecosystem of today is situated and embedded in a specific ecological and cultural context. Unfortunately, in many cases, agency practitioners and Western scientists have assumed that this "knowledge" of how to burn the forest or how to manage the fisheries can be described by Karuk people, shared in various agency processes and then applied by multiple actors in different contexts. Leaf Hillman describes the frustration of working with agencies when they try to blend these distinctive perspectives in the absence of understanding that they reflect such different epistemological orientations:

> You do a paper on TEK and we talk about specific practices, you write them down on a piece of paper and then the Forest Service thinks that they can take that. "Okay, we paid for this under a contract for you guys to develop this, so now we are going to take this and apply it." Just the notion that they can apply those things, within their structure—within the boxes that they have—as if they just knew what they were. "Tell us what they are, and if you describe them well enough then we can apply those things." But they can't just apply those concepts, because what they require is cultural practices of a land-based people. They must be used by people who are on the land, not people who are separate from the land as part of a government agency.

While non-Native agency practitioners and Western scientists have assumed that traditional knowledge can applied in generalized ways, a substantial literature now details the pitfalls of such efforts.[9] Despite their real differences, each epistemological framework is powerful and collaborations can be enormously productive. Many tribal scientists and practitioners within the Karuk Department of Natural Resources have not only welcomed collaboration with Western academics but also successfully deployed these two knowledge systems strategically side by side to communicate tribal perspectives to agencies and the general public. The gap between holistic Indigenous cosmologies and Western academic systems has, however, meant that conducting this research has required collaborators across a wide range of academic disciplines from biology to chemistry. Ron Reed and other Karuk DNR employees have worked closely with fisheries biologists, geographers, medical doctors, and especially scholars who are trained in the interdisciplinary social sciences. But it is Indigenous ideas, observations, and understandings of the world that have driven the framing, content, and process of their cutting-edge projects on questions from fire behavior to the nature of food security. And at every juncture, relationships and interconnection have been critical themes.

As a sociologist, I have benefited enormously from Indigenous perspectives. One of my favorite concepts in sociology is the sociological imagination coined by C. W. Mills in 1959 (Mills [1959] 2000). Mills describes how, in this "Age of Fact," what we needed most was not more information but a "quality of mind" he called the sociological imagination. Mills wrote that "it is by means of the sociological imagination that men and women now hope to grasp what is going on in the world, and to understand what is happening in themselves as minute points of the intersections of biography and history within society" (7). With such an imagination we bring in context, make connections between the personal and the political, and are thereby better able to see the operation of power. Attending to this broader context and emphases on interconnection are hallmarks of Karuk and other Indigenous worldviews. Indigenous cosmologies make visible many layers of power and relationships that are (not coincidentally) obscured by Western frameworks. In the case of the altered diet study, our collaborative research approach benefited enormously from our ability to set multiple knowledge frameworks and social and environmental realms in conversation with one another. And Indigenous perspectives regarding the role of the state in the creation of environmental harms, as well as identification of these harms on a continuum with past genocide, pointed to the lens of colonialism. It is precisely this ability to identify power that Pulido (2016, 2018), Pellow (2016, 2017), and other commentators of the environmental justice and environmental health movements have called for.

Yet while taking Indigenous experiences into account will certainly improve any theory, it is important to keep in mind the words of Rima Wilkes (2017), who reminds us that Native people must not be taken as one more "case" to add to existing theories. Dwanna Robertson (2016) describes how Indigenous scholars came together at a convocation in March 1970 and called for "the development of an academic discipline that was developed by Indigenous intellectuals and that centered on Indigenous cultures and tribal sovereignty, while simultaneously accomplishing the decolonization of the educational system" (249). To accomplish the goal of decolonization, the scholars called for "an interdisciplinary body of Indigenous knowledge that no longer focuses on Native peoples as 'exotic others' but rather uses this knowledge to influence U.S. Indian policy" (249). Robertson emphasizes that "the value of Indigenous research is more than knowledge production: it is about the being and the doing of that knowledge and its impact on our communities. Indigenous research is transformative. Consequently, Indigenous researchers do social research for more than the profitability of intellectual property or freedom. Indigenous research is an act of subversive resistance because the knowledge produced must be practical and applicable in our movement toward collective liberation, particularly within the academy" (248).

TRADITIONAL ECOLOGICAL KNOWLEDGE AND ENVIRONMENTAL HEALTH RESEARCH

Our approach of centering Indigenous knowledge, perspectives, and values alongside Western science and developing causal explanations that turned attention onto the state not only allowed us to achieve significant traction in our particular case but also may be useful for developing theoretical perspectives within the fields of environmental justice and environmental health. For example, while much important work has been done on the health and cultural effects of the *contamination* of tribal traditional foods (e.g., Chan et al. 2006; Kuhnlein and Chan 2000; Laird et al. 2013), the Altered Diet Report was the first study to link increasing tribal diabetes rates to the loss of a traditional food in the context of generalized environmental degradation, species decline, and the loss of traditional food sources. Indeed, scholars in the field of environmental health emphasize that policy-driven work often enriches existing theory (Brown 2013; Krieger 1994, 2011). In his 2013 Reeder lecture, Phil Brown reflected that "combining medical sociology and environmental sociology have brought me ways to address pressing spheres of personal, professional, institutional and global importance. The linkage is more obvious each day that we discover additional connections that must be addressed together as with potential connections between obesity, diabetes, neighborhood walkability, and community gardens" (160). When Ron Reed first approached me about documenting the heightened rates of diabetes and heart disease that he was seeing in the Karuk community, the relationship he hypothesized between the loss of salmon and increased disease came from his direct experience. Salmon are a healthy food source, high in protein, iron, zinc, an omega 3 fatty acids and lower in saturated fats and sugar. Yet until we combined this direct experience with medical and social science research, it was dismissed as "anecdotal evidence." Without this three-part combination of firsthand community experience, Indigenous epistemologies of interconnection, and Western science, our research could never have made such a substantial contribution.

When it comes to environmental health research specifically, few health studies draw upon the framework of colonialism to understand health outcomes, and fewer still draw upon Indigenous conceptions of health and wellness. In contrast to biomedical models where researchers look for one variable that will change health outcomes, Indigenous perspectives on health are refreshingly holistic. Even with the "social determinants of health" frameworks, the tenacity of colorblind ideologies, overemphases on biology, and latent cultural racism lead to very individualized foci on how to improve health. Furthermore, Indigenous people often have sophisticated environmental and social knowledge relating to specific foods, medicines, and practices. Take, as one small example, the recent explosion of literature on the downsides of gluten (especially bread) for human health. Today books on the health hazards of refined flours and gluten intolerance are hitting the bestseller lists and reframing how health practitioners think about everything from weight loss to blood chemistry and brain function. Consider that in the 1930s, Karuk elder Phoebe shared these words with anthropologist J. P. Harrington. Her original words are recorded in Karuk; I provide here the English translation. Maddux explicitly describes bread and other white foods as "poison" and "world come to an end food":

> All did the same, the way that the Ikxareyavs used to do. And what the Ikxareyavs[10] ate, that was all that they ate. They told them: "Ye must eat this kind." The Ikxareyavs ate salmon, they spooned acorn soup, salmon along with acorn soup. And they ate deer meat. And they claimed that the Ikxareyavs had two meals a day, and they also did only that way. When the whites all came, then they said: "They eat poison, poison food, world come-to-an-end-food." The middle-aged people were the first to eat the white man food. When they liked it, they liked it. They told each other: "It tastes good." They said: "He never died, I am going to eat it, that bread." But the old men and old women did not eat it till way late. We are the last ones that know how the Ikxareyavs used to do, how they used to eat, the way our mothers told us. And even we do not eat any more what they told us to eat. And what will they who are raised after us do?

Indigenous worldviews center social, political, cultural, and environmental dimensions of human health that are enormously beneficial for advancing policies to promote human health (Colomeda and Wenzel 2000). In particular, researchers would do well to better understand the complex ways that colonialism structures not only health outcomes but also the research frameworks and methods of academic practitioners. Krieger (2011) argues that the act of engaging ecosocial context of disease enables the greatest theoretical advancement for the field of epidemiology. Sociological approaches to epidemiology have contributed important insights to understanding the patterns of health inequalities—specifically regarding the importance of social structure and racism on health (Williams and Strenthal 2010). These emphases on social context have been important additions to the dominant biomedical framework on health and disease that emphasizes individual biology and behavior (Wing 1994). The inclusion of how the natural environmental and colonialism shape a community's health circumstances—as well as the interactions between them—provides further valuable context. There is an extensive literature on the benefits of community-based science in the context of environmental health research (Brown and Mikkelsen 1997; Brown 2013), but little of this work occurs in Native communities or employs Indigenous cosmologies. By contrast, ecologists are beginning to describe in detail the scientific and applied benefits of Indigenous-Western science collaborations in fields from restoration ecology (Charnley and Poe 2007; Hummel, Lake, and Watts 2015) to forestry (Kimmerer and Lake 2001; Lake et al. 2017) and fisheries management (Verma et al.

2016). I encourage scholars to consider the many benefits of collaborations with Indigenous communities as a means of improving community health, advancing theoretical scholarship, and working to "unsettle" academic knowledge structures.

DECOLONIZING ENVIRONMENTAL JUSTICE

Last, lessons from the Karuk altered diet study hold both practical and theoretical implications for the field of environmental justice. Robertson (2016) writes, "Within the discipline of Indigenous studies, the term decolonization generally means to first comprehend how oppressive structures of colonialism (such as individualism and patriarchy) color our everyday worlds and then to apply that knowledge in ways that disrupt colonized ways of thinking, doing, and being" (248). What can such perspectives add to studies of environmental justice? In her poignant critique of the environmental justice movement, Pulido (2017b) notes, "While the environmental justice movement has been a success on many levels, there is compelling evidence that it has not succeeded in actually improving the environments of vulnerable communities. One reason for this is because we are not conceptualizing the problem correctly.... If, in fact, environmental racism is constituent of racial capitalism, then this suggests that activists and researchers should view the state as a site of contestation, rather than as an ally or neutral force" (524). Pulido notes that in the absence of such awareness, the state itself is racialized and capitalism directly benefits from the state's production of inequality: "Numerous problems stem from not conceptualizing the problem accurately, including not giving sufficient weight to the ballast of past racial violence, and assuming the state to be a neutral force, when, in fact, it is actively sanctioning and/or producing racial violence in the form of death and degraded bodies and environments" (524–525). When Ron Reed was first working in this area, he asked about the framework of environmental justice but was told that this term did not apply to him. Instead, he was told that this term referred to racial minorities facing toxic exposure in urban areas and latinx farmworkers dealing with pesticide exposure at work. Indeed, environmental justice is commonly understood as emerging from the fights of communities of color against disproportionate siting toxic facilities, highway redevelopment projects, and other urban communities of color who have used a civil rights framework.

But what if rather than tracing environmental justice as an outgrowth of the black civil rights movement (Pulido and De Lara 2018) or a response to white environmentalism, we traced its origins to Indigenous conquest and resistance? Why have we not done so, and what does this absence imply? Environmental justice activists and scholars have explicitly identified the state as "neoliberal" and "racist" (Pulido et al. 2016; Kurtz 2009; Harrison 2011, 2014; Ducre 2018) but rarely use the term *colonial*.[11] Karuk and Indigenous perspectives on environmental justice in general reframe the dominant environmental justice discourse in (at least) two important ways. First, Indigenous notions of desired movement goals point to a deep reframing of prevailing notions of health, relationships, and "the other worlds that are possible" beyond either capitalism or colonialism (Whyte 2015, 2016d, 2018d; Grey and Patel 2015). The environmental justice movement is most commonly understood as emerging in the late 1970s and early 1980s as pioneering scholar-activists and legal voices such as Dr. Robert Bullard, Benjamin Chavez, Charles Lee, and Luke Cole began identifying and challenging the disproportionate exposure to toxins and regulatory discrimination in African American communities in the American South. Certainly, despite the presence of Indigenous leaders at the time, the predominant initial legal strategies and communities' conceptions of desired outcomes reflected a civil rights discourse that emphasized appeals to the state regarding unequal burdens of environmental harms. In contrast, the goals, values, and worldviews of Karuk and other Indigenous communities emphasize relationality, kincentricity, responsibility, and the notion of nature as animate (Alfred 1999; Martinez 2011; Salmón 2000; Simpson 2017; Watts 2013). Rather than language about "equality" or "rights to clean water or air," Karuk visions for an "environmentally just" and good world are framed as caretaking responsibilities that are disrupted by natural resource policies of the settler-colonial state. "Nature" in the form of salmon, eels, or acorn trees is much more than a platform for human action or a space within which "environmental resources" might be distributed unequally. These species are part of the human community—treasured relatives to whom people have real responsibilities.

Second, when it comes to the process of environmental justice activism, Pulido (2017a, 2017b), Pellow (2016), and others note a mysterious emphasis on appeals to the state for protection (Harrison 2014, 2015). Pulido (2016) observes how "for the most part, activists have not only prioritized engaging with the state, but have done so with the expectation of positive results. They have believed that by working closely with regulators, through regulatory attention, judicial action, and the implementation of the EO [Executive Order 12898 on Environmental Justice], the conditions in their communities would improve" (12–13). This is a tendency that Pulido and coauthors (2016) surmise arises in part from the successes of this strategy during the civil rights movement. Yet "environmental justice activists' reliance on state regulation has inhibited their ability to achieve their goals" (12). By contrast, Indigenous environmental activists coming from the perspective of colonialism tend to be very clear that the state is not an ally. Leaf Hillman, director of the Karuk Department of Natural Resources, describes the adversarial relationship with the state in explicit terms: "Every project plan, every regulation, rule or policy that the United States Forest Service adopts and implements is an overt act of hostility against the Karuk People and represents a continuation of the genocidal practices and policies of the US government directed at the Karuk for the past 150 years. This is because every one of their acts—either by design or otherwise—has the effect of creating barriers between Karuks and their land." Thus, the notion of "decolonizing environmental justice" is an opportunity to expand understanding of the origins of the environmental and environmental justice movements, whether the state is conceived as a potential ally or explicit foe, and especially the desired goals and outcomes of social action. While useful, emphases on frameworks of equality and rights nonetheless signal equal participation in a mainstream democratic project as the desired outcome. Such frameworks tend to apply more to dynamics within settler communities of color. Indigenous visions for environmental justice may include such desires as equality or access to participation in a particular process but ultimately center on the desire to continue existing in one's own cultural, economic, and political system. And if we are to do more than tinker with the existing system, paying closer attention to Indigenous cosmologies, ethics, and understandings of power will be necessary for deeper understandings of sustainability, community, ecological relationships, and the other worlds that are possible beyond the capitalist or colonial imaginations.

In contrast to the failure of environmental justice activists to effect change identified by Pulido and others, Indigenous political movements have made remarkable gains in the past forty years (Gilio-Whitacker 2015; Steinman 2012, 2016; Whyte 2017b; Wilkinson 2005). On the Klamath, the Karuk Tribe's federal recognition has been reinstated, and the Tribe operates its own health, transportation, and housing programs and a thriving Department of Natural Resources with active programs in fisheries, fire, watershed restoration, environmental education, and water quality.

REIMAGINING FOOD JUSTICE AND FOOD SOVEREIGNTY

This larger framework on food, health, and the social and environmental relationships that support them relates to another dimension of environmental justice: that of food justice and food sovereignty. Interest in food studies has produced an explosion of academic and popular interest in "food" complete with a new vocabulary of terms like *slow food, local food, food security, food deserts, food sovereignty,* and *food justice.* Books on these topics abound and include popular bestsellers from Michael Pollan's *Botany of Desire* and *The Omnivore's Dilemma* to dozens of books about the social histories and nutritional merits (or lack thereof) of particular foods from butter to bananas to bread (Bobrow-Strain 2012). My own university is one of many in the country that has recently launched a thriving food studies program. Yet in contrast to the generally commodified understandings of "food" where there are concerns about inequalities in the "production" and "consumption" of food, Karuk and other Indigenous people speak of the foods they eat as relations. They speak of a longstanding and sacred responsibility to tend to their relations in the forest and in the rivers through ceremonies, prayers, songs, formulas, and specific stewardship practices they call "management." Rather than doing something *to* the land, ecological systems prosper because humans and nature work together. Working together is part of a pact across species, a pact in which both sides have a sacred responsibility to fulfill. Traditional foods and what the Karuk call "cultural use species" flourish as a result of human activities, and in return, they offer themselves to be consumed.

While "food" no doubt holds significant individual and cultural meanings for all peoples, the profound connection that Indigenous peoples draw between "food" and identity, community, and spirituality are of another order. Colonialism and its resistance remain almost entirely invisible within this new field. Karuk cultural practitioner and traditional fisherman Kenneth Brink describes how cultural capital is reproduced along with ecological knowledge within the family, and on the land, "there is a cycle that goes around." When he reflects on how this cycle was taught to him, he states,

> It might have been through picking mushrooms, it may have been through going and gathering acorns. It's certain, you know, family values like that, you know, family adventures like that you know. And this is like, next to morals, I've thought about this. They get passed down from generation and family to family and on down the line. And it just ... if you don't have some of these key elements that bring us together to learn morals and traditional values, which may be the mushroom or the acorn on the salmon, then we are like missing our gathering, we are missing our food. We are like missing the teaching right there.

Brink emphasizes that ethics and cultural values are taught alongside skills and experiences: "by picking the right type of mushroom," you "teach your kids to do things you know, how to respect your Elders and how you treat your land." Both social ties and the presence of specific material ecological conditions are central to how young people come to internalize gender identities, as will be the focus of chapter 4. Notions of food justice or even so-called food sovereignty that circulate in most public conversations today do not even begin to capture these relationships. Indeed, to comprehend and acknowledge Native relationships with what gets called "food" would require non-Indians to recognize not only the depth of the human scale of Native American genocide but also the fact that this genocide has been an assault on a spiritual order that nourished and governed an entire field of sentient beings and ecological relationships. Hence, the powerful notion of justice from Kyle Whyte and colleagues (2018): "Injustice is a form of domination that works to undermine Indigenous peoples' capacity to have moral relationships with nonhumans and the environment, which are crucial to their resilience" (144).

Indeed, the food justice and food sovereignty movements are one place where underlying tensions between competing understandings of power and movement goals have played out in the United States. Local food movements often stress the pitfalls of a corporate-controlled food system and the need for "food security' in the face of large-scale forces that take power away from community hands. Many such efforts to secure local foods have, however, unwittingly enacted economic and racial inequality (Guthman 2008; Alkon and McCullen 2011; Reynolds 2015). In contrast, the concept of food justice emphasizes the broader ways that racial and economic inequalities are deeply entrenched within and reproduced by today's system of food production, distribution, and consumption. In reviewing two U.S. food movements that define themselves in terms of food sovereignty, Alkon and Mares (2012) observe that although the food security and food justice movements in the United States have drawn much inspiration from La Via Campesina, they have "not wholly embraced a food sovereignty approach that would explicitly oppose neoliberalism": "Both of our US cases can easily coexist with industrial agriculture, and in some ways even serve to relieve the state of its duty to provide basic services. As such, they fail to challenge a neoliberal political economy in which services that were once the province of the state—such as the provision of food to those who cannot afford it—are increasingly relegated to voluntary and/or market-based mechanisms" (348). Instead, Alkon and Mares describe how an emphasis on citizen empowerment in the United States, "while of course beneficial in many ways, reinforces the notion that individuals and community groups are responsible for addressing problems that were not of their own making." As a result, the authors note that "neoliberalism constrains the ability of the West Oakland Farmers Market and Seattle's urban agriculture projects to provide fresh, healthy food to the low-income citizens they seek to serve" (348). Yet even within frames that use the term *sovereignty* to emphasize how the globalization of agriculture undermines the abilities of agrarian peoples to produce food for self-reliance, food still appears as a commodity or resource to be controlled. Food sovereignty as articulated by the Karuk testimonies on these pages is all about fixing relationships on the land such that ecosystems and humans may thrive together. As Grey and Patel (2015) write,

> Indigenous food sovereignty is about more than the familiar bundle of rights that attach to production and consumption. Here, a "right to define agricultural policy" is indistinguishable from a right to be Indigenous, in any substantive sense of the term. Upon being told there was no word that translates directly as "health," medical anthropologist Naomi Adelson was given the closest Cree equivalent, miyupimaatisiium, or "being alive well." For the Whapmagoostui Cree, among whom Adelson was working, miyupimaatisiium entails the ability to hunt, to have shelter, to eat iyimiichim ("bush food," or "food that was meant for the Cree"), and to engage in other traditional land-based activities (2000, p. 103). This makes "being alive well" about food sovereignty, and food sovereignty about land, identity, and dissent—and not just for the Cree. (439)

Food sovereignty as Karuk people have articulated in these pages is about resorting those relationships and the ability to carry out one's responsibilities—in particular, the ability to carry out one's responsibilities to one's kin that are not in human form but in the form of other species. And if activists and scholars listen to what Native people are saying here and take food justice and food sovereignty to be about fixing relationships, any understanding of what is wrong with those relationships would do well to look to the long arc of tribal knowledge and achievement with the natural world on this continent for the last 10,000-plus years (Trosper 1995, 2003). The fact of Indigenous management is testimony to the need for food movements today to have a stronger critique of what is wrong even more than that of the globalization of either commercial agriculture or racial capitalism.

4 · ENVIRONMENTAL DECLINE AND CHANGING GENDER PRACTICES

What Happens to Karuk Gender Practices When There Are No Fish or Acorns?

> It's been a way of life for as long as I can remember. I mean, my grandfather and his parents and on and on have been fishermen for so many years. I mean I can't even imagine life really without having a whole line of fishermen. I don't even know how to compare it to someone else because this is what is known to me, I just can't even imagine my life without it, you know, it's just such a big part of who I am. Our whole culture... it's such a huge part of it.
> —Jennifer Goodwin

> In managing for, tending, harvesting—all of that—we work together. As women, we work together as a team. We take care of each other and of each other's kids. We listen to each other's stories, each other's woes. We offer advice and listen to the advice of others. Sometimes we scoff, but the fact remains that we learn how to be women from each other.
> —Lisa Hillman

For nearly all of North American human history, gender constructions have looked very different than they do today. By all accounts, the diverse Indigenous notions of gender that organized human communities in this place from which I write on the Klamath, and across what is now called California and the United States, have long been more fluid, less binary, and organized around caring and stewardship rather than hierarchy and domination. Gender constructions have also been intimately interwoven with ecological activities and responsibilities. For Karuk men from fishing families, fishing comes with a set of responsibilities to family, community, and the fish themselves. The acts of harvesting, preparing, and sharing fish organize important material and symbolic features of Karuk social life still today. Fish are a gift from the Creator and an important food source in an impoverished rural community. The "right" and responsibility to fish at specific sites is an honor passed down to men through family lines.[1] The absence of fish resulting from ecological damage affects both food availability and the quality of social connections, which in turn affect individual gender practices and represent a genocidal act to the community. What happens to this form of Karuk masculinity when there are no fish? How are "traditional" male gender constructions and practices simultaneously about racial resistance and resistance to colonialism? What options do men have for restructuring masculinities in the face of environmental decline?

Karuk women hold particular responsibilities regarding burning meadow habitat, digging brodiaea (also known as Indian potatoes), and tending and harvesting tanoak acorns. Activities traditionally considered the domain of females have also been affected by ecological decline—both directly (see, e.g., Baldy 2013; Norgaard 2007; Norgaard, Reed, and Bacon 2018) and indirectly. As Lisa Hillman, Karuk tribal member and program manager for the Karuk Tribe's Pikyav Field Institute, explains,

> Whole areas which were traditionally managed with fire—by female fire practitioners, that is . . . there used to be a lot of meadows where there were seed-grasses, like wild oats and blue rye, and women used to take care of these places so that the forest didn't take them over. We used to eat a lot of different grass seeds . . . we used to have basket catchers to harvest these seeds . . . I don't know anyone who really knows how to do that anymore—either harvest the seed or make that kind of basket. There aren't those meadows anymore because not only has fire been banned from our landscapes, but some of those open areas were planted with conifers and others have just grown over. (2018, personal communication)

The majority of the grasses now found in the mid-Klamath region are not Native, having been planted postcontact in order to feed non-Native animals (Anderson 2005). Ecological decline also indirectly affects activities traditionally practiced by women when their absence from the landscape over time erases the living memory of these practices from the collective. What happens to this form of Karuk femininity when there are no pine nuts, no tan oak acorns, no Indian potatoes to dig? How are "traditional" female gender constructions and practices simultaneously about racial resistance and resistance to colonialism? What options do women have for restructuring their particular gender roles in the face of environmental decline? How does one understand the role of femininity as an honorable role—caregiver, resource and landscape steward, and provider for the community—when one feels shame in the face of its Western expression? Land management policies that criminalize traditional Karuk activities such as burning forest and meadow habitat, as well as fishing, gathering, and hunting in a traditional manner, are thus powerful disruptions to the possible realm of individual and collective gender identities. Colonialism is thus enacted through ecological violence that is deeply gendered (see also Meissner and Whyte 2017).

Just as chapter 1 examined the process through which race came to be real, there was a process through which white, capitalist, colonial constructions of gender and sexuality came to be the standard on this continent. This forced alteration of the gender structure was so necessary to the material access to resources and the symbolic normalization of the new order that without it, what we now know as California would not exist. The process through which gender was forcibly reorganized has had everything to do with the natural world. Not only did it take place for the explicit purpose of settler access to Indigenous lands, but the process through which it occurred also relied upon tautological binaries of mutual associations between gender and nature for its symbolic construction. It required that people believe that both a gender binary and a nature/culture binary were reality. The forced alteration of gender structures was enforced through extreme physical and psychological violence, just as it continues to be maintained by physical and psychological violence today. As my Karuk colleague Leaf Hillman emphasizes,

> Colonialism has favored Western definitions of the masculine as exerting power, dominance, and authority, and has projected that onto indigenous society from the very first contact, where you have non-Natives interpreting Native cultures and lifeways. The first ethnographic accounts have assumed that the main characters—especially those that have dominant roles—are males. That's just not true. So even the translations of our origin stories are already perverted—projecting settler-colonial ideologies onto our own Karuk culture. Our culture is not about dominance or control. We don't see ourselves *above* nature. Just as the Western concept of femininity fails to capture the cultural underpinnings of what it means to be a woman, likewise Western notions of masculinity cannot be imposed or easily inserted when one talks about what it means to be a man. Many of the social ills we are experiencing today in Karuk country have their origins in this mistaken substitution of the cultural construct of gender roles. When men put "women in their place" as they say, they're accepting the colonial definition of masculinity as their own, but it's *not* a part of their own culture. It's not just a question of "male" or "female": it's the culture. (2018, personal communication)

In contrast to emphases on masculinity as hierarchy, voices in this chapter will describe how fishing, participating in ceremonies that regulate the fishery, and distributing fish to the community are each gender accomplishments that serve ecological functions, unite communities, and perpetuate culture in the face of settler-colonialism. And it is not just constructions of masculinity that serve to assert colonial domination. In contrast to emphases on femininity as subservience, managing with fire, participating in ceremonies that perpetuate the species understood to belong to the domain of the feminine (or support the role of the "new" woman as provider and healer), and gathering, preparing, and sharing food, fiber, and medicinal plant resources with the community are fundamentally structured by the presence and absence of other species in what the dominant society calls "nature." Powerful and autonomous forms of Karuk femininity have been erased as well, as Lisa Hillman describes how today, "Even the very word 'femininity' smells of roses and baby powder, evokes images of high heels and lace." In such a framework, "The epitome of a threat to femininity is the bra-burning feminist, who deign to want to destroy the idea of the subservient woman in the male-dominant society. Karuk women were never any of those things. My great-grandma came and went as she pleased, and whether or not all of her children were my great-grandfathers was never put to question. Lots of women had multiple partners, owned gathering sites, homes and regalia. They weren't known for being nice, either." Her husband Leaf concurs: "Yeah, my grandma had a husband on each side of the river" (2018, personal communication).

I make several assertions in this chapter. I describe how colonialism has forcibly altered Karuk gender arrangements, imposing European gender constructions and hierarchies onto people as a mechanism for accessing Indigenous lands. In so doing, I draw upon the work of Leanne Simpson and others who describe how Indigenous gender constructions across North America have been forcibly transformed through colonialism. Second, I describe how this trajectory of gendered colonial violence continues through environmental degradation today. I then place Karuk and other Indigenous scholars' observations about gender and gender violence into conversation with those of non-Native feminist sociologists and gender studies scholars.

Feminist sociology has had an uneasy relationship with the notion that the natural world would structure gender. There are significant reasons for this aversion. We can trace the dualistic associations of women/nature/body and men/social/intellect back to the fields' founders in the work of Durkheim and Weber. Indeed, Ann Witz (2000) observes, "The woman in the body served as the foil against which a masculine ontology of the social was constructed" (2). Witz describes how, for Durkheim in particular, the very conception of the social involves "the simultaneous exclusion of the corporeal and of women" (13): "Durkheim's sociological project centered around establishing the social as a form of life *sui generis* distinct from the biological, the natural or the corporeal—'The more elevated it [civilization] is, the more, consequently, it is free of the body. It becomes less and less an organic thing, more and more a social thing' Durkheim 1964, 321). The For-itselfness of the social is defined in contradistinction to the in-itselfness of 'nature,' 'the body' the 'instincts' and (more latterly in The Elementary Forms of the Religions Life), the 'profane'" (13). It is hardly surprising that feminist sociologists refocused gender as social and distanced gender from the natural or the biological. Indeed, the notion of gender as social became the cornerstone of feminist theory on gender:

> The distinction between the corporeal material of 'sex' and the sociality of gender—however problematic and unstable this has since become—was the enabling moment of feminist sociology. Feminist sociologists sidelined body matters and foregrounded gender matters. Precisely because they were sociologists they did latterly for women what sociology had done formerly for men: they retrieved them from the real of the 'biological' the 'corporal' or the 'natural' and inserted them within the realm of the 'social.' They began to release them from the theory of biological or natural determinism and subject them to a theory of social determinism. The feminist sociological turn *to* gender entailed a corresponding turn *away* from corporeality. (Witz 2000, 3)

A parallel critique of the natural-social dualism shaped the formation of the subdiscipline of environmental sociology where, not surprisingly, environmental sociologists have been at the forefront of seeking to bring focus in sociology back on material conditions. Here too scholars have returned to early texts to trace the assumptions leading to this omission. Dunlap (2002, 2010) and others have argued that a disciplinary bias toward "social facts" at the exclusion of the material world has limited the scope of sociological work on the environment and contributed to a marginalization of environmental sociology within the discipline.

Yet we can use the term *nature* or think about what is "natural" in many different ways (Sturgeon 2009). Biological determinist theories of sex and gender have certainly drawn upon association with "nature" to imply a fixed inevitability or "naturalness" to gender arrangements, but not all references to nonhuman species, biology, or materiality need be deterministic.[2] Feminist sociologists have achieved enormous movement in the past twenty years—moving from a position on the fringe of the discipline to the very core of sociology (the American Sociological Association [ASA] Section on Sex and Gender is among the largest), yet if it is true that as Casas (2014), Latour (2004), and others have argued, the notions of separation between the hierarchies between the social and natural have been foundational to the projects of modernism and settler-colonialism alike, then it would appear that feminist sociologists have spent too much effort in arguing that gender belongs in the realm of the social, rather than dismantling the dualism itself. Indeed, such a strong need to separate the social from the natural is only necessary in the context of the nature-social dualism that also underpins constructs of race, colonialism, and the very notion of modernity. Feminist sociology has yet to examine the ongoing dynamics of colonialism or genocide as ongoing processes in "modern" societies in North America, much less their entanglements with gender performances, gender violence, or gender scholarship (Glenn

2015; Simpson and Smith 2014; Smith 2012). But our failure to do so makes it impossible to see or theorize many forms of gendered violence taking place in Indigenous communities today, thereby continuing the legacy of academic collusion with Indigenous erasure and settler-colonialism. For non-Native feminists, there are fertile possibilities for both learning and solidarity with communities for whom alternative gender structures are not abstract theoretical utopias—communities that have been engaging in over a century of gendered resistance as part of cultural and political survival (Arvin, Tuck, and Morrill 2013; Ross 2009; Jaimes and Halsey 1997).

While there is much prospective for generative engagement across Indigenous and non-Indigenous feminist standpoints, the potential for solidarity has yet to be realized. And the stakes are higher than ever. Now in the face of today's rapid environmental degradation, the importance of the natural world for social dynamics of power and inequality, as well as possibilities for and limitations to gender expressions, gendered forms of resistance, and gendered experiences of violence has become both more visible and more important to understand. To see any of this as feminist, sociologists and gender studies scholars need more nuanced and specific ways to think about the mutually co-creative relationships between human experiences, social outcomes, and the natural world. For one example of how this can look, we turn to the case on the Klamath.

GENDER, GENOCIDE, AND LAND

Colonialism in the form of genocide and forced assimilation has obscured our knowledge of past gender structures, but we know enough to see that Karuk gender arrangements have long been very different than their non-Native counterparts and that Karuk conceptions of gender have long been intimately interwoven with ethics of responsibility and care for community and the natural world. A great number of traditional Karuk stories feature women and their importance as a food provider and healer, centering their success or failure as overarching themes and serving to socialize female listeners in their current or future roles and males in their assessment of their female counterparts. The responsibility of the woman as a food provider is symbolized in the crown of brodiaea flowers that traditionally adorn the head of an Íhuk maiden (i.e., a young woman who, as a result of having had her first menses, is the center of attention at the puberty ceremony held for women). The manner in which these young women sustain the arduous activities required of her over the course of this ten-day-long ceremony are considered representative of her ability to perform activities that are her responsibility, including managing for, harvesting, storing, preparing, sharing, and serving Indian potatoes to her extended family and community. In the traditional Karuk culture, women taken into marriage were "bought" from their families. The value of the woman depended largely on her ability to produce and care for children but, more important, on her ability to produce food and baskets—both for the immediate family and the village community as a whole.

The responsibility and etiquette involved in managing for and harvesting a wide variety of food, fiber, and medicinal plants are important topics in creation stories. Like salmon and acorns, Indian potatoes are also considered First People. As explained in the "Teacher Background" to the Karuk Tribe's Nanu'avaha (Our Food) Curriculum:

> The Karuk way of life is known to us through origin stories that have been told and re-told, heard and re-heard for countless generations. Listening to the stories together, we again re-imagine "the times before" human existence, "when the animals, plants, rocks were people." ... These First People understood their responsibility to figure out how the yet-to-come humans should live. Hearing the stories, we learn that they fulfilled that responsibility through repeated sequences of contemplation, discussion, inspiration, and both collaborative and random experimentation. As human re-hearers, we inherit the same ancient responsibilities of the First People, each of us in our own way trying to figure out "how people should be living."[3]

Not only are they the relations of the Karuk people, who also stem from the First People, but their use also carries specific responsibilities: how these are enacted reflect the honor and value of the performer. Participation in the system of tending, harvesting, and consuming is part of a pact across species. Bob Goodwin explains this responsibility:

> It comes back to me being responsible for what the Creator has entrusted me to do. That responsibility goes from me to my sons, my sons are gonna give that to their sons, and hopefully, within a period of time we will be able to get back to what our responsibility is, and bring the health of the river back to the way it should be.... We are a people that were created for our environment. And we were, or have been instructed as to how to keep that environment intact. What we do each year is to make it new again so that the next year we would have everything we need to sustain life. And that is the primary reason that we have the religious ceremonies that we do, to renew who we are as a people.

Another younger fisherman in his early twenties explains, "I've actually dipped my first time last year, and its ... and it feels good ... it felt good and I want my son to do that ... and it's an important role in being a man in the tribe ... you know ... you fish for your family, you fish for the people ... and there's fish days, and the ones who owned those fish days were responsible for feeding the community." Sociological conceptions of masculinity emphasize the *social* in socially constructed and performative aspects of gender identity but have not theorized how the natural environment may serve to structure the accomplishment of gender. Much present theory emphasizes masculinity as domination, but are there circumstances under which masculinity can be an enactment of racial resistance, resistance to colonialism, or cross-species reciprocity? Along with other Indigenous feminists, Sarah Deer (2015) observes that "patriarchy is largely a European import. Native women had spiritual, political, and economic power that European women did not enjoy. That power was based on a simple principle: women and children are not the property of men. I am guarded about pan-Indian essentialisms suggesting tribal nations were all 'matriarchal' and therefore rape free—Plains Cree Metis scholar Emma LaRocque cautions that 'it should not be assumed that matriarchies necessarily prevented men from exhibiting oppressive behavior towards women' . . . still there are some common themes in tribal histories and epistemologies that serve as counterpoints to patriarchy" (18). In a later passage, Deer notes that "power and equality are not the same thing; women may have had power, even as gender was used to prescribe the division of labor; balance between male and female can be described as non-binary complementary dualism, wherein binary gender lines are fluid without fixed boundaries" (19).

Just as chapter 1 detailed the mutual coproduction of race and the natural world, the construction of today's normative heteropatriarchy and gender binaries relied on symbolic notions of what is natural and on the material reorganization of the natural world. On the material level, the life-sustaining activities of Karuk women and men had to be reorganized into Western or European-identified gendered practices. Ideas of Karuk masculinity and femininity had to be symbolically altered as well. Through this process, a new gender order came to be understood as "real," on one hand, and came to have a new form of material power for its enforcement, on the other. The roots or the basis of the genocidal project is the erasure of the Indigenous people. Indigenous feminists have articulated this process across North America. Leanne Simpson (2017) describes how "missions were intense sites of assimilative education designed to make Michi Saagiig Nishnaabeg devote Methodists, the men and boys farmers and carpenters, the women and girls managers of effective British households and patriarchal nuclear families in village-like settings, thus removing Indigenous peoples from the land completely and erasing those who did not conform to the colonial gender binary completely. . . . This is sexual and gendered violence as a tool of genocide and as a tool of dispossession. It is deliberate" (98–99). Thus, what Bacon (2018) calls "colonial ecological violence" is deeply gendered. Hawaiian anthropologist Ty P. Kāwika Tengan (2008) writes that "the touristic commodification of culture and land in Hawaii proceeds most notably (and profitably) through the marketing of a feminized and eroticized image of the islands as the hula girl; meanwhile men are either completely erased from the picture, relegated to the background as musicians for the female dancers, or portrayed in similarly sexualized fashion as the surfer beach boy or Polynesian fire-knife dancer whose body and physical prowess are highlighted in an economy of pleasure" (8).[4]

Across North America, the alteration of Indigenous gender structures was forced through a variety of overt mechanisms that included the separation of children from socializing adults through so-called boarding schools. Leanne Simpson (2017) describes how "Indigenous forms of gender construction and fluidity around gender had to be replaced with a rigid heteropatriarchal gender binary and strict gender roles. Indigenous peoples had to be removed from educating Indigenous children. The sexuality of Indigenous women and 2SQ people had to be removed from the public sphere and from the control of Indigenous women and 2SQ people, as normalized within Indigenous societies and contained within the white heteropatriarchal home" (110). Forced alteration of forms of masculinity organized around community and cross-species responsibility to those of domination occurred because, according to Simpson (2017), "hierarchy had to be infiltrated into Indigenous constructions of family so that men were agents of heteropatriarchy and could therefore exert colonial control from within" (109). Simpson emphasizes how "dismantling the power and influence of Indigenous women became important to the destruction of Indigenous nations" (111).

> Everywhere in the world Indigenous women and our sexual agency provided a dilemma for the colonizers.... The more Indigenous women exercise their body sovereignty the more we were targeted as "squaws" and "savages," subjected to violence and criminalized. A large part of the colonial project has been to control the political power of Indigenous women and queer people through the control of our sexual agency because this agency is a threat to hetero-patriarchy, the heteronormative nuclear family, the replication and reproduction of (queer) Indigeneity and Indigenous political orders, the hierarchy colonialism needs to operate, and ultimately Indigenous freedom. (107)

In contrast to the earlier examples of Karuk women as sophisticated land managers who were politically and sexually autonomous, legal structures and cultural conceptions of gender were set in place to create hierarchy. This targeting of the gendered practices and identities was fundamentally directed toward settler acquisition of land. However, early manifestations of the project on the Klamath utilized two separately branched strategies to arrive at the same goal: those used to deal with men and those to deal with women. Men were viewed as a threat and impediment to land and resource acquisition, and the policy translated into a practice equated with the murder of all adult males—manifested first in the county, state, and federal (political and fiscal) support of bounty for heads—incentivized murder of adult male Indians. Indeed, the general for the Pacific theater's order from U.S. Army directing military forces to kill the males and spare the females was reissued under the commanding officer at Fort Humboldt. Next, this targeting and erasure of Karuk men took place by forcing them into indentured servitude through a number of ways, including charging them with the criminal offense of "loitering," or paying for their work with hard alcohol and then having them arrested for drunkenness. And third, it was carried out by rendering Karuk men impotent or powerless in the newly forming gender order by not employing them or by assigning them to hard or disrespectful labor paid minimally—sometimes the only paid wages were earned by doing things to the land and resources that went against all tribal code, honor, and value systems.

At the same time, the females were treated quite differently—nonetheless with the same "end game" in mind: taking the land and the resources. Karuk women were considered quite useful in, or at least harmless to, fulfilling settler goals. Filling the female void in the settler population, women were "taken" by the newcomers. And while listed as "wives" in early censuses, the ages of the "married couples" belie yet a completely different picture: the age difference between white husbands and Indian wives often averaged thirty years, and there is not a single recorded listing of an Indian husband with an Indian wife. In fact, to use the term *marriage* here serves again to erase the coercive and genocidal nature of these alleged unions: the implication is that the girl/woman had agreed to this relation, and therefore the results would lead to the "natural" and inevitable assimilation of Natives into white culture. In the erasure of Karuk males, the white settlers could move right into their shoes and enjoy the benefits of their land and resources with an Indian woman who knew how to live from them. Next, the children of this alleged "union" were listed promptly in later censuses as "white," leading to the appearance of the Natives' disappearance. While this practice may be interpreted as being an act of benevolence toward the Natives (i.e., acceptance of the halfbreeds into the white society), it was an explicit act of Karuk erasure. The forced reorganization of the family structure in order to access Karuk lands was violent and deliberate, with longstanding social and ecological consequences that continue to play out in the present. Lisa Hillman describes how, "Since contact, Karuk femininity has suffered by the fact that many women were 'taken,' otherwise known as raped or forced into servitude disguised as 'marriage.' Looking at the census' published throughout the early decades of the late nineteenth and twentieth centuries, the fact that tribal families were accounted for—including the number of children and how many of them could read and write, but that there was always a list of young girls and women who were 'orphaned.' There weren't any young boys or men, apparently. Only women. It just makes you sick" (2018, personal communication). She draws a connection between women's loss of status as active members of the Karuk community and the intergenerational trauma that men (and women) attribute to the damage to Karuk masculinity.

> It's like, now we don't have a say about anything. We can only cook and clean, but can't participate otherwise in tribal ceremonies or other active roles in traditional culture—besides making baskets—because we are "just women." People don't

think about the "why" anymore. A tribal elder told me that women aren't supposed to hunt because of their role as caregivers—they are shielded from the powerful medicine at play so that their role as mothers and community provider can be protected for the greater good. Not because they are somehow deficient, but because they are the carriers of our tribal continuity. That's not the message we hear now at all.

Instead, negative notions of femininity are inscribed in the very lands to which Karuk people are connected, where they continue to do the work of settler violence. Across California from the Squaw Valley ski resort near Lake Tahoe to less-known places like Squaw Creek and Squaw Canyon, place names contain this obscene term for Indigenous women.[5] In Karuk country, the use of the term *digger* as a derogatory term for women inscribes the new hierarchical gender order into the landscape. This pejorative term for California Indians, *digger*, is attached to the *Pinus sabiniana* tree, commonly known as the "digger pine." Historic accounts of Indian "digger" women abound in which the act of digging for various roots, corms, and bulbs was equated with acts of an "uncivilized" and "dirty" race of people. Such terms continue to exert gendered colonial violence today. Karuk Elder Janet Morehead recounts, "Once I heard this woman talking about 'digger Indians' at a party and I recognized that it made me feel bad. I told her I was offended, and then she got mad about it and said that she wasn't trying to be disrespectful. She sure acted like she was" (2015, personal communication). Today Karuk women are reclaiming once stigmatized traditional gender presentations and practices (Baldy 2017, 2018). Hillman points to the so-called "one eleven" (111) mark that traditional Karuk women have tattooed to their chins as young women: "This was done to protect women in times of war. It marked the women so that they would not be targeted by other warring tribes in order that the families would be preserved. Nowadays women are beginning to do this again after many years of suffering derogatory remarks by non-tribal people—as an act of resistance and outward connection to their tribal identity."

FEMINIST SOCIOLOGY AND CONSTRUCTING GENDER

These Karuk voices engage several areas of uneasy terrain within the sociology of gender. The first concerns the relationship between constructions of gender and the natural world. Although sociology was borne of the modernist view that humans had "risen above" nature and that human society is most appropriately understood through a focus on "social facts," the functionalism of early sociological work justified unequal gender relations through a determinism that naturalized what we now call "gender." Feminists responded by asserting that gender is socially, not naturally, constructed. Thus, rather than addressing the modernist notion of a nature-social dualism, sociological scholars (including sociologists of gender in particular) have theorized two distinct categories: sex as biological and gender as specifically *social*. As Anne Witz (2000) aptly notes, "The very term 'gender' as it has come to be used within sociology and anthropology, linguistically denotes that which is not natural, although it is presumed to be" (7). Against this backdrop, references to "nature" continue to be associated with biological determinism or essentialism. Yet "nature" in the form of nonhuman plant and animal species, rocks, rivers, and places are central organizing features of Native cosmologies, culture, and social life, including gender (Anderson 2005; McGregor 2005). Furthermore, the Western discursive negation of the natural world has been a building block of modernity and central to the legitimization of North American colonialism. The discipline of sociology has played its own part in setting up the modernist frameworks that divide humans from the so-called natural. In his historical review and analysis of the epistemic implications of inequality and marginalization within sociology, Go (2017) reminds us that "the very notion of the 'social'—as a space between nature and the spiritual realm—first emerged and resonated in the nineteenth century among European male elites to make sense of and to try to manage social upheaval and resistance from workers, women, and from so-called natives" (195). And that the purpose of doing so was explicitly elitist. Go continues, noting that sociology

> first emerged in the United States and Europe as a project for the imperial metropoles and the white, straight, middle- to upper middle-class males in those metropoles (Connell 1997; Steinmetz 2013). And its emergence was not purely an intellectual matter. We know that Auguste Comte first used the term sociology in 1839, theorizing "the social" as a space distinct from the political, religious, and natural realms. But recall that a key part of his larger project was to create a group of a technical elite experts, armed with knowledge of the social realm, whose ideas could help manage and control society. Sociology was to be the "science" of the social, and it was to serve the powers that be. (195)

As a result, the move away from materiality in much present gender scholarship obscures colonial power relations and the ever-associated environmental degradation that continue to be highly salient for Native peoples, making their experiences invisible and irrelevant within dominant discourse.

Feminist sociologists have struggled to identify and navigate the artificial construct of the nature-social that many believe is foundational to many of the social sciences. This conversation has recently played out in work on the body (Witz 2000; Rahman and Witz 2003; Crossley 2001; Malacrida and Boulton 2012; Jaggar 1989) and queer theory (see, e.g., Mortimer-Sandilands and Erikson 2010). It is exciting to note that within some areas of gender scholarship such as queer theory and trans studies, these binaries are beginning to change. Just as Messerschmit

(2009) and others call for more adequate theorizing about the presence of material bodies as part of "doing gender," I argue that theorizing the symbolic and material importance of the natural environment is necessary to understand both the constructions of traditional Native masculinity and femininity, and the operation of gendered and racialized colonial violence in the form of environmental degradation today. In fact, there is a difference between biological determinism per se and theorizing a place for the natural world in the structure of social outcomes. Part of what is needed to do this is to open up a wider discussion of "nature" and "the natural" by providing specifics and language to interrogate the workings of the natural world in shaping gendered understandings, practices, inequalities, and possibilities for resistance. It is my hope that this chapter contributes to movement in this direction.

We encounter a second area of uneasy terrain when it comes to dominant focus on masculinity as achieving and maintaining one's position within hierarchies of power. Raewyn Connell (1990) writes that "masculinity is socially constructed and has a material existence at several levels: in culture and institutions, in personality, and in the social definition and use of the body. It is constructed within a gender order that defines masculinity in opposition to femininity, and in so doing, sustains a power relation between men and women as groups" (454). Masculinity as presently theorized is socially produced and about power, as Candice West and Don Zimmerman (1987, 126) assert in their foundational essay. Masculinities are understood to change over time, yet again one key impetus for such change in the dominant framing has been the ability to sustain power relations (see Pascoe and Bridges 2015). But can masculinity be equally about racial and colonial resistance? Is it possible that the prevailing constructions of masculinity as domination merely privilege colonial masculinities and normalize the colonial project? Indigenous feminists and feminists of color have articulated how dominant gender norms are tied to white supremacy and likewise, in a settler-colonial context, to colonial domination (Collins and Bilge 2016; Coulthard 2014; Fenelon 2015a, 2015b). Although understanding masculinity as hierarchy has been critical in the context in which it was developed, voices in this chapter speak to the possibility that Indigenous masculinity can be equally about what Kyle Whyte (2013a, 2018d) calls "collective continuance," responsibility, and resistance to racism and colonialism (Vinyeta, Whyte, and Lynn 2016b).

Finally, people in the Karuk community place much importance on notions of *tradition* and *traditional roles* that have—for good reason—become intensely unfashionable among anyone thinking seriously about gender and power. Here I use the term *tradition* to refer to those aspects of values, norms, worldview, and social practices that, while not unaffected by outside influence, biologically inherent, historically fixed, or socially static, are nonetheless intensely meaningful *to Karuk people themselves, especially in a context of genocide and forced assimilation.* According to Lisa Hillman, "People always tend to say that traditional women 'gathered, cooked and raised children.' Past tense. The fact remains that we are still here doing 'traditional' things—and it's a whole lot more than that. Women manage oak stands, meadows, and certain plant communities with fire; disturb soil to enhance corm and bulb growth; seasonally gather hundreds of food, fiber, medicinal plant, and regalia species; produce tools, clothing materials, and a beautifully crafted array of food-related utensils; process, store, and prepare foods, fibers, and medicinal plants; care for young children and more" (2018, personal communication). And the responsibilities do not end there. For women from traditional medicine families—especially herbalists, for example—the practice of healing comes with a long list of land management skills and responsibilities to teach their eventual successors, serve their community, and steward the plant communities themselves. "These are our relations," notes Hillman, "and we care for them as they care for us in return" (2018, personal communication). It has also been pointed out to me that while roles matter and they are linked to gender expression, the Indigenous counterpart of "roles" is less ridged than in non-Native contexts. Instead, there are many options for how to be traditionally male or female, many options for how to fill such gender "roles."

Indigenous conceptions of tradition may be highly varied, but in the context of genocide, colonialism, and forced assimilation, their importance cannot be underestimated. Here there is some room to expand upon existing theory in race.

Vasquez and Wetzel (2009) describe how, in the context of racist social stigma, an emphasis on tradition is used to reassert social position and collective dignity: "Mexican Americans and Native Americans establish their social worth—that is their group's social position and collective dignity—by making discursive comparisons with, and drawing distinctions from, the American mainstream. They evaluate themselves using a metric that highlights their traditions, specifically their roots, values and cultural toolkits" (1557). But traditional gender structures matter not only in their ability to reinscribe positive self-conceptions in the face of racial stigma as highlighted by Vasquez and Wetzel. Here they are about physical and cultural survival.

The next section draws upon the words of Karuk people to describe how the natural environment has been a central feature of how traditional Karuk people construct gender, shaping important social practices and structuring their internalization of gender identities and power structures and their resistance to racism and genocide. In this chapter, an initial focus on masculinity and fishing came about as a result of the earlier work on declining fisheries due to the dams and the fact that my long-term collaborator is a traditional fisherman. Yet the focus on femininity has been brought to my attention as one of the essential factors missing in this discussion, as well as lacking in depth in the sociological study of Indigenous perspectives on race and its projection through environmental degradation.

CONSTRUCTING KARUK MASCULINITIES AND FEMININITIES: RESPONSIBILITIES TO RELATIONS, FAMILY, COMMUNITY, AND COLLECTIVE CONTINUANCE

Throughout my time working in Karuk country, people have impressed upon me that humans have vital and substantial responsibilities to the natural world. These responsibilities include caring for the other species as well as utilizing them. Ron Reed explains that the reason for this is that, otherwise, they won't come back:

> People sit there and say, "How can sit there in good faith . . . and eat a coho salmon, they're endangered!" . . . I say you know what, I look at it the other way, the Karuk way: "That's the reason why they're endangered! Is because we can't eat them! Our people can't eat them!" That the stretches of the river where they go—our people are supposed to be up there eating them! And taking care of them! And managing them! And we're not able to . . . so it doesn't surprise me one bit . . . the only reason why they're still around is because we're eating them, we're holding that ceremony where we provide them a place to go, spiritually . . . if we don't eat them, they're taught to not come back.

Fishing is thus one of the responsibilities people have to the fish. Other traditional activities that people do to fulfill this responsibility include ceremonial management of the fishery to ensure "escapement" and burning of the forest to enhance runoff and maintain other ecological conditions beneficial to fish as described in chapter 2. Karuk fishermen provide salmon both to their own families and the entire community. As Jennifer Goodwin, a woman from a prominent fishing family, explains, "And my family when they go fishing, they go fishing for all the elders, not just our grandparents or our great aunties and uncles. They go up to the Tribal council even and give out fish to the Elders. They take that responsibility, you know, very seriously."

In the same vein, harvesting Indian potatoes and huckleberries is one of the responsibilities women have to these species, as the very acts of managing for and harvesting stimulate greater production and encourage healthier results. As the custom with all "relations," harvest is only done when adequate resources for regeneration are provided for and the first harvest is sacrificed. Wasting resources is strictly prohibited, and, for example, acorns, fish, and so on fed to the ceremonial priest that are not finished in their entirety are "fed" to the sacred oak tree, which represents one of the First People. Prayers are said when the first of certain seasonal species are spotted. All harvests are shared among community or family members. Leaf Hillman explains, "The first harvest of the year or season is never kept and consumed by the person who harvested. It's required to be shared outside the immediate family of the harvester. One of the purposes of that is to

FIGURE 14. Ron Reed and family packing fish up to elders at Ishi Pishi Falls. Photo credit: Achviivich.

discourage selfishness, greediness—it's to remind us of our obligation not to be greedy." The activity of community provider is also a crucial aspect of fishermen's identities and a source of pride. My research collaborator, fisherman and father Ron Reed, reflects on his pride in providing fish, as well as on how it connects him to both the earth and his people across time—an experience of collective continuance: "Being able to fish for people, for the ceremonies. You know, there's a great deal of pride being able to deliver fish to people. And you know, it connects you to the earth, it connects you to *pananahouikum*, the people that walk before us." Because it is men but not women who fish, and it is women but not men who gather plant resources, carrying out these responsibilities is deeply entwined with gender identity. A female Tribal member described the importance of the huckleberry for what it meant for her to be a woman:

> My earliest childhood memory is picking huckleberries. I remember it was hot and my hands were sticky, my clothes were all dusty from sliding down a bank. I remember wanting to go home and being thirsty, but that I was stuck there with all my cousins and sisters, my grandma and aunties, picking and picking. I remember wishing I could use the nippers that my mom had, and listening to her as she explained why we took branches instead of trying to pick off the berries. "Do like

the deer do—see how they've nibbled on this branch last year, and see now how many berries are on it this year?" And even though I can remember hating being there, it's one of my fondest memories. That was the day I started to learn what it was to be a woman, a Karuk woman.

Another female Tribal member described the importance of managing resources for men's sense of what it means to be a man: "And so for them it's like that is what is what makes them a man, it's part of who they are, and so I think that one of the things that fishing and hunting and forest management . . . those are what makes a man a man . . . you know and I so I think that they're really excited to get to participate in those kind of things when they do, you know, and take great pride in it." Being a woman in the tribe carries other crucial aspects with regard to fish, and those are mainly centered on preparing, drying, cooking, storing, and serving. Only an older girl is allowed to use the traditional quartz or contemporary metal knives to filet salmon, as the taboo surrounding "wasting" food carries a great deal of weight. Similarly, that one's first time "dipping" and "cutting" fish, for example, serves as an informal rite of passage for boys, girls, and young adults further underscores the importance of fish to gender identity. David Goodwin describes the strong community in his youth when many people gathered at the falls: "You'd get up on the big rock, it was called, and watch the older men down there fishing, and you know, ordering each other around, like: 'Come over here! Grab these fish out!' You know, clubbing them up, and it was just always happening. It was always exciting, you know. You always felt important, because everyone was involved. It was a step process of learning, but yea it was still good." Ron Reed explains how as families fish, boys learn a host of skills and traditional values.

> Originally, you had family rites down at Ishi Pishi Falls. Now kids are just going through life and living, but actually they're being trained at a very young age. You teach them how to go down and, first of all, just being able to walk down the falls. And the first issue is carrying them down there as a baby and getting them familiar with the falls, and pretty soon, the next thing you know, they're walking. You take them down and let them go as far as they can. The next thing you know, now they're walking all the way down. Pretty soon, they're packing little Jack out. Now they're packing the salmon out. Now pretty soon, they're clubbing for you out there on the rock a couple of the fish that you're catching. Then all of a sudden, they're cleaning fish and packing them out and then actually giving them to elders and then learning how to process fish and learning how to gather the materials for your fish poles and how to do all those type of things.

Daughter of a ceremonial leader, Sophie Neuner, shows how the same can be said for the socialization of women in the framework of Pikyávish, also known as the Karuk annual World Renewal Ceremonies:

I'm always surprised at how much the girls learn each year. First, it's about fasting and not complaining. Then, when the girls are about 4 or 5, they can start serving at mealtime, put away some of the dishes and pound acorns. Easy tasks at first, right? As they get older, they learn to prepare food, gather wood, stoke the fire and keep it the right temperature for what they [the cooks] need it for, thread salmon and eels on sticks, leech the acorns, heat the cooking rocks—and so on. All the while they are learning language, culture—values like respect for our relations, sharing with the community, generosity, you know? It's a lot of work being a woman, but they all look forward to it each year.

And as ceremonial leader and DNR Director Leaf Hillman notes, "Like the Ikyávaan,[6] who prepares the meal for the Fatavéenaan[7] when he breaks his fast, the girls learn that the purity of their body and mind is critical to the overarching success of the ceremony . . . or the preparing the daily meal, while fasting, reinforces the important connection between their good thoughts and . . . the importance of that what you're feeding the people . . . that you maintain a good state of mind" (2018, personal communication). Present gender scholarship emphasizes how social context and settings structure the ways people "do gender." Here the natural world is also fundamentally central for how men and women construct Karuk gender identities. Constructing this form of Karuk masculinity is not possible without the presence of very particular *ecological conditions* (a river full of fish) that are central to men's ability to perform masculinity. Likewise, because the role of the woman is to prepare and serve food for her family and community, the presence of food resources is central to her ability to enact this form of traditional Karuk femininity. Wife of a Karuk ceremonial leader, whose role is to lead the "kitchen" throughout the ten ceremonial days, explains her frustration during a year when her husband was unable to harvest a deer: "Okay, yeah it was a bad year and I know he feels bad. But I'm not going to stand here and not serve any *púufich*[8] for our guests. That would look really bad on me! So I'm not sorry I got the meat from some other guy. He's [the husband/ceremonial leader] got to deal with it." In this case, both the Karuk identities of masculinity and femininity were threatened by the lack of critical Native food resources, whose population number was reported to have been at least fivefold prior to Euro-American contact and whose habitat has suffered tremendously under environmental degradation. Moreover, the tribal leader's wife could only preserve her own sense of identity by further increasing the "emasculating" effect of not being able to harvest a deer and accepting the gift of another—here more successful—man.

Whereas colonial notions of masculinity center domination, Karuk constructions of masculinity and femininity are not only about the acts of gathering, hunting, or fishing, or even the act of actually harvesting resources, but about participation in a set of *responsibilities* to these resources that they know as their relations and to family, community, and the future, each of which can in turn only be enacted if ecological conditions are right. For example, the presence of specific material ecological conditions is central to how boys and men from fishing families come to internalize masculine identities. Traditional fisherman and cultural practitioner Mike Polmateer reflects on how he came to understand his importance in the community and sense of responsibility in the course of providing for his people:

> I killed my first deer when I was about twelve years old. It was a big four pointer and one of the things that really, that really sticks in my mind right now, is that after my brothers cut it up, gutted it, at that time, I realized that I was an important part of our society, because I was able to give meat to my aunts, my great-aunts, my uncles. Even today when I catch fish in the falls, I don't just bring fish home. I give it away to Elders. People who don't have kids who can go down and pack fish up for them. I take fish to these people.

Leaf Hillman explicitly underscores this difference: "The domination over nature is at the heart of Western notions of the masculine. Contrasting these notions are the gender roles embodied in traditional Karuk society that emphasize working along with our relations in a reciprocal manner." And while notions of tradition and gender look very different than in Western systems, they are also less binary, as reflected in earlier quotes from Indigenous feminists, and this Karuk elementary school teacher's explanation:

> I think the curriculum is really important for these [tribal/Karuk] kids. They're learning about who they are, and one of the ones I think is *really* important—and I say that because we've got a bullying problem—is the lesson on gender roles.⁹ It's important that kids *get* that guys to this and girls do that—but that at the same time there are exceptions. People did and do what they gotta do. And every time I've taught that lesson, the kids are really quiet: they're listening hard, because they want to know how to act. Then they get really quiet when it comes to the part about the traditional acceptance of transgender and homosexuality. I think they really appreciate hearing about how they should be respectful, I guess.

Being "Karuk" is a racialized as well as gendered experience. That is, racial and gendered identities are interwoven. The act of hunting, gathering, or fishing simultaneously confirms one's gender identity and *Karuk*, as this Karuk mother in her mid-thirties surmises: "I think that that's one of the things we end up with today is because we have a limited view of roles. It's like OK, you're either a fisherman or . . . if you are a guy, you gotta be a fisherman. You don't want young boys to think, 'I've never been to the falls to fish' you know, 'so maybe I'm not quite the Indian that someone else is who goes to the falls and fishes.'" For Karuk people, both ethnic and gender identity are constructed through the fulfillment of responsibilities to their relations and collective continuance. Karuk gender practices become synonymous with enacting these responsibilities. Nature is necessary to how individual men perform masculinity, how individual women perform femininity, and the gender structure of the community. The presence of food resources is an *environmental* context that links sense of self, social interactions, and social structure. But as these speakers describe, these constructions of gender and race are vulnerable in the world today.

CHANGING ENVIRONMENTS, CHANGING GENDER PRACTICES: WHAT HAPPENS TO KARUK MASCULINITY WHEN THERE ARE NO FISH? WHAT HAPPENS TO KARUK FEMININITY WHEN THERE ARE NO ACORNS?

If achieving Karuk masculinity or femininity is about performing ecological responsibilities to care for fish, foods, families, and community, as well as about participation in collective continuance, what happens to these gender constructions when there are no fish? What happens to these gender constructions when meadows with important bulbs have disappeared, when basketry materials cannot be used because the forest has not been properly managed with fire? Many of these ecological changes are quite recent. Historic aerial photographs and forest data illustrate drastic reductions in the total extent of open grassland and meadowland habitat over the past century as a result of fire exclusion as described in chapter 2. Despite the significance of earlier environmental impact such as hydraulic mining and overfishing in the early 1900s, the testimony of adults and elders about river conditions and foods they ate until recently indicates that very damaging impacts to the abundance of salmon and other important riverine species such as sturgeon, steelhead, and eels occurred within the lifetime of most adult Karuk people today. Whereas many people reported eating salmon up to three times per day during fishing season in the 1980s, in 2005, the entire Tribe caught fewer than 100 fish at their main fishing site of Ishi Pishi Falls. In 2017, the Karuk Tribal Council imposed strict limitations on even ceremonial fishing due to record-low numbers of returning adult species. This environmental change has drastic implications for the entire community. The way Jennifer Goodwin understands it,

> I think that traditionally you know there is are pretty strong gender roles and roles for individuals within the community and when you can't fulfill those roles, that's when we start to have problems like drug and alcohol abuse even domestic violence and child abuse because I think . . . this is just my perspective, but I think if fishing and hunting and providing for your family in that manner is what makes you a man so to speak, then if you are not able fulfill those things, how do you prove that? How do you show that?

In the Orleans Maiden creation story mentioned in chapter 3, the plot revolves around the harvest and associated etiquette hereto of the brodiaea species—locally known as the Indian potato. There are at least three types known to have been traditionally harvested on a regular basis in the mid-Klamath region, and a number of other stories feature them.

> People hardly ever mention them anymore. Now it's only "salmon and acorns," as if we only ever ate *them*, but we used to eat a lot of other things. It's just that the landscape has been literally assaulted by settlers, and we haven't been able to come after them and repair the damage. Take Indian potatoes, for example. We used to go to certain places like over there [pointing to an area across the Klamath River] and spend all day digging for potatoes and packing them back home at night. It's a woman's thing, and we all had our own digging sticks—I have one that my son made for me out of yew wood. The digging was good for the patch, and they grew up again year after year. They used to be bigger because of it—because they had room to grow, you know. Now they are pretty small and really hard to dig up, the dirt is so dry and hard. (2018 anonymous, personal communication)

The enactment of traditional Karuk masculinities and femininities has both material and symbolic significance in the projects of family survival and Tribal collective continuance. Lisa Hillman emphasizes that

> the tan oak acorn is such an important staple—we used to eat it all the time. Now we try to make sure we have at least enough for ceremonies, although we had a fairly decent crop a couple of years back. The Forest Service let the whole country grow over with Doug fir, but sometimes we can still get a good harvest. The idea that sudden oak death is just a question of time before we'll see it here is debilitating. What will we do when the trees die? Who's going to remember how to make acorn soup when there's none to practice with?

In this case, what Goffman would call the "dramaturgical task" of performing gender is fundamentally dependent upon the presence of a river full of fish or a forest full of acorns. And these gender performances are not merely of individual consequence as so many sociologists have emphasized, but also fundamentally necessary for Karuk culture to continue. Karuk Department of Natural Resources Director Leaf Hillman describes how

> I can say from a professional point of view: at a policy level at the Department of Natural Resources, certainly when I first became aware of this, I thought: "oh, this is yet another challenge that we have to address in our advocacy." But then to attend an informational "sudden oak blitz" in the community and you hear from

the leading Western scientists from the field who report on their lack of success in controlling, containing, slowing the spread, or otherwise altering what is described as a "when" our forest will become infected and hearing the prognosis of lack of control once it arrives, is like a punch in the gut. It is an emotional trauma that is in some ways even worse than the typical environmental degradation, like fire exclusion. Those are things we think we can cure, but sudden oak death? This horrible infestation could completely wipe out the most heavily relied-upon resource we have. Acorns are no less potent a symbol for tribal identity than salmon, and no less critical to our continued survival as Karuk People.

Ron Reed describes how people who are unable to uphold their responsibilities feel not only a sense of shame (which can be understood more in the individual sense of status loss) but also guilt in relation to community responsibilities.

> Elders are always trying to give you money, and you know, you respectfully deny it, and then some elders respectfully demand that you take it. And so there's prayer back. And so when you do that, you're living the good life, you're giving people fish, you're doing what you're supposed to do. But when you're not able to do that, there's a sense of emptiness, there's a sense of not meeting your responsibility, and there's a sense of emptiness when you have different functions. When people tell you, they hadn't had fish in five years. Or three years, or two years, or I haven't eaten eel in ten years. And I might have just eaten it last night. Your elders yearning for the food, and you know, it's a scary situation to be in. All of traditional fishermen back in the day, you're supposed to take care of your family first. Your family was taken care of first, then you reached out to the other community members, and now it's impossible to do that.

In addition, people describe stress and disgrace due to their inability to fulfill responsibilities to the Creator, the First People, to particular species in the ecosystem, and to the human community. "If an Elder is asking upon you to go fish for him, that's like an honor. If you can't fulfill that honor, it's kind of degrading inside. You know, it's hard on the spirit." Similarly, "So, I was really embarrassed [laughs]—I coined a new term: 'tribal shame.' I was really busy that fall and didn't get my shit together, so when it was time for Jump Dance, I had to get on Facebook and beg for acorn meal. I even offered to buy it." Connell (1990) writes, "It follows that any particular form of masculinity can be analyzed as both a personal project and a collective project. Conflicts over a form of masculinity similarly have two levels, in which rather different things are at stake. On one level, alternative transformations of personal life are at stake; on the other, alternative futures of the collective gender order" (454). As in the case mentioned above describing the plight of the ceremonial leader's wife, this situation is analogous to the concepts of role stress and role strain from the mental health literature. Mirowsky and Ross (1989) define, "Role stress is a disjunction or inconsistency in the system of roles, so that normal obligations cannot be met.... Role stress produces role strain, which is the frustrating sense of not being able to understand or meet the normal expectations of one's roles" (15). Yet while ideas like role stress and role strain to which these Karuk speakers implicitly refer describe part of what is going on here, these concepts fail to capture a situation in which an entire moral system is threatened by an outside force. The symbolic violence entailed in the emotional experiences of environmental violence is expanded in more detail in chapter 5.

Anger is also a response men and women describe to the changing circumstances and inability to perform appropriate activities. During the original interviews that Ron and I did in relation to the dam removal, a number of men voiced sentiments similar to this traditional fisherman: "You know when I get pissed off, you know what I do? I go out and start drinking. But what if I had salmon? If I had a fucking, if I had a sweat lodge in the back of my house . . . or if I had a fishery that we had enough fish and if I could go give to my mother and my kids and the way we need to live . . . that's what we're looking for." Women too expressed anger at the gendered dimension of ecological violence: "So, it's like . . . you're really a good catch if you're a good basketweaver. And, heh—guess what? I suck at it, but I think that's really bogus. There hasn't been good quality materials for forever, and I didn't have anyone to teach me that. Now I could, I guess . . . but it's too late for me to be any good." People interviewed made connections between threatened self-concepts and negative behavioral outcomes from alcohol and drug use to violence in the Karuk community in the face of declining food, fiber, and medicinal plant resources. As Jennifer Goodwin notes, "Just seeing my brothers and what they've gone through and you know being able to go down every summer to the falls to fish, and now not being able to do something like that, I mean they're finding other things to occupy their time, and you know, it's not always constructive!" Another man in his forties made reference to job loss and unemployment in attempt to explain what was going on:

> They would fish and that was their role and they distributed that fish to people, and that was, kind of their capital, again, of networks. When that's not there, where's your value? What's your role as a human? What's your role and function if you don't have the ability to exercise that? Yeah, you're going to have a sense of loss. It's no different from somebody who had, you know, a good-paying job, and, you know, was respected and well-known in the office and blah blah blah, and then they get laid off, there's that loss of like, well, what do I do now?

In the course of our conversations and interviews, many people made connections between men's unemployment and job loss in the formal sector and the experience of Karuk men being unable to provide fish to their families. As traditional dipnet fisherman Kenneth Brink explains,

> You know, a fisherman that's able to catch fish and provide for his family is just as good as anybody else in the professional world, you know . . . so if this is your way of living . . . your way of living . . . if it revolves around the fish, you know providing for your family . . . now the fish are gone but now you can't provide no more, so like you're changing your whole life around . . . some people don't have time in their life to wholly switch their life around.

People in the community are keenly aware that men are unable to catch fish as a result of "outside" and involuntary interference with gender order, in this case as a result of state environmental regulations that have ordered the natural world around the goals, values, and economic activities of non-Native society. As Karuk practitioner and fisherman Jesse Coon describes: "You know the federal government came in here a long time ago and started to take everything over, which they damn near did, you know? We can fish at the falls. Dipnet and that, you know, that's the only place we can fish really. But we're not able to go out and go hunting anymore, without getting in trouble for it or something, you know." Ron Reed's brother, Mike Polmateer, vividly describes the agonizing interference of the state in the gender practices and family responsibilities of everyday life:

> I fish at my family's hole up here at Dillon Creek every single day during the winter, and I'm checked for my license no less than six times per year, by the same game warden, by the same two game wardens over and over and over. They're trying to catch me keeping fish. They sit up here on a point with binoculars watching me catch fish, and they watch me return them to the water. Because I'm—I'm afraid. But when my great aunt Marge, when she wants fish, I bring her fish, regardless of what the consequences might be. I'm going to suffer that consequence because an elder wants fish.

Thus, it is through *the declining river conditions and through impacts to men's ability to construct desired gender identities* that individuals come to understand their positions in a system of power relations. Power is internalized and experienced via disruptions to individual gender identities and to the community gender structure at large. People across the community are aware that fish are declining, and they explicitly associate both environmental decline and the inability of Karuk men and women to be able to carry out what they describe in their own language as "traditional roles" with colonialism, racism, and the end of the Karuk way of life.

ECOLOGICAL DECLINE AS COLONIAL VIOLENCE: "DOING MASCULINITY" AS A PROJECT OF RACIAL RESISTANCE AND COLLECTIVE CONTINUANCE

Better theorizing of the natural world is not the only suggestion from Karuk country pointing to how Native voices can enhance understandings of power within gender studies. Present theorizing on masculinity, for example, emphasizes its social origins and purpose in domination. Schrock and Schwalbe (2009) write, "All manhood acts as we define them are aimed at claiming privilege, eliciting deference, and resisting exploitation" (281). West and Zimmerman's key (1987) essay describes how, "rather than as a property of individuals, we conceive of gender as an emergent feature of social situations: both as an outcome of and a rationale for various social arrangements and as a means of legitimating one of the most fundamental divisions of society" (126). We can think of individual Karuk men engaging in "gender performances" that solidify self-esteem and gender identity, but we miss something important if we fail to understand that for both these men and the broader Karuk community, to see young men bringing fish to elders is fundamentally an expression of the continuity of culture. Thus, it is critical to understand that for the Karuk men interviewed, individual experiences of shame, anger, stigma, or loss are not only in relation to self and the inability to provide for a nuclear family but also interwoven with the sense of responsibility to care for a larger community that extends beyond the human and to fight against racism, genocide, and colonialism. Conversely, the inability of Karuk fishermen to carry out these gendered social practices is understood as a manifestation of cultural genocide. Ron Reed describes how

> as a traditional fisherman you clean your fish immaculately. Pretty fish, you know your relation, that you're preparing food that you're giving to somebody, and you clean that fish with great reverence, with great understanding, with practically prayer. Because what you get back is prayer in return.... And so when you do that, you're living the good life, you're giving people fish, you're doing what you're supposed to do. But when you're not able to do that, there's a sense of emptiness, there's a sense of not meeting your responsibility.... You know, our fisheries at risk. Our fishing identity is at risk. There's maybe a handful of fishermen left, so therefore there are a handful of families that are associated with that fishery, and it's important that those families continue to fish there and can carry on and lead you know, the traditional ways.

These experiences of a degraded river and forest are associated with a long-felt awareness of Karuk culture and life under attack, as well as a longstanding sense of their imminent destruction. A sense that Karuk life and culture could come to an end provides a grim background cadence to people's everyday activities. In the course of our interviews, many people shared statements such as Ron's cousin Binx emphasized: "The Karuk people actually believe that if the salmon quit running, the world will quit spinning, you know, maybe the human race as we know it may be nonexistent... if the river quits flowing, it's over... if salmon quit running, it's like the sign of the end." Here changing environmental conditions become the leading edge of genocide, and the anxiety that underlies these accounts forms a backdrop to the many layers of struggle associated with the degraded river. Bacon (2018) coins the term *colonial ecological violence* to underscore this dynamic. Whereas intentional resource destruction was used as a tool of genocide and forced assimilation in North American colonization during the 1800s (e.g., the destruction of the buffalo is classic example), the ecological destruction of today continues to be the leading edge of forced assimilation, genocide, and colonialism for those Native people who have managed to retain relationships with ancestral landscapes and species. Degradation of the natural environment is a central vector continuing the transfer of power and material resources from Native to non-Native people today.

Colonial ecological violence is racialized and gendered. It is racialized in that it targets people of particular racial categories and operates through disruptions of group identities that are central to individual self-esteem, on one hand, and collective meaning systems and cosmologies, on the other. Like colonialism (Smith 2005; Smith and Kauanui 2008), colonial ecological violence is also gendered. Not only does it affect women and men in specific ways, but *it also targets them as men or women*, disrupting the process of gender identity and the gender structure of the community.

Colonial ecological violence is both symbolic and material. It generates material outcomes of wealth and poverty at the same time as inflicting symbolic violence onto individuals through shame and powerlessness when people internalize power structures of racism and colonialism. In the present context of environmental decline and community resistance to genocide, continuing to engage in "traditional" masculine activities such as fishing is also understood as fighting back against the intertwined forces of colonialism and racism (not to mention poverty) that have deeply structured individuals' lives through the appropriate enactment of authentically Karuk gendered selves. Men's individual struggles to understand themselves as men are interwoven with resistance to ongoing racialized ecological violence in the river basin. Continuing to engage in traditional activities asserts that "we are still here" in the face of a hegemonic social discourse of extermination (Smith 2012; Steinmetz 2014). Potawatomi philosopher Kyle Powys Whyte (2013a) describes how environmental degradation threatens the "collective continuance" of Tribal peoples: "These challenges lead many tribes to remain concerned with what I call collective continuance. Collective continuance is a community's capacity to be adaptive in ways sufficient for the livelihoods of its members to flourish into the future" (518). Thus, within this context today, Karuk constructions of masculinity and femininity are not only about responsibilities to community, fish, and family, but they are centrally also about participation in collective continuance via resistance to colonialism.

Such conceptions of gender practices are, however, invisible in current theory. Fishing, participating in ceremonies that regulate the fishery, and distributing fish to the community can each be understood as "manhood acts." But these manhood acts can only be enacted if ecological conditions are right. Furthermore, while the field acknowledges multiple and subordinate masculinities, the overall framework negates the possibility of male gender identity unrooted in domination. In the theoretical orientation of the field away from sex roles and toward performative masculinity, contemporary understandings have centered the notion of masculinity as achievement and maintenance of one's position within hierarchies of power. The enactment of traditional Karuk masculinity requires particular ecological conditions in the natural world (a river full of fish) and fulfilling a set of responsibilities to the natural world and to the human community. Definitions that emphasize domination leave little room for conceptions of masculinity in the form of carrying out responsibilities to the natural world or to community. In this case, the "manhood acts" performed by Karuk men serve ecological functions, unite families and communities, and perpetuate culture in the face of genocide. To account for Karuk masculinity as an enactment of racial resistance, resistance to colonialism, and cross-species reciprocity, we must extend these definitions.

RESTRUCTURING MASCULINE IDENTITY AS ACTIVISM AND FISHERIES MANAGEMENT

As society changes, people work to create new gender identities and continue to find ways to express gender constructions and cultural values in the present. As Sherman (2009) found with the loss of the timber industry in rural California, men may restructure ideal forms of masculinity in order to make them attainable. This is how unemployed men in Sherman's Golden Valley restructured masculinity through an emphasis on fatherhood and engagement in activities of hunting and fishing. Judith Large describes how in conflict zones where daily life is disrupted and there are few options for pride as a breadwinner, young boys are prone to images of violent masculinity and may be drawn into war (1997, 27).

Salmon and other riverine species have been central to social arrangements on the mid-Klamath. Their material absence cannot simply be replaced through symbolic reconstructions. Yet while the use of alcoholism, violence, and drugs appears to be a response to the absence of salmon and corresponding disruption of gender order, some Karuk men also describe engaging in Tribal activism regarding the removal of Klamath river dams and working for the Karuk Tribal fisheries program as realms through which they are able to transfer traditional cultural responsibilities to fish, community, and collective continuance to new settings. When Ron and I asked how they coped with the present ecological situation, many men named their work on fish crews:

> I feel like we're making a difference, yes, you know, the numbers will start to come up and ... they've already, you know, they're already coming in the tributaries that were blocked off and did a little field assessment and made pools for them to make it up there ... now they're up there and they're spawning in there ... that's the proof right there ... is where our work ... we go in here and we try to better this creek to make it better for the fish so that they can spawn.

Men expressed pride in their work and a feeling of connection to the larger cause of caring for place and community. Here a man in his early twenties describes his experience:

> I've been doing this [working on a fish crew] for about seven months now, and six of them were volunteer, and that for me personally in itself was awesome ... I see a lot of my peers running around and they don't know what to do, and I've been there before and running around drinking alcohol ... I think it's very important for people my age and even younger to get involved with this. I mean it's ... ultimately important, so that you know, they get in touch to what's going on and it's good, and you feel like you're a part of something, and you know, it's a strong connection.

The Karuk Tribe is part of a regional and national effort to remove the four dams on the mainstem Klamath. Many people in the Karuk community have actively participated in rallies and marches and even direct action protests. In the act of engaging in protest, one can simultaneously assert "traditional" values of caring for the fish, caring for the community, and engaging in collective continuance. For example, my collaborator Ron Reed explicitly translates the notion of traditional responsibilities to fish and community to a responsibility to speak out against what is happening. He even describes speaking on behalf of the fish and for his people as the "fisherman's duty" of today.

> Before, it wasn't easy. Ceremonies, subsistence, those, you know, those aren't easy things to accomplish. There's a great deal of responsibility and pride involved in those activities, but we never had to go speak for the fish, we never had to go talk about our values, our cultural ways, our traditional values. As long as we followed them, we were taking care of them. But now the fisherman's role is also to speak for the fish. Speaking publicly isn't a common trait of the Karuk people. The people who speak on behalf of the fish or resources are people that have taken that responsibility and have been able to utilize what God has given us, what the Creator has

given us to be able to speak for the resource, and manage for the resource in a way that is foreign to us. . . .

When I heard farmers talking about it you know, I heard farmers talking about bankruptcy, traditional values, culture, that's when I decided that I needed to be able to start speaking on behalf of the fish. On behalf of the fishermen. On behalf of the basketweavers, on behalf of the people who walk before us, and on behalf of the people who walk after us. I asserted traditional values, traditional core values into my way of thinking. That's the fisherman today. It's a burden, it's a responsibility, it's what I cherish, and I wouldn't do anything else. I mean, this is, God put me, the Creator put me on this earth for a reason. I think I'm fulfilling that reason. That's what it is to me being a fisherman today.

Karuk masculinity is not static. Men seek to remake their identities in the face of environmental decline, and to some extent, they succeed. In these new forms, masculinity remains interwoven with fulfilling responsibilities and resistance to racialized and colonized power structures. The natural world remains central to each aspect of these constructions. But these forms of masculinity that have emerged in the context of environmental decline are different. Nature plays an increasingly symbolic role, and the many social interactions associated with harvesting and distributing fish do not occur. These constructions of masculinity look to feel less "steady." They are tinged with worry. Ultimately, they rely on the possibility of a future return of the salmon. How long they can be sustained in the absence of continued declining salmon runs is unclear.

These are but a few of the multitude of examples of changing gender dynamics in the context of environmental decline for Indigenous communities (Vinyeta, Whyte, and Lynn 2016a, 2016b). The ongoing presence of Indigenous people is testament to the many daily forms of creative resistance people have employed in the face of the onslaught of settler-colonial structures and the shifting gender arrangements described here are among them (see, e.g., Eriksen and Hankins 2014).

UNSETTLING FEMINIST SOCIOLOGY AND GENDER STUDIES

I have suggested here that in the course of distancing from environmental materiality, gender studies and sociology of gender have refocused in a manner that obscures power relations particularly important for many Native people—those associated with colonial violence and environmental degradation (Bacon 2018; Vinyeta et al. 2016a, 2016b). Only by theorizing the natural can we understand the construction of traditional Karuk femininities, masculinities, or the ongoing operation of gendered and racialized colonial violence in Native communities today. On the Klamath, the fight to retain traditional forms of masculinities and femininities is part of a larger collective resistance to assimilation by an advancing racialized state that attempts to force inculcation to a non-Native social order. This struggle has a gendered dimension. It is experienced through the struggle of individuals to attain a positive gendered sense of self, at the same time as it is a collective effort for cultural survival.

I have argued that whereas sociology of gender has emphasized *the social* and *social construction* in order to theorize power, it is here through the material environmental circumstances (environmental decline) *as well as* their symbolic structures that people experience gendered violence. Just as environmental justice activists point out how the unequal health effects from environmental contamination are a form of racism, just as Indigenous people experience the absence of traditional foods as a mechanism of forced assimilation, the reorganization of the natural world violently restricts possible gender identities, expressions, and arrangements.

It is much easier to follow a claim that the natural world shapes gender for Karuk people on the Klamath where individuals have such nuanced interaction with "nature" than it is to visualize how the natural world shapes gender constructions for urban dwellers. Racism and essentialized notions of Indigenous people can easily cause readers to doubt more general applications of this concept. Do we really need to account for how the natural environment may influence gender constructions more broadly? Such questions are not even on the table in geography where feminist political ecology has long examined similar issues, gender and development literature where "less modern" women's lives are understood as connected to nature, or anthropology where the flashy new concept of "interspecies ethnography" has recently been developed. Existing sociological analyses of changing gender dynamics rarely theorize the natural environment as an explanatory factor in changing gender relations—even in cases where there are direct connections to environmental degradation. For example, while the declines in the timber industry in the Pacific Northwest emerged from a complex of social and environmental factors, sociologists have often confined their explanatory variables to the social fact of "changing labor market practices" and "labor market transformation." Others examine changing masculinities in relation to the social and political factors of environmental regulations such as commercial fishing net bans and refer to "changes in the industry" rather than the changing material environmental conditions that underlie them (see, e.g., Smith et al. 2003).

Yet many sociologists have quietly begun theorizing a place for the environment in their work. Sanyu Mojola's (2011, 2014) study of the spread of HIV around Lake Victoria places the "disrupted lake ecology" at the center of a complex of factors that reshape gendered behaviors and lead to the spread of HIV. Peek and Fothergil (2006) examine gendered practices of parenting in the aftermath of Hurricane Katrina. Bryson, McPhillips, and Robinson (2001) show how, for women living near a lead smelter in Australia, intense efforts to prevent children from ingesting toxins add to unequal gendered burdens of household labor. But as any word search will reveal, the term *environment* in sociological literature rarely refers to anything beyond the social environment, and literature on gender and environment is especially sparse.

Today we are witnessing large changes in social arrangements that result in part from environmental decline. Environmental decline in the form of species loss, toxic contamination, energy shortages, and now climate change is literally reshaping the baseline conditions around which human social, economic, political, and cultural systems are organized. Especially in the face of climate change, scholars across the natural and social sciences have begun to theorize the concept of the "Anthropocene"—described as an entirely new geological epoch in which human activity is fundamentally reshaping the ecosystems of the earth (Steffen, Crutzen, and McNeill 2007). But exactly how this level of environmental degradation translates into specific social outcomes is a complex process that few sociologists are tracking. We must theorize the importance of the natural world in social action more generally, both in order to keep analyses of gender and power front and center and especially now to understand the gendered nature of the challenges people face in light of the so-called Anthropocene.

Can we imagine a feminist sociology that theorizes nature without compromising an understanding that power operates through the sociality of gender? Although still in its infancy within feminist sociology, the explosion of work in Native studies is now beginning to gain traction in women's and gender studies. Feminist sociologists hold clear, longstanding commitments to analyses of power. Can this agenda be enacted without theorizing how colonialism shapes not only our notions of gender but also our disciplinary frameworks? There are relatively few Native academics, and relatively few feminist sociologists in particular have engaged Indigenous perspectives on the world (Wilkes and Jacob 2006). Perhaps this fact explains why neither the power relations surrounding colonialism or the natural world have been significantly theorized in my field. Yet cross-disciplinary Native studies programs are on the rise, and the outpouring of work within Native studies—and by Indigenous feminists in particular—has much to offer sociology and the broader field of feminist theory. Native feminists, including Cutcha Risling Baldy (2018), Luana Ross (2009), Stephanie Teves (2015), Kim TallBear (2014, 2015), Leanne Simpson (2013, 2017), Leilani Sabzalian (2018), Angie Morrill (2017), Mishuana Goeman (2009, 2013), J. Kehaulani Kauanui (2008), and many more, assert that both colonialism and decolonization are gendered processes.

It matters that sociologists engage Indigenous experiences of and perspectives on gender. Failure to do so makes it impossible to "see" or theorize the dynamics of power described in this chapter, and theory working within the nature-culture binary continues to inscribe gendered colonial violence onto Indigenous communities and constrain theory in many ways. Not only does the case of Karuk masculinity underscore assertions by Native feminists that present conceptions of masculinity as domination in gender studies are privileging colonized masculinities, but Indigenous feminists also have many important critiques of and contributions to non-Native feminist orientations (see Arvin et al. 2013; Barker 2017; Deer 2015; Smith and Kauanui 2008; Maracle 1996; Goeman and Denetdale 2009; Ross 2009; Sabzalian 2018). It has been pointed out that non-Native conceptions of "rights" and "power" are complicit with imperialism and alienated conceptions of the self and community that are themselves a product of colonialism. When it comes to the feminist discourse on rights, voices from the Klamath point to the multiple ways that notions of tradition and responsibility operate to sustain community and resist colonial encroachment. Such concepts are not new to Native scholars and have resonance in other communities of color as well. These observations are similar to those of Patricia Hill Collins (1994), who describes fighting against eradication by the dominant culture as one of central themes of ethnic women's struggles. Collins (1994) writes that for ethnic women, "The locust of conflict lies outside the home as women and their families engage in collective effort to create and maintain family life in the face of forces that undermine family integrity. But this 'reproductive labor' or 'motherwork' goes beyond ensuring the survival of one's own biological children or those of one's family. This type of motherwork recognizes that individual survival, empowerment, and identity require group survival, empowerment and identity" (47). Evelyn Nakano Glenn (1985) describes that while the family is generally seen as the site of gender conflict for white feminists, women of color often experience their families as sources of resistance to the racism of the broader society.. Lisa Udel uses Patricia Hill Collin's term *motherwork* and describes its importance in resisting colonial oppression. "Native women's motherwork, in its range and variety, is one form of this activism, an approach that emphasizes Native traditions of 'responsibilities' as distinguished from Western feminism's notions of 'Rights'" (43). Udel further notes that "Native women thus articulate their responsibilities in terms of their roles as mothers and leaders, positing those roles as a form of motherwork" (54). Udel broadens this assertion, describing how maintaining and asserting traditional gender practices become part of how families and communities resist colonial oppression. In the context of colonialism and racism, maintaining "traditional" gender practices is understood as a central part of cultural resurgence and survival, and they are of upmost importance for collective continuance right alongside and indeed fundamentally interwoven with the maintenance of political and economic systems.

Until feminist social scientists theorize across the nature-social dualism, our theories themselves perpetuate colonialism through Indigenous erasure. Until we engage Indigenous perspectives and experiences, we will miss critical opportunities such as understanding how the framework of masculinity as domination is itself a colonial structure. The alternatives, however, are quite promising. Many of these insights from Indigenous scholars about self, community, and the natural world can help gender studies move beyond the alienated cosmologies of modernism and colonialism alike.

For most of the history of human habitation in this place from which I am writing today—that is to say, some 10,000 years or more—gender relations have been very different and more life sustaining than they are now. What difference could it make for all of our lives—Native and non-Native alike—if we could think, live, and theorize from a place of awareness that the current problematic gender structures have only been around for a relatively very short period of time? The gender structures most of us have internalized to various degrees feel so very real because of the material and symbolic violence used to enforce them. Today, that same symbolic and material consolidation of power underpinning heteropatriarchy and gender binaries has a parallel violence in the natural world. Male domination is being enacted not only onto human communities but also onto the land. Indeed, the heteropatriarchy and gender binaries that non-Native feminist sociologists find so problematic today were made real for the purposes of the ecological violence that now threatens all humanity. Indigenous erasure is a significant tool of that violence. Social theory that fails to imagine the natural world cannot help us understand these present circumstances. Let all of us as feminist sociologists, Native and non-Native alike, reject that violence and erasure. Let us listen to, read, cite, and otherwise engage Indigenous scholars and peoples on their terms.

5 · EMOTIONS OF ENVIRONMENTAL DECLINE
Karuk Cosmologies, Emotions, and Environmental Justice

There are so many plants, and each watershed has lots of different ones that are known by the families that lived there—they have specific medicinal qualities. These families cared for the plants, and the plants cared for them in return. They healed. They healed each other. Plants have feelings, too, you know. But now that most of these families had to leave the land...were kicked off the land one way or another...a lot of villages were burned, you know. Then they kept extracting all kinds of shit from it, and the plants suffered. Now they're sick—so how are they supposed to heal us?
—Sophie Neuner

There are spirits in every living thing, and the rocks and the soil and the river. So [salmon] you know, it's not like it's just a piece of food and it's not a big deal. It's definitely something that has that value, that it is a living spirit, like the spirit of a person, really.   —Jennifer Goodwin

Up Red Cap [road] there used to be a place where you could get a bunch of elderberries. We used to get all we wanted, but then the Forest Service came along and mowed them all down...completely. They never came back. I still think about that...every year. It makes me really sad...all those elderberries. It still makes me mad.   —Janet (Wilder) Morehead

Concepts of "environmental racism" and "environmental justice" first articulated by academics and activists in the 1980s have moved to the center of environmental sociology, environmental studies, ecocriticism, food studies, and many related fields. As the field of environmental justice developed, definitions of "environment" and the types of claims for justice expanded from an initial emphasis on proximity to toxic sites to consideration of more and more dimensions of environmental inequalities (Agyeman et al. 2016; Schlossberg 2013; Brulle and Pellow 2006). Yet still today, environmental justice work continues almost exclusively to address unequal *physical health* impacts from environmental degradation. There is little development within this literature (or the movement) of the notion of unequal *mental harms*, how the psychological impacts of racism might be part of environmental justice, or symbolic dimensions of how power operates through environmental degradation more generally. While a solid disaster literature exists on the negative psychological consequences of environmental degradation and their unequal distribution along the lines of race, class, and gender, mental health impacts are sparsely covered within the environmental justice framework.[1] Most important, there is only a beginning of interrogation of the concept of emotional harm, or the relationships between emotions, environmental change, other features of social structure, and the process of inequality formation. There is little engagement with theories of emotions. In this chapter, I argue that neglecting the natural world as a causal force for "generic" social processes (Prus 1987; Schwalbe et al. 2000) has limited not only work on Native Americans but also work in sociology of emotions and masked the theoretical significance of environmental justice.

Yet as Ron Reed and I worked on the altered diet study that is the focus of chapter 3, our interviews with people about impacts to the river and forest were rich with emotions. People raised their voices, shed tears, and sighed in resignation. Sociology tells us that emotions lie at the heart of social organization. Emotions animate meaning systems and structure power relations. Some emotional experiences harm, as when soldiers experience posttraumatic stress disorder, or a child grows up stigmatized or shamed as a consequence of racism (Brown 2003; Thoits 2010). Despite the profound effects of environmental decline on communities worldwide, the emotional dimension of ecological destruction has not been taken up as an issue of environmental justice. Even more surprising, given the particular strength of Native ties to land and species, and the central importance of Native traditional management for the culture and daily life of many Indigenous people, little existing research examines the emotional dimension of ecological destruction in Native communities.[2] What role do emotions play in the embodiment of power, oppression, and resistance? In what sense can emotions be part of a racialized experience of the environment, or might emotional experiences constitute an occurrence of environmental injustice? What role do emotions play as environmental degradation inscribes racialized power relations, advances assimilation and genocide, or does the work of colonial violence? What can Indigenous cosmologies teach academic disciplines about the environmental context of emotions?

While work in environmental justice has yet to engage emotions, sociological work on emotions has likewise yet to theorize the role of the natural environment in social action. Whereas many social scientists regard emotions as personal or "private" experiences, sociologists of emotion describe them as deeply embedded in both social structure and culture (Bonilla-Silva 2018; Collins 2004; Hochschild 1983; Schwalbe et al. 2000; Wilkins and Pace 2014). One area of this literature describes emotions as the link between micro-level social interactions and the macro-level reproduction of social structure (Scheff 1994; Schwalbe et al. 2000). The people Ron and I spoke with illustrated this link with their words, especially as emotional distress confirms structures of power. As the link between individuals and power structure, emotions matter in part for their role in cognition. As Arlie Hochschild (1983) notes, emotions serve a "signal function." They are part of how people make sense of their place in the world. For Collins (2004), cognition and emotion are linked to social structure as people engage in "interaction ritual chains." Here shared emotional experiences are part of the production of group reality and meaning and the construction of social order. Taken together, scholars in this area illustrate how emotions are fundamentally related both to the cognitive processes of interpretation and meaning construction that undergird important social processes such as framing, identity formation, the maintenance of ideology and social order, and "cognitive liberation" (Jasper 2011).

Emotions literature also conceptualizes emotional harm and points to relationships between emotions and mental health (Scheff 2014). Racism and other forms of oppression are understood to manifest as negative mental health outcomes (Brown 2003; Thoits 2010). There is a significant gap, however, between literature that considers emotions as socially constructed embodiments of power linking micro-level agency to macro-level social structure, clinical scholarship on emotions and mental health as mentioned above, or work on how emotions function to reproduce colonialism or racism (see Thoits 2012). If negative emotional states can embody oppression, what exactly is the "harm" of these experiences? What might we understand about the production of inequality or the dynamics of settler-colonialism by examining disruptions to relationships between nature, emotions, and society?

Symbolic interaction details how people attach symbolic meaning to objects, behaviors, themselves, and other people and then develop and transmit these meanings through interaction (Scheff 1994). Emotions are central to this process. Emotions link micro-level interactions that connect both identity and emotions to the larger reproduction of social structure, in part by taking social context into account. In her work on "ecology of interaction," Lynn Smith-Lovin (2007) describes social settings influencing interpersonal encounters. Between their signal function in cognition and their link with social settings and individuals, emotional experiences may thus represent a three-dimensional "embodiment" of power relations, or terrain of resistance (Jasper 2011). As South African anti-Apartheid activist Steven Biko pointed out decades ago, "The most powerful weapon in the hands of the oppressor is the mind of the oppressed."

I concur with Smith-Lovin and other symbolic interactionists that the social context of emotions matters. Yet *environmental* contexts also link sense of self, emotional experience, social interactions, and social structure. Existing theory is only beginning to engage this terrain.[3] I offer here details of the mechanisms through which the natural environment structures emotions. Voices in this chapter will illustrate how the natural environment is at minimum part of the stage of social interactions and a central influence on the emotional experiences of the people interviewed, including their internalization of identity, social roles and power structures, and their resistance to racism and genocide. Environmental decline thus removes the stage upon which social interactions involving emotions would otherwise occur and serves as a signal function regarding racial inequality, colonialism, and forced assimilation. Furthermore, racialized emotions norms negating the relevance or even possibility of intimate emotional connections between people and the natural world form an added layer of racism and assimilation by situating Karuk experiences as invisible, "abnormal," and even potentially subject to categorization as "mental illness." By contrast, taking seriously the experiences of Native people and the role of the natural environment offers an opportunity to extend sociological analyses of power and move sociology toward a more decolonized discipline. In his review of the subfield of emotions, Jasper (2011) laments the conceptual limitations that emerge from dualisms between emotion-reason and body-mind. I agree as to the limitation of these dualisms and aim here to bring attention to a third: the dualism between nature-society.

This chapter continues the themes of environmental influences on social action by examining the emotions experienced by Karuk Tribal members in the face of environmental degradation. In chapter 3, I described how the altered diet study was conducted as an effort to illustrate the physical health impacts of declining salmon runs. Yet as Ron Reed and I framed that work, we incorporated Karuk understandings of health that are very much interwoven with social, cultural, and mental health. These emotional dimensions of health were so significant that we launched a second set of interviews specifically focused on these impacts (see Norgaard and Reed 2017. Here I use interview data from that project, historic source material, and more current ethnographic interviews and personal communications that illustrate the expansion of an earlier focus on salmon runs to a broader theme that encompasses Native foods, fibers, and medicinal plants. In all of these sources, the natural environment figures centrally as a driver of emotional experience for those we interviewed.

EMOTIONS ANIMATE RELATIONSHIPS WITH OTHER BEINGS AND NATURE

People members vividly expressed emotions of *joy* from being out in nature and *grief, anger, hopelessness, and shame* with the decline of the Klamath River. These emotions were not discrete but related to each other in the lives of the individuals with whom Ron and I spoke. To better understand the significance of the emotions associated with environmental decline, it is useful to first describe emotions people experience interacting with the intact river and landscape. In our interviews, people emphasized the central connection between daily life, their identity as Karuk people, and the Klamath River. In the words of traditional fisherman and cultural practitioner Robert Goodwin, "When I was a young child, my first conscious memory is being at the falls. That's you know to say that's where I've always been. That's where my life source comes from. That's who I am, that's what identifies me as a person, as a Karuk person. Being on the river." Bob emphasized how the river is a central orienting point for life: "Everything within our culture surrounds the river. The river is the center of our way of life. You're up the river, down the river, up big hill, on this side. Everything comes back down to the center, which is the river. The river is our life course. When that river is no longer healthy, we're no longer healthy." People spoke of intimate connections with the river, with important sites and species, especially salmon but also eel, trout, sturgeon, and steelhead. In a conversation while on the fish crew, Rabbit and David Goodwin, two traditional dipnet fisherman then in their thirties, offered vivid descriptions of a sense of oneness and joy while being on the river.

> I come out here ... come out to these places, you know, and get that connection back. Just that silence and the liveliness of everything surrounding us ... everything is alive when you're out here and you can feel it. It's a bliss that you can feel—it's indescribable ...

> You know, my first time I went down to the falls, it was almost like being in heaven ... that's our ceremonial fishing grounds and it's right at the base of our mountain that we pray to ... and it's medicine ... and to be at both those places, you know, to be there and the falls right there is just magical. To hear the raw power of the river ... it's like you're on earth but you are in a different place at the same time.

Other Karuk people vividly expressed emotions of *grief, anger, hopelessness, and shame* with the decline of the Klamath River. The impact of each category of experiences is underscored by their invisibility and corresponding lack of legitimacy within the dominant culture—a phenomenon Ken Doka (1989) calls "disenfranchised grief." This disenfranchisement operates as a vector of racism and assimilation, as will be discussed below. In the next sections, I have organized people's words to underscore the specific ways each emotion operates in connection to identity, social interactions, power, and meaning systems.

GRIEF: "JUST LIKE TEARING MY HEART OUT"

In our original interviews about the impact of the Klamath dams, the most frequently expressed emotion in the face of the degraded water quality and diminished quantity of fish in the river was grief. Nearly everyone with whom Ron and I spoke conveyed this emotion with intensity. For Rabbit, a father and traditional fisherman then in his early thirties, "It gets pretty emotional for me, you know when I see salmon *dying* because of the algae or the river is too low." He went on to elaborate how "it saddens me . . . to see all the algae and all the toxic, it just saddens me that they continue to allow this to happen. They know the long-term effects it's going to have . . . it's going to be devastation . . . you know for everything in that river, not only the salmon, for everything in that river." In addition to grief in the form of direct pain on behalf of others, for many people, experiences of grief are bound up with other important social experiences: disruptions to identity, disruptions to social interactions, and the association of environmental degradation with both cultural and physical genocide. Table 4 outlines these. While I discuss these emotions in relation to these seemingly distinct causal outcomes, in people's lived experience, there is no clear line between the experience of grief on behalf of dead salmon, grief on behalf of one's son who cannot go fishing, or grief that Karuk people may come to an end. The same can be said for many other plant and animal species: grief on behalf of the casualties of sudden oak death, grief on behalf of the loss of those acorns to the next generations, or grief that without acorns, the Karuk people will lose their identity. This interconnection is evident in the source materials as speakers move from one aspect of the emotion to another. Indeed, these aspects are not only interconnected but also *compounded* by one another.

Continuing with the example of grief, the inability to catch and share fish is a deeply painful experience for individuals who carry this responsibility and for the community as a whole. Rabbit relates the intense grief of not being able to provide fish to Elders: *"If I couldn't be able to dip for my Elders, it would just break my heart, you know, if I couldn't go down there and gather up some eels for them, it would just be like tearing my heart out."* As this quote from Karuk descendant, traditional practitioner, and research ecologist Frank Lake shows, the angst surrounding his inability to provide fish and eels for Elders blends into fears about the future. The inability to "make things right" both for the fish and human community weighs heavily on those who carry this responsibility. As Frank Lake describes,

> And, I think I see this with my own family, people in the community, they feel responsible in part for what's happened with the fish. Not that they did it, that caused the degradation, but more of that loss of—<crying> But what else can you do? You can only do so much. You know, you can make your prayers, you can rally, you can go down there and pray at the falls, you can go to the ceremonies, but at some point you know, it's just beyond your ability to do something and when you reach that point where you can only do so much, and you can't do anymore, that's when it's lost <crying>. You know? And, when you see your role as a young man

TABLE 4 Grief in Relation to Identity, Social Interactions, and Social Structure

Identity	Social interactions	Social structure
Like tearing my heart out	Sadness because "quiet down at the Falls"	Sadness that Karuk people may disappear

or as a person who's supposed to get fish and go feed these people so they can have that ceremony so they can fix the Earth, fix the world, or the salmon are just—you go down and you look below the bridge or you look at the falls and there's more dead, you know <crying>. You can't be the one who makes the water ten degrees cooler like it needs.

Here grief serves a signal function indicating both awareness of the degraded environment and one's location in a system of power. In her revision of Sheldon Styker's classic work on structure within symbolic interaction, Smith-Lovin (2007) argues for more emphasis on social settings in the construction of identity. Given the necessity of fish in the river for Karuk people to perform the activities that uphold positive identity and re-create culture and meaning systems, it is not only social environments that constrain behavior but also the condition of the natural world that profoundly shapes the "menu of opportunities" (Smith-Lovin 2007, 108) available to individuals. Throughout the source materials used here, speakers describe threats to social roles (8). Vera Vena Davis, former Karuk council member who passed in 2003, drew the connection between the threats to fiber materials and her role as a basketweaver, as well as the connection between the threats to water quality and fish survival: "I remember my old people saying basket materials weren't like they used to be. They were kind of scared to put them in their mouths. Scared if there was something on them. So they heated them up before they did anything with them. We would get willow, grape, and blackberry roots in the spring. You know there is poison in the river so how is a fish surviving?" Traditional practitioner and fisherman Kenneth Brink put it this way: "Just the whole sense of being able to be a provider . . . you know, a fisherman that's able to catch fish and provide for his family is just as good as anybody else in the professional world, you know . . . so if this is your way of living . . . if it revolves around the fish, you know providing for your family . . . now the fish are gone then now you can't provide." The sadness, shame, and frustration described here can be linked to grief experienced on behalf of others, as well as how it affects social interactions—especially the dynamics of upholding gender expectations, as expressed in the following comment from a Karuk woman who wished to remain anonymous: "So no, it's not really okay for me to go eeling, go hunting . . . whatever. But what the fuck am I supposed to do? He's all upset about not having anything for ceremonies, but I've got nothing to cook. It's my responsibility to provide the meals for ceremonies, so it's my shame to put beef stew on the table. So what am I supposed to do? It's hard on both of us, and it's totally not good for our relationship." Frank Lake describes the stress in losing the social capital that comes from participation in the barter economy that continues to operate around traditional foods:

> Sometimes, when there's low economy and there's no other jobs to do it's just tough—you drink it away because, well, you know, what the hell, there's nothing that you can really do that's going to be good anyways. So you pass the day by numbing the senses. You know when things aren't good with the fish people take it out because they're stressed, right? Normally, that salmon would be that role of building that capital when you don't have that capital, it's not a reservoir of, either monetary or even, kind of like, "I owe you one," type of thing to draw from. Just like people in a contemporary sense would get stressed for not having financial security, when you don't have salmon security, it adds all those other dimensions of stress to it.

As Frank notes, one can understand parallels between the inability of people to carry out cultural responsibilities and provide subsistence foods and the effects of unemployment on identity, gender practices, or drug use (Segal 1990; Sherman 2009). Again, parallels to non-Native communities are also evident. In Newfoundland, the collapse of the cod fishery in the 1980s resulted in complex changes in identities, health outcomes, household practices, and gender arrangements (Ommer 2007).

Furthermore, the grief that people experience can also be on behalf of others and on behalf of the many social interactions that can only occur when there are cultural species to steward, harvest, process, or even eat. Important social interactions traditionally happen during seasonal harvests. Grief over the decline of the health of the forest and the river is inseparable from the sense of loss of social relationships that occur with the harvesting of food. Leaf Hillman speaks to the loss of cultural and gender identities that are connected to the harvest of sugar pine nuts:

> It's a community . . . it's a family event. Every piece of the process is embedded with gender roles, cultural and community norms. Umm . . . it involves the entirety of the community . . . the children, you know, the old people, the adults, you know . . . they all have a role. They all hold an important role. Boys climbed the trees and the old men told them what to do . . . how to build their hooks. Girls were on the ground gathering up the cones. Women were in charge of the fire pits and worked all night to release the nuts. The old folks cared for the little ones and sang the songs.

Everyone had an equal part in the success of a harvest. So, umm . . . the loss of pine nuts from our diet represents not only a loss of culture, identity, subsistence, food . . . it represents [draws a circle in the air with his finger] a loss of everything that is Karuk.

Ron Reed describes how things have changed at the falls: "It's not only just a fishery. It's a social area. So people come from all over the place still today. They go to Ishi Pishi Falls, to mingle, to get their fish, to share their wisdom, their knowledge about when they were kids. People come down to the falls because they know there's something to come down to. And this year was awful quiet at the falls because they knew there wasn't anything to come down to. It was very quiet down there this year, and it was very sad." People voice sadness when describing how their children may not have the same experiences eating the Native foods they have had as children. When learning about the plans to establish a Native plant demonstration garden in Happy Camp, one Karuk Elder lamented, "What are those potatoes called? Are you going to put in those kind in with the blue flowers? I remember my mom used to get those for us when we lived out Salmon River. They weren't really my favorite, but it's so sad that I can't get them for my own grandkids. They just aren't there anymore. Maybe I'm not looking in the right spot."

An anonymous survey respondent noted, "I think it's remarkably sad that in my teenage years I ate a tremendous amount of salmon or deer meat, and now it's hardly ever eaten. . . . There is just a TERRIBLE shortage in salmon now that when my little Indian daughters eat it they think it's a treat . . . it saddens me to have my children not enjoy the same simple happy memories of eating salmon with all the old Indians and hearing stories of catching them, dipping them, and packing them out." Smith-Lovin (2007) writes that "the person we become depends profoundly on the networks in which we are embedded" and "the actions we take and the emotions we experience depend on these networks. These networks are, in turn, shaped powerfully by the social settings that we occupy" (106). But for theory in sociology of emotion, all of this activity is conceived as taking place in a vacuum. Here changes in the natural environment alter the quality of social relationships. A parallel argument to the above can be made in terms of the missing ingredient of the natural world in the interplay of emotions, interaction, and identity. To draw specifically from the cases cited above, when pine nuts and salmon are present, a host of family and community interactions associated with harvesting, processing, and distributing these traditional foods will occur. People gather to harvest, to watch others harvest and process, to see friends and family, and to distribute food. These interactions have profound meaning for individuals, families, and to the Karuk community. When these traditional food sources are absent, these interactions do not occur, and a very different set of emotional and social experiences is set into motion. As one Karuk teacher explains, *"When*

I try to teach kids why, in Karuk, you are calling someone 'lazy' by calling them 'assaxvuh'—turtle . . . when I tell them the story about how turtle takes so long to get to the pine nut gathering event that he misses the whole harvesting party, they can't understand it really. It's so sad . . . those social events don't occur anymore, since there aren't many sugar pines left at all." As one man in his mid-thirties put it, *"You got to have fish to teach them how to fish. You got to have fish to teach them about fish . . . with the numbers dropping like they do, it'd be hard, you know to tell your son how to dip when there's nothing in there to dip."* Socialization and identity-building processes are also intrinsic to the traditionally social activities of digging "Indian potatoes," gathering acorns and huckleberries, processing deer and elk meat, winnowing for a variety of seeds, and so on.

Grief from the association of the degraded landscapes and waterscapes with genocide forms a final category of how emotions from the environment shape social action. Witnessing the degradation of the forest and river is associated with genocide in multiple ways. On one hand, because the absence of cultural resources, such as acorns, salmon, deer, and pine nuts, makes impossible the social and cultural practices described earlier, the decline of the landscapes and waterscapes literally becomes the vector of forced assimilation and structural genocide. Geena Talley, a young Karuk woman, begs the question: "Let me ask you this: if I can't collect enough acorns . . . I mean, good ones that don't have worms or mold in them. If I don't get enough acorns, I have to buy bread. If I don't have enough berries, I have to buy oranges. So what does that make me? Am I going to feel like an Indian at the supermarket picking out oranges? [laughs] Maybe I should have the check-out lady get me a fifth [of alcohol] from under the counter, aayyeee?" Leaf Hillman, director of the Karuk Department of Natural Resources and a prominent cultural practitioner, explains, "How do you perform the Spring Salmon Ceremony, how do you perform the First Salmon Ceremony, when the physical act of going out and harvesting that first fish won't happen? You could be out there for a very long time to try to find that first fish and maybe you won't at all and then of course in the process you'd end up going to jail too if anybody caught you. So, will that ceremony ever come back? Well, I don't know. But, once again, it's a link that's broken. And restoring that link is vital." Former Karuk council member and traditional practitioner Bob Goodwin describes the sense of human cultural genocide with decline of the river:

> I am always going to be identified as a Karuk person . . . and looking at it from the standpoint of our native community, every aspect of it is affected by the unhealthiness of what is going on with the river. Not only does it have an impact on present day life, but it is going to affect the future if this isn't changed. And now is the time to make these changes. If much more time goes by, we're not going to be a fishing people, because there are going to be no more fish. We have to do something now.

In 1916, the Forest Ranger Jim Casey responded to the earlier mentioned letter from "Klamath River Jack" who had asked why the Forest Reserves did not understand that low-intensity fire was essential for forest health. Casey's reply was also reproduced in the Del Norte Triplicate, much to the mirth of its readership. Casey explained to Jack that if his "acorns are wormy, don't blame it on the white man for keeping fire out of the country." He also made explicit recommendations for assimilation: "Anyway, there are other things that make better flour than acorns. Why not plant some grain and vegetables and fruit trees on that flat back of your cabin? That's white man's grub, but it's pretty good." Karuk People have long been aware of the connection between cultural genocide and forced assimilation with the transition from Indian food to "White-man's food," as expressed here in the dying words of one of the last Karuk people who were born before the influx of Euro-Americans:

"yavík ikuppítihe'ᵉsh.
"Be good,
    koovúra yáv ikupeekyâatiheeshik.
    be good to everybody
    véek táay vúra paxuntáppan íffike'ᵉsh,
    and pick lots of acorns
    káru vúra xúun ikyávish.
    and make acorn soup.
        xáyfaat ík pamíshpuk ikvár pa'apxantich'ávaha—"
        Do not throw away your money on white-man grub—"
        ittáayvar pamíyav.⁴
        (lest) you spoil your good.

On the other hand, experiences of a degraded river and forest are connected to genocide because the experiences are associated with a long-felt awareness of Karuk culture and life under attack and a longstanding sense of their imminent destruction. In their work on cultural trauma and substance abuse in Native communities Les Whitbeck and coauthors (2004) underscore how many aspects of daily life become triggers for past events: "American Indian people are faced with daily reminders of loss: reservation living, encroachment of Europeans on even their reservation lands, loss of language, loss and confusion regarding traditional religious practices, loss of traditional family systems, and loss of traditional healing practices. We believe that these daily reminders of ethnic cleansing coupled with persistent discrimination are the keys to understanding historical trauma among American Indian people." As cultural practitioner Renee Stauffer put it in 2002, "The Karuk people have survived, managed their land for thousands of years. And how long has it taken the White man to come in and destroy it? What does that say about their land and water management? They come in and they try and play God and they've ruined everything, threw everything out of balance. And I don't see any way to fix it because there's too many of them." Chapter 1 described how, from the early 1800s to 1880, the Karuk population went from about 2,700 people to about 800 people largely due to state-sponsored genocide (McEvoy 1986, 53). Many Karuk families carry specific stories of these events. Emotional responses to the destruction of the environment reanimate these histories of physical genocide. That Karuk life and culture could come to an end is the grim background cadence of people's everyday lives. Many people have shared stories such as the following words from cultural practitioner Vikki Preston: "Sudden oak death, it's like the next wave of cultural genocide. Just the next chapter. When I think about sudden oak death, I think about people and places . . . the connection between them. It is a real fear, besides the fact that it is part of your identity, your ceremonies. There's a thought that people aren't going to help until it's gone. You have to fix it every year, not just when it's gone." Explicit connections to genocide were also made in a conversation between traditional fishermen "Binx" (Kenneth Brink) and "Scrub" (Earl Aubry) as we discussed the impacts of the Klamath river dams one afternoon on Scrub's outside porch.

> The Karuk people actually believe that if the salmon quit running, the world will quit spinning, you know. Maybe the human race as we know it may be nonexistent. . . . If the river quits flowing, it's over. If salmon quit running, it's like the sign of the end. (Binx)
>
> My grandma she said the deer would probably go first, according to what the medicine people talked about when she was real small. She said from what she could understand the animals were going to let us know when the end is here. Because they'll disappear. (Scrub)

The fear and dread interwoven in these stories form a foreboding backdrop to the many layers of struggle associated with the degraded landscapes and waterscapes as changing environmental conditions become the driving vector of genocide—an emotional manifestation of what Bacon (2018) calls "colonial ecological violence."

ANGER: "I GET PISSED OFF"

Although grief is the emotion people most frequently conveyed when talking about the degradation of the Karuk ancestral homeland, many also expressed anger. Like grief or sadness, James Jasper and other sociologists of emotion consider anger a reflex emotion. Yet like grief, here anger also serves a signal function and operates in complex ways in relation to cognition identity and moral understandings (see Table 5).

In a 1928 letter to the forest supervisor of the Klamath National Forest, Finn Jacobs expresses his frustration and anger about the non-Natives' disrespect to

TABLE 5   Anger in Relation to Identity, Social Interactions, and Social Structure

Identity	Social interactions	Social structure
Anger that cannot fulfill expected roles	Anger that children do not have same opportunities anymore	Anger at agencies that arrest people for fishing according to tribal custom

many aspects related to the World Renewal Ceremonies, as well as the disregard of tribal rules: "All this time the people that are at the Pick-ya-wish should feast on acorns, salmon, and deer. But the whiteman will not allow us Indians to have our food that is salmon and deer. We want our food, and our rules carried out. We do what the whiteman commands us, so why can't they do as we say?" In relation to the loss of salmon, Ron Reed emphasized that "you might be pissed off. You might be really super angry. You'd be super angry like I was and not really know what the hell you was angry about." Another traditional fisherman from the same community added, "When you don't have something that you feel like you have a right to have [fish, access to a healthy life], you're disenfranchised. You're angry." The anger voiced here is not just about what is happening to the landscapes and riverscapes but how anger and frustration are intertwined with a sense of "denied access" to a variety of important activities and responsibilities that lie at the very heart of being Karuk as in this anonymous man's words referenced in the last chapter: "You know when I get pissed off, you know what I do? I go out and start drinking. But what if I had salmon? If I had a fucking . . . if I had a sweat lodge in the back of my house . . . or if I had a fishery that we had enough fish and if I could go give to my mother and my kids and the way we need to live . . . that's what we're looking for." Here note that in alluding to the inability to provide for his family, this man's identity as a both traditional Karuk man and fisherman has been disrupted and social relationships have been disrupted as well. Again, note how anger at the condition of the river serves a "signal function" locating the speaker in a system of racial inequality and ongoing colonialism at the hands of non-Native land management projects such as the dams. More specifically, people express anger toward specific land management policies and then state actors who enforce them. Ron's older brother, traditional fisherman and practitioner Achviivich, put it this way:

> We still do not have a "right to fish." We are fish people. That is another thing. Rocks, mushrooms, and *áama* [Karuk word for salmon]. You want a line around them? Hey, acorns, yeah. They can arrest me for that. You know. I've gone to jail for some stupid-ass shit, so I don't mind going to jail for something I believe in. So before

the White Man came here what part of this river do you think the Indian could fish? . . . Every single square inch that they could fish they fished because the fish were there, the fish were everything. You know. If you didn't get no fish you know, you didn't make it to the next year. You know they say this was a land of plenty. Well it was a land of plenty at one time, before they started catching all the fish out you know, 200 miles limit or wherever you know [referencing commercial offshore ocean fishing] . . . I mean just about any Indian around here has been in trouble with the law for killing deer.

Later in this same interview, the speaker's anger toward the agencies that arrest people for fishing and gathering according to tribal ways blends with anger at denied ability to perform traditional management (here the speaker refers to burning to keep brush down and provide forage) and for the degradation of the environment that has occurred from non-Native management. Each of these links between emotion and cognition forms visceral understandings of present conditions of environmental decline and colonial violence:

Mushrooms is one area I draw the line in. I don't give a shit what anybody says about mushrooms you know. But the rest of the times, do I want to go out there and be hassled about it? Why? I go down to the damn store and buy that stuff a lot, you know, it is going to cost you more to go hunt, to go out into the woods and get it. It is not like it is readily available no more. It is not like you have a gathering spot like we used to have a gathering spot. You know, you used to have a gathering spot to gather something and you would go there and gather. Now you don't. Now you can't burn there. You can't burn there every year and every other year or however often you need to burn it in order to make your crop come up good. You can't do that. You can't burn. And you have to have a permit to get everything. Everything. You have to get a permit to get rocks off the goddamn river bar out here. Did you know that?

This sentiment is inferred through the 2016 Klamath Basin Food System Assessment (KBFSA) data collected from 286 Tribal households representing 843 Karuk people. Here, 68% reported that the *heavy degradation* of traditional gathering sites and *climate change* were the greatest barriers to accessing Native foods, followed by 60.91% reporting the *limited availability* of Native foods, and 56.11% noting the barriers posed by *rules* related to hunting, gathering, and fishing (KBRSA 2016, 26–27). Analogous to how Smith-Lovin (2007) attends to the importance of interpersonal encounters as the "link between macro-level community structure and the micro-level experience of self-conception, identity, performance, and emotions" (106), these passages describe how encounters with the natural world are a key link between micro-level experiences of identity and emotion and macro-level power relations of racism, colonialism, and structural genocide.

SHAME "THAT PUTS YOU IN THIS LITTLE DOWN FEELING"

Shame has been considered one of the most important emotions in the formation of social structure and stability (Scheff 1994, 2000). Here too, shame operates to inscribe ongoing racism and colonialism across spheres of social action from relations with individual identity to social interactions and structure. According to Thomas Scheff (2000), "Shame arises when subjects fail to achieve social ideals, or diverge from certain social standards." But Scheff and other important scholars writing about shame also define shame more fundamentally—in terms of seeing oneself lacking as other(s), real or imagined, see you. That the non-Indian community and government agencies have perceived, or perceive, the Indians on the Klamath as lacking is evidenced in numerous documents, such as the following quote from a 1949–1950 General Integrating Inspection Report for the Six Rivers National Forest: Native Americans in this region *"in reality, have simple minds. They have inferiority complexes and are more or less confirmed in their thinking that the land should be theirs and that incendiarism is one way of retaliation towards the white man for various controls, disciplines and laws."* As Pamela Conners comments on this report in her *History of the Six Rivers National Forest* (1998), "This ignorance of Native American cultural practices coupled with equating Klamath River 'incendiarists' with 'Indian,' and the prepossession against an entire racial group had the effect of clouding problem-solving and of poisoning relations for years to come."

People describe shame in relation to personal identity in several ways. First, there is a sense of direct identification with the contaminated entity articulated poignantly here by Frank Lake: "I think particularly for indigenous people social, cultural, and community wellness reflects the ecological quality of their environment. So when the river's degraded, and it's liquid poison in some ways, and you're supposed to draw all your sustenance and your identity as a river Indian or a river person, then of course that weighs on you. It's like, you want to be a proud person and if you draw your identity from the river and the river is degraded, that reflects on you." Identity and shame are also at play in relation to people's inability to perform social and cultural responsibilities, as Rabbit describes: "If an elder is asking upon you to go fish for him, that's like and honor, and if you can't fulfill that honor, it's kind of degrading inside, you know, it's hard on the spirit." In one of our recorded conversations, Ron Reed emphasized how,

> when you're not able to go upslope and manage, you're not able to go up and reap the harvest of that management. If you're not able to go produce for your children and give things for each other for the well-being of life, then all of a sudden, that puts you in this little down feeling. You're down casting yourself. I think that's where a lot of the people in Karuk Tribe are because of our inability to get to these

resources that have been given to us by the Creator. We understand very much that we're a proud people. We're here for a reason, but a lot of us struggle trying to figure out how do we integrate into modern society.

In the face of environmental decline, two types of responsibilities were threatened: those to cultural species and those to the human community. Director of the Karuk Department of Natural Resources Leaf Hillman describes Karuk responsibilities to tend and care for the natural world through traditional management:

> We believe that we were put here in the beginning of time, and we have an obligation, a responsibility, to take care of our relations, because hopefully, they'll take care of us. And it's an obligation so we have to fish. They say, "Well, there aren't that many fish this year, so I don't think you should be fishing." That is a violation of our law. Because it's failure on our part to uphold our end of the responsibility. If we don't fish, we don't catch fish, consume fish, if we don't do those things, then the salmon have no reason to return. They'll die of a broken heart. Because they're not fulfilling their obligation that they have to us.

Traditional management refers not only to care for the environment but also to specific social and cultural responsibilities people hold to their families, elders, and the Karuk community. One mother in her thirties who wished to remain anonymous underscored the importance of these duties: "To be a fisherman . . . it's an important role in being a man in the tribe . . . you know . . . you fish for your family, you fish for the people . . . and there's fish days, and the ones who owned those fish days were responsible for feeding the community." This dimension of management is embodied in the very word for the World Renewal Ceremonies, *pikyávish*, which literally translates to "the time to fix it." This intention to "fix it" is understood to be with regard to relationships to oneself, between human relations, and between humans and nonhuman relations as described here by Leaf Hillman:

> Responsibility and management are both tied to *pikyávish*. Those specific ritual practices carried out by the World Renewal priest, such as ignition of fires on the sacred mountain, which would burn until the rains would put them out, are carefully timed to produce conditions which will trigger the onset of the fall Chinook salmon migration. . . . Prayers calling the salmon home correspond with the ignition of these ceremonial fires. We do this for our people, our land, our river—all of the inhabitants of this land. We do it every year, for this is our responsibility as Karuk people.

Finding one's role and identity in the present context can also affect youth, as well as their caregivers as they imagine how they must feel. Another Karuk mother recounts,

> I know when my first daughter . . . before she started menstruating . . . I really wanted to understand what to do for her Íhuk [Karuk coming-of-age ceremony]. That came and went, and when the next daughter was getting closer to her time, I thought I'd try again. I never did give a ceremony for one of my three girls, and I never taught them how to weave. So I say to myself, "Don't feel bad. People shamed that practice, that knowledge out of our family way back when. You couldn't teach them how to weave 'cause grandma was ashamed of being Indian, and now the materials are bad." But I just can't let it go. I just can't feel okay about it. And now they feel bad about not knowing, not being Karuk enough. A tribal card and number just doesn't cut it. (Anonymous)

Conceptualizing the nature of physical harms from environmental decline due to cancer or lead poisoning appears more straightforward, but what exactly is the nature of emotional harm? The ability to maintain a coherent meaning system is considered a vital component of mental health and psychological well-being (Mirowsky and Ross 1989; Thoits 2010). By contrast, Ron Reed described the sense of shame and emptiness that he sees in the community:

> When we don't have a way of life, you're left with emptiness . . . if you don't know the creation stories, if you don't know tribal philosophy—there is a big void in your life. I think there is a level of embarrassment with the lack of knowledge, with the lack of presence in the culture. With all that, you end up with a low self-esteem. You can't fish. You can't hunt. You don't know how to pray. People aren't really eager to talk about something that embarrasses them. And it must be an embarrassing moment not to be coming down to the falls, or going to the ceremonies.

As Ron's words reveal, it is the emotions of shame related to one's inability to carry out valued social roles in the face of the degraded environment that inscribe racism and ongoing colonialism. The emotional "harm" is a function of their cognitive dimension in the inscription of social power. Table 6 summarizes how emotions of shame in relation to environmental decline structures people's experiences of identity, social interactions, and social structure on the river today. As with earlier emotions, past and ongoing genocide are bound up with these emotional experiences. And with shame in particular, the failure of the dominant society to recognize either genocide or the relationships and social processes that people so vividly experience becomes part of the problem. Sociologists have done Native people few favors in this department; rather, as Bacon (2017) and Huyser (2017) have documented through their content analyses of sociological scholarship on Native peoples, the dominant characterizations are pathologizing.

TABLE 6　Shame in Relation to Identity, Social Interactions, and Social Structure

Identity	Social interactions	Social structure
If you draw your identity as a river Indian but the river is contaminated, that reflects on you.	Shame that cannot provide for elders or family, cannot perform responsibilities to other species	Shame that cannot find way in "modern society"

"THE NATURAL THING IS TO FEEL HOPELESS"

Grief, anger, and shame were also mixed with feelings of powerlessness in the face of institutional forces working against the health of the Klamath River. Leaf Hillman put it this way:

> People say, "Do you really think they are going to take out the dams on the Klamath River? You'd be out of your mind to think that." Well I don't know. Do I really think there is justice in the world? No. That's an easy one. Do I ever think that they'll be justice? No. Do I think there is any hope? I don't know. People say, "How can you be even the slightest bit optimistic?" It's not easy to be optimistic about any of these things that I'm talking about. The easy, and I think the natural thing, is to feel hopeless.

Frank Lake vividly described the experience of living with the degraded river as "enduring an assault on one's relations" yet being powerless to fully stop it:

> You know, that spiritual tie, kind of more like kinship or family type of relationship. That's where I think the grief comes in. It's like, a sense of powerlessness. You know, and yet what can you do? . . . You basically see this assault or this attack on your family, either directly as humans, but also the extension of your family relationship and the tribal perspective of seeing that with salmon, you see this attack. You see this, you know, and there is this constant, I guess the only word I can think of is assault on them. And there are certain things you can do within your capacity, and then some things are so broad outside of the influence, that it's hard to comprehend what's going on.

Ron's older brother Achviivich described an analogous loss of control in relation to cultural activities. As do others, he clearly associates genocide with both literal killing of people in the past and the ongoing structural genocide. "*Our way of lives has been taken away from us. We can no longer gather the food that we gathered. We have pretty much lost the ability to gather those foods and to manage the land the*

TABLE 7 Hopelessness Operates across Identity, Social Interactions, and Social Structure

Identity	Social interactions	Social structure
Feeling unable to fix problem, powerlessness	There are only a handful of fishing families left.	Maybe this is a sign of the "the end."

way our ancestors managed the land." People's lived experiences of hopelessness are interconnected, but as sociologists, we can also attend to the specific operation of this emotion in relation to identity, social interactions, and social structure as presented in Table 7.

In contrast to the experiences articulated here, decades of research from sociology and psychology indicate that vital components of psychological wellbeing include a positive sense of self-worth and self-efficacy, coherent meaning systems, and sense of personal and cultural identity (Mirowsky and Ross 1989; Thoits 2010). The experiences Karuk people articulate are similar to those Downey and Van Willigen (2005) found in their work on proximity to environmental contamination as generating personal powerlessness and the work of Shriver and Webb (2009), who describe how an "endless battle to validate health and environmental concerns, along with the constant assault on Native American values, has fostered a sense of apathy and hopelessness among some tribal members" (282).

Karuk descriptions of grief, anger, and hopelessness provide important opportunities to think about the role the natural environment plays as a source of emotional experience in multiple ways. First, these emotions are a direct response to environmental decline. Grief, anger, and hopelessness are experienced in response to changes in a treasured riverine system and the loss of species that are considered relatives. Second, these emotional experiences shape how people understand their own identities, including, in the case of hopelessness, their sense of efficacy in the world. Third, changes in the natural environment alter the quality of social relationships. Smith-Lovin (2007) writes that "the person we become depends profoundly on the networks in which we are embedded" and "the actions we take and the emotions we experience depend on these networks. These networks are, in turn, shaped powerfully by the social settings that we occupy" (106). But for theory in sociology of emotion, all of this activity is conceived as taking place in a vacuum. A parallel argument to the above can be made in terms of the missing ingredient of the natural world in the interplay of emotions, interaction, and identity. When salmon are present, a host of family and community interactions associated with catching and distribution of the fish will occur. People gather to fish, to watch others fish, to see friends and family, and to distribute food. These interactions have profound meaning for individuals, for families, and to the Karuk community. When salmon are absent, these interactions do not occur, and a very different set of emotional and social experiences is set into motion. Finally, individual grief matters because it serves as a "signal function." In our data, the emotions of both grief and anger in the face of environmental decline are the means through which people understand their location in a racialized power structure and an ongoing process of genocide and colonization.

The reduced ability of Karuk people to participate in traditional management negatively affects both individual mental health and generates chronic community stress. Chronic community stress occurs when long-lasting psychological stressors are present across a community (Gill and Picou 1998). Such community stress is more than the sum of individual parts because the simultaneous disruption of many people's lives affects social structure and the maintenance of day-to-day activities, creating an overall normlessness or anomie (Edelstein 2004; Gill and Picou 1998).

DISENFRANCHISED GRIEF

As I have alluded to periodically throughout, impact or "harm" of all these emotional experiences is also underscored by their invisibility and the corresponding lack of legitimacy within the dominant culture, including the discipline of sociology. Ken Doka's (1989) term *disenfranchised grief* refers to grief that is experienced but cannot be openly acknowledged or publicly mourned. Because the dominant non-Native society does not recognize the deep emotional ties described between humans and the natural world, Karuk grief and other emotions described here over their loss are invisible. Brave Heart and DeBruyn (1998) describe how this "disenfranchisement" of emotional experience produces "unresolved grief" that itself becomes a significant "harm." They note that American Indians face particularly high rates of mental health challenges ranging from suicide, homicide, and accidental deaths to domestic violence, child abuse, and alcoholism, and they argue that the lack of social recognition of this grief is key to explaining these challenges: "These social ills are primarily the product of a legacy of chronic trauma and unresolved grief across generations. It is proposed that this phenomenon, which we label historical unresolved grief, contributes to the current social pathology, originating from the loss of lives, land, and vital aspects of Native culture promulgated by the European conquest of the Americas" (56). Indicative of this legacy are the results of a 2016 Karuk Needs Assessment for K–12 Education. Of the total 72 Native American parents, Karuk tribal employees, and local school staff respondents, the overwhelming majority reported that documented academic underperformance of Karuk students was caused by poverty, domestic violence, substance abuse issues, and low self-esteem. Whitbeck et al. (2004) similarly link discrimination and historical loss to subsistence abuse in

Native communities. From a sociological perspective, we can apply concepts of emotion norms and feeling rules to understand this situation. Theoretical work in sociology clearly states that emotion norms vary by social context, including racial and ethnic contexts (Mirchandani 2003; Wingfield 2010), and describe a complex interplay between these racialized feeling rules and systems of oppression (Wilkins 2012). In this case, dominant cultural emotions norms (non-Native, white), which situate Karuk experience of deep connection with the natural world as "abnormal" and even subject to categorization as "mental illness," form a mechanism of racism, assimilation, and colonial violence. On the one end, stigma over aberrant" emotional experiences are associated with potential substance abuse as described by Brave Heart and DeBruyn above. Multiple community members underscored the importance of this connection:

> I've talked to a lot of guys and their dream is to be able to fish and hunt and take care of their families and you know, be able to do that. They have all this guilt about not being the person you had wanted to be, so you avoid it. A lot of Karuk guys avoid feeling that, so I think that's why you have a lot of drugs and alcohol. They feel really bad about not being able to participate or to provide in a manner that they felt they should, so just they do drugs to avoid having to deal with that feeling. (Anonymous)

On the other end of the spectrum, this disjuncture in emotion norms becomes a mechanism of forced assimilation in non-White expressions of grief and can be sanctioned by force through the mental health system when people are institutionalized for inappropriate emotional presentations. Although there is not space to elaborate upon this here, Native studies scholars have illustrated how attempts to assimilate Native cultural norms into a hegemonic normative structure of emotions has been central to the mechanisms of colonialism and racism (Alfred 2005; Bacon 2018; Simpson 2017).

Taken together, emotions of grief, anger, shame, and hopelessness each work across multiple social categories to inscribe racism and ongoing colonialism. Environmental decline is understood and manifested via the emotional experiences that translate into threats to identity, disruptions to systems of social interaction, and structural outcome in the form of racism and ongoing colonialism. Emotions serve as a signal function in that it is the cognitions about these experiences that confirm structures of power. Thus, one unique offering of this chapter is to build an additional layer to Goodwin, Jasper, and Polletta's (2004) and Jasper's (2011) typology of emotions by illustrating how features of the natural environment cause negative emotions that in turn form a three-part relation between cognition and identity work for individuals, social interaction, and broader enactment of social structure (see Table 8).

TABLE 8 Emotions of Environmental Decline Confirm Structures of Power

	Identity	Social interactions	Social structure
Grief	Like tearing my heart out	Sadness because "quiet down at the Falls"	Sadness that Karuk people may disappear
Anger	Anger that cannot fulfill expected roles	Anger that children do not have same opportunities anymore	Anger at agencies that arrest people for fishing according to tribal custom
Shame	If you draw your identity as a river Indian but the river is contaminated, that reflects on you.	Shame that cannot provide for elders or family, cannot perform responsibilities to other species	Shame that cannot find way in "modern society"
Hopelessness	Feeling unable to the fix problem, powerlessness	There are only a handful of fishing families left	Maybe this is a sign of "the end"

Thus, while drawing upon Goodwin et al.'s (2004) and Jasper's (2011) framework, this chapter also emphasizes how emotions can simultaneously operate in different ways. While the authors categorize some emotions as affective emotions (these concern themselves with social bonds and loyalty) and others such as those that are considered central to advanced moral reasoning, all emotions described here have moral dimensions. Taking, for example, anger—categorized as a reflex emotion—here anger operates in relation to identity when the person is unable to fulfill social roles on behalf of other family members who cannot experience fishing, gathering, and/or hunting and in relation to systems of oppression. Anger may well be a reflex emotion operating along different brain pathways, arising and receding quickly, as the authors note, but in our data, this does not necessarily preclude the ability of anger to simultaneously signal important social experiences in relation to loyalties, social ties, and moral reasoning. Similarly, hopelessness fits the categorization of a mood operating across multiple settings, but it too operates across categories of individual identity, social ties, and moral perception (see Table 7).

EMOTIONS OF RESISTANCE

While some of the original research behind this project was intended to document social impacts from the Klamath River dams as part of legal and policy process to redress harms, research focusing solely on negative impacts and oppression in Native communities can itself be profoundly damaging (Tuck 2009). As much as emotional experiences of the degraded environment inscribe racism or form a mechanism of assimilation and colonialism, one would profoundly misunderstand what is happening in this setting without mentioning the wide variety of forms of resistance undertaken by Karuk tribal members on a daily basis. Resistance to colonialism and assimilation has been continuous, pervasive, and diverse in the range of forms that people employ. People resist by engaging in direct action at protest events and legal actions against federal and state agencies via natural resource policymaking, testifying in public hearings, and participating in dances and ceremonies. People use personal prayer, continue to hunt and fish according to tribal law despite personal risks, learn and teach the Karuk language to their children, and develop tribal cultural heritage-centric educational curriculum. Here too the relationship between emotions and the environment is fundamental to any valid understanding of resistance. Anger is an emotion that is often associated with political agency, but the dynamic between fear and hope and other emotions associated with group solidarity are part of resistance as well. Traditional fisherman and fish crew technician Mike Polmateer alludes specifically to the dynamics between fear and hope as he moves from reflecting on things that are lost to referencing hope for the future as his motivation for traveling outside the area to engage in protest events, such as when the Tribe went to the shareholder's meetings of the companies that owned the dams first to Edinburgh, Scotland, and later to Omaha, Nebraska: "But we never give up hope. That's why I went to Scotland. That's why I went to Omaha twice. Because we have hope to fix this river. Which is one step to getting the Karuk people back to where they once were." Ron's cousin Binx also describes how work on the fisheries crew keeps him going and serves as an avenue to fulfill cultural responsibilities in today's world: "My job, for one ... bringing the salmon back and restoring the health of our river, if everybody reaches out and does their own little part ... that's kind of how I'm feeling right now with my job, you know, I'm trying to restore the river, doing these fish surveys, and create a positive effect for the fish." A mother in the community who wished to remain anonymous notes the importance of activism against the dams for the identity of young people: "Now there is such a big push because we see an opportunity. I think in a lot of ways that this provided people with a role. These younger folks have a purpose, knowing that they can make a difference in the Klamath Campaign. I think that that makes a difference." While the emotions implicit in these passages are more subtle, the natural environment clearly is a motivator for social action, and emotions in response to environmental change operate here in the form of what Jasper (2011) calls "moral batteries" whereby "fear, anxiety and other suffering in the present" are combined with "hope for future change" to motivate action (14.7).

SO WHAT CAN NATIVE COSMOLOGIES TEACH ABOUT EMOTIONS AND ENVIRONMENTAL JUSTICE?

Emotions associated with environmental degradation shape social experiences, serve as a signal function inscribing racialized power relations, and function as a mechanism of assimilation and genocide. But to understand such things, the term *environment* as used within sociology must be broadened to refer not only to the "social environment" but also to the material, beyond-human environment that includes other species, rivers, landscapes, and rocks with whom humans carry out multiple activities and interactions that give rise to personal and social identities, cultural meanings, and a range of emotional experiences.

It is somewhat disheartening to try to add my voice to the long list of attempts to challenge the nature-society dualism within sociology (see, e.g., Brulle 2015; Dunlap 2002, 2010; Salleh 1984). The field of environmental justice makes a particularly significant intervention into the nature-society divide by expanding the concept of "nature" to include more and more feathers of the "social," although the wider disciplinary implications of these literatures have yet to be realized. For example, inclusion of work in urban areas and on the body each has blurred the boundaries of what is perceived as "nature" and "social," and environmental justice scholars and activists have critiqued the idea of "nature" or "environment" as a wilderness that is separate from humans and refocused attention on "the environment" as human inhabited spaces and even human bodies (Alaimo 2010; Sze 2007). Yet like environmental sociology, the field of environmental justice has taken a primarily positivist approach, focusing almost exclusively on material descriptions of harm in the form of pollution exposure or lead poisoning, whereas much important work on gender and race includes an interpretative dimension to power and oppression, including concepts such as symbolic violence and social suffering.[5] Indeed, the field has in many ways upheld the "nature" side of the nature-social—just as some believed that negation of the natural was necessary to politicize power in gender relations for feminist theory (Rahman and Witz 2003), environmental scholars have tended to downplay social construction to legitimate claims of environmental contamination.

Just as chapters 1 and 2 detail the workings of environmental and natural resource policies in the process of racial-colonial formation and ongoing violence, environmental justice scholarship can benefit from incorporating symbolic conceptions of power in the form of structural violence from ethnic studies, sociology of race and ethnicity, and sociology of emotions. As the field of environmental justice expands to include more nuanced attention to symbolic dimensions of power and injustice, Indigenous perspectives have much to offer. The most recent theories of settler-colonial oppression are becoming increasingly specific about the idea that settler-colonial erasure physically alters environments cultivated by

Indigenous peoples (Whyte 2018c, 2018d; Whyte et al. 2018). These environments served to support Indigenous cultures, mental health and well-being, and political and economic sovereignty (including food sovereignty). For example, mental and emotional health are harmed through structures of domination that invent national and state borders and distinctions (e.g., urban versus reservation) on Indigenous landscapes (Goeman 2009; Tamez 2016), create false forms of Tribal or First Nations governance or multiculturalism (Simpson 2001, 2004; Lomawaima and McCarty 2006; Richardson 2011), dissociate people from having place-based identities (Lawrence 2003; TallBear 2013; Whyte 2016a, 2016b, 2016c), disrupt Indigenous systems of responsibilities (McGregor 2009; Bang et al. 2014; Whyte 2013a; Coombes, Johnson, and Howitt 2012), or have adaptive capacities coupled with ecological systems (Whyte 2015, 2016d) that underwrite the possibility for having a positive emotional attitude toward the possibility of there being a future (Tuck and Gaztambide-Fernández 2013; Whyte 2016d) or of having a sense of place (Hoover 2013; Johnson and Larsen 2013; Watts 2013). Hence, theories of anticolonialism or decolonization—traditional lands as well as reclaimed urban spaces—are not just about land reclamation but about community relationships. These relationships also include those with nonhumans and ecosystems that advance individual and community mental health and wellness (Bang et al. 2010; Tuck and Yang 2012; Million 2013; Coulthard 2014; Miner 2014; Simpson and Coulthard 2014; Todd 2014; Whyte 2017a). While emotion is not invoked as a specific term, these theories are very much about emotions using other terminology. Hence, it can be argued that they connect emotional life to the destruction of ecological relationships through structures of settler-colonial oppression.

The emotional experiences of environmental decline will influence and mediate social action in complex and unique ways for diverse communities (Jacobson 2016; Willox and Ellis 2018; Willox 2012; Cunsolo and Landman 2017; Davidson 2018a, 2018b). I have written elsewhere about how, for privileged social actors, emotions of grief and hopelessness can work to solidify political inaction on climate change (Norgaard 2011, 2012). Certainly, in the face of widespread and ongoing environmental decline, relationships between emotions, power, and environmental decline will become both more apparent and more important for our understanding in the structuring of inequalities. Given the unfortunate facts of climate change and expanding environmental degradation, connections between changes in material environments and social outcomes will become both more significant and more visible in the years to come. Greater interplay between existing theories of emotions and environmental sociology will be necessary for understanding these dynamics.

CONCLUSION
Climate Change as a Strategic Opportunity?

[Mainstream political ecology frameworks] ascertain how indigenous people are politically, economically, culturally, and ecologically marginalized, but rarely provides a way forward from this plight. Indeed, indigenous people and struggles are only visible within Western framings—resulting in a situation whereby they are seen but not heard.
—Beth Rose Middleton (2015)

We are trying to get back to an intact world. Climate change can be a vehicle for that because of the awareness it brings to so many about limitations in the current management practices. We believe there is genuine interest in Karuk perspectives about how to care for the land, we offer these explanations in the hopes that this is true.
—Ron Reed, cultural biologist and traditional dipnet fisherman, Karuk Tribe

The traditional land and resource management practices of the Karuk people is based on traditional ecological knowledge. It's adaptive in nature. It's not about a certain date when you do something. It's not about a measurement that tells you that you're right or wrong. It's about listening, observing, feeling, remembering… and communicating. We have it all over the Western scientists when it comes to adapting to climate change. We've been adapting to a changing climate for thousands of years.
—Lisa Hillman, Píkyav Field Institute program manager, Karuk Tribe

Climate change as a strategic opportunity? Climate change as a vehicle back to an intact world? These are rather different perspectives from the dominant doom-and-gloom narratives of climate change and the Anthropocene within academic circles. Are these merely the words of a naive optimist? From what standpoint could this statement possibly be made?

Throughout these chapters, I have sought to bring what I have learned from my Karuk colleagues, collaborators, and friends into conversation with the thinking across the social sciences and humanities and with the discipline of sociology specifically. That these worldviews are so different—and hence that there is so much to be said—is a function of the profound epistemic exclusion wrought by colonialism itself. At some level, the omission of Indigenous perspectives might seem surprising—after all, sociologists have centrally concerned ourselves with questions of power and inequality. Critiques of capitalism, racism, and sexism have been at the heart of sociology and many other disciplines, yet when it comes to colonialism, fewer Western academics or sociologists in particular have been able to see the air we are breathing. Rather, as Kemple and Mawani (2009) write, sociology's "ontological moorings, categories, and modes of analysis have been fundamentally structured by imperial pursuits and formed within cultures of colonialism" (238). We may have thriving American Sociological Association (ASA) sections on Race and Ethnic Minorities, Race/Gender/Class, Asian Americans, Latina/o Sociology, and the Association of Black Sociologists, yet there is no section on either Indigenous peoples or colonialism. As I have argued throughout the preceding chapters, a central component of the logic of colonialism is that social relations can be analyzed without attending to dynamics in the natural world. Now in the face of climate change, this façade has begun to crack. The ecological conditions around which our contemporary societies are organized are now changing. Whether it is through an increase in the frequency of large-scale wildfires, collapsing energy grids, or disrupted transportation systems, the fundamental relevance of the natural world to the social—relationships that were masked through technological capacities—is now becoming visible once again. Can climate change be a vehicle to eradicate the nature-culture dualism within sociology, thereby allowing for a fuller understanding of social dynamics and power?

In these pages, it has been my aim to add my voice to those of the many others working to show how Indigenous perspectives can powerfully invigorate and expand a wide range of sociological questions and theories from the discipline's undue fixation on assimilation (in both sociology of immigration and sociology of race) to the ways that gender practices may be structured by and in relation to other animals and places in the natural world. Throughout these chapters, I have sought to illustrate how taking seriously the experiences of Native people and the importance of the natural environment offers an opportunity to extend sociological analyses of power and move sociology toward a more decolonized discipline— one that serves and supports Indigenous survival and self-determination rather than Indigenous erasure and demise. Chapter 1 makes a case for the importance of the "natural environment" for racial formation. In addition to detailing the importance of the natural world for the development of the racial categories of White and Native on the Klamath, I described how the changing ways that groups interact with natural environments shape the production of race and racism. I described how racialization is not only about "the elaboration of racial meanings to particular relationships, social practices or groups" (Omi and Winant 1994, 91) but also to particular environmental practices and places in the landscape. White supremacy is enacted upon not only people but also the natural world. Chapter 2 built upon these themes tracing the history of human-fire relationships on the Klamath to illustrate the relevance of the framework of settler-colonialism for the social sciences, particularly my own discipline of sociology, and to highlight the importance of manipulation of the natural world for the success of the settler-colonial state.

Chapter 3 examined colonialism as it relates to food and the links between environmental and human health. Here I situated the negative physical health conditions so many Karuk people currently experience, including diabetes, strokes, and heart disease in a context of social power, colonialism, racial formation, and environmental degradation. Health studies rarely start with Indigenous perspectives on health, theorize the impacts of colonialism on Indigenous physical or mental health, or account for how health practitioners and researchers themselves enact colonial violence on communities. Chapter 3 also described the vital need for Native-led research programs and the many benefits of incorporating Indigenous environmental knowledge into social science research to counter deeply colonial notions of knowledge and agency that stigmatize Indigenous peoples, on one hand, and restrict them to the roles of passive research subjects, on the other. I described as well the importance of Indigenous collaboration in the areas of environmental justice, and the erasure of Indigenous presence within the food sovereignty movement, arguing that both movements can benefit greatly from Indigenous knowledge and perspectives on health, food, justice, and more.

Chapter 4 theorized the particular ways ecological relationships structure gender, emphasizing that gendered and racialized colonial violence operates via environmental degradation. Because the absence of abundant and healthy Native food, fiber, and medicinal plant species affects individual gender practices and socialization methods, as well as symbolize genocide to the community, Karuk people's individual process to understand themselves as women and men is interwoven with struggles against racism and ongoing colonialism. Chapter 5 engaged the field of emotions, bringing attention to the prolonged trauma of long-term environmental decline as a dimension of environmental justice. Here I sought to extend work in sociology of emotions by describing how the natural environment is part of the stage of social interactions and a central influence on people's emotional experiences, including individuals' internalization of identity, social roles and power structures, and their resistance to racism and ongoing colonialism. While environmental justice has focused on physical health impacts, I detailed how emotions of grief, anger, shame, and hopelessness associated with environmental decline serve as signal functions confirming structures of power, as well as how hope and anger inform ongoing, sustained resistance.

Across these chapters, what unites seemingly disparate topics has been an emphasis on Indigenous epistemologies of interconnection, on the relevance of the natural world for social happenings, and on the critical need to attend to the settler-colonial nature of state structures and actions in North America. In each case, bringing in settler-colonial understandings of power dynamics into existing critiques of capitalism, sexism, and racism significantly changes theoretical conversations. If readers find even some of these arguments useful, then it would seem that indeed, the traditional knowledge that shaped this continent remains relevant today in the halls of the academic world and social science classrooms and the pages of sociological theory.

FOUR THOUGHTS ON CLIMATE CHANGE

> We [American Indians and Alaska Natives] have been on the receiving end of Western ideas, opinions, and colonial institutions for five hundred years. Now we face a situation on the planet where Native voices must be heard in order to avert or hopefully minimize the deadly events emerging.
> —Daniel Wildcat (2010)

Whether they come in the form of emphasizing the necessity of paying attention to colonialism and Indigenous understandings of resistance or in the form of emphasizing the importance of community and responsibility, I believe Indigenous insights and perspectives are necessary now for human survival. The accelerated alternation of the biophysical systems around which modern capitalist societies are organized and upon which all humans and existing life forms on earth depend represents a crisis of rather unfathomable proportions. As Daniel Wildcat asserts, Indigenous peoples have long been outspoken concerning climate change. What might happen if the non-Native settlers of this land began to listen to what Native people have been trying to say?

If climate change does represent the proverbial crisis as dangerous opportunity, what do non-Indian academics and activists need to better understand? In the remainder of this chapter, I will reflect on the sociological implications of four main ideas that Karuk and other Indigenous peoples have been emphasizing regarding climate change. First: the situation is extreme, but it is not new. Climate change needs to be understood as the latest intensification of dynamics on a long continuum of colonialism. This assertion will resonate with other deeper systemic critiques, including those of capitalism. Business-as-usual responses to climate will not only be ineffective but also serve as a vector of the further displacement and erasure of Indigenous peoples. Second—relating to the notion of the Anthropocene—humans have always been a part of and related to the other beings that make up what we often call "the natural world." Just as the "discovery" of the Americas is an absurd concept, the "discovery" of the "Anthropocene" or the impact of humans on the natural world must equally be understood to be an event occurring for only a segment of the world's population. Third: the natural world, including other beings with whom humans share this earth, is a source of power and agency. Fourth: humans have a responsibility to act now. I offer these reflections with some caution, especially regarding the encouragement toward action. As Whyte and colleagues (2018) highlight with their essay titled "Indigenous Knowledges Are Not Just for All Humanity," there is both a long history of the appropriation of Indigenous ideas and of white settlers jumping much too easily into actions that only reaffirm their/our own social position. At the same time, listening and understanding always makes sense, and under the right circumstances, settlers may be valuable—even essential—allies within Indigenous-led organizations and social movements (Bacon 2017; Grossman 2017; Steinman 2018; Whyte 2017b).

FIRST IMPLICATION: OUR SITUATION IS URGENT, BUT IT IS NOT NEW

The day after the Trump election, I had the students in my climate change class move their chairs into a circle. Students were devastated regarding the potential consequences of the election for the recent climate agreements, for the security of their communities and families in the face of rising white supremacy, and for war and peace around the world. We needed to talk, be together, and listen to one another. Students shared fears of economic and social insecurity, as well as a sense of bewilderment that any of it was really happening. When one Inupiaq student spoke, she expressed similar sentiments to the group as a whole, but also some measure of frustration with the others. Her comment was something to the effect that "none of this is really all that new" and "by the way Native people had to travel all the way to Standing Rock in order to be able to sit in a circle like this with like-minded individuals who shared their concerns and perspectives."

Whether it be the consequences of the Trump election, the concept of "the Anthropocene," the attacks on science, or the challenge of climate change, the Native people in my life and whose words I regularly read frequently express some version of the sentiment that "this is nothing new."[1] Certainly, if we take the view from Karuk country, what people experience as climate change is only the latest on a continuum of failed non-Native environmental policies that threaten human and ecological survival. We can see this in a concrete example through fire policy as discussed in chapter 2. In the mid–Klamath Basin, the most immediate impacts of climate change come in terms of the increasing frequency of high-severity fires and the associated fire suppression politics. Rather than a "new" event, the increased frequency of high-severity fires manifests as an intensification of the same longstanding and highly problematic social relations and fire policies that have been at play for well over a century: as larger hot fires are more frequent due to a combination of changing patterns of precipitation and temperature, on one hand, and a fraught legacy of fire suppression, on the other. Indeed, the Karuk Tribe's Climate Vulnerability Assessment emphasizes that the extent to which this pattern of fire behavior is a result of climate change or the past management practices that have led to high fuel loads cannot even be untangled. Furthermore, the ways that natural resource management plays out when fires occur continue these dynamics.

My colleague Kyle Powys Whyte has been writing specifically about climate change as a case of déjà vu. In Whyte's (2017a) words: "Climate injustice, for Indigenous peoples, is less about the specter of a new future and more like the experience of déjà vu" (88). Whyte elaborates, "Indigenous peoples face climate risks largely because of how colonialism, in conjunction with capitalist economics, shapes the geographic spaces they live in and their socio-economic conditions. In the U.S. settler colonial context . . . settler colonial laws, policies and programs are 'both' a significant factor in opening up Indigenous territories for carbon-intensive economic activities and, at the same time, a significant factor in why Indigenous peoples face heightened climate risks" (88). Chapters 1 and 2 detailed how the wealth that made the new state of California into one of the most powerful capitalist economics in the world was taken from Karuk and other Indigenous peoples. Karuk traditional knowledge may have shaped an abundant forest and riverine ecosystem, but Karuk people were killed and displaced to create and sustain the economic and political order that now generates enormously high emissions of climate gases. Just as Grey and Patel (2015) flag how the concept of food justice "calls attention to the tremendous economic and ecological debt owed indigenous Peoples, which remains unacknowledged (never mind unpaid)" (442), the situation is equally evident for climate change.

Karuk and other Indigenous people face heightened climate risks today not only because the causes of climate change are a continuation of colonialism but also because the dominant society's responses to climate change themselves produce further harm to Indigenous communities. Norton-Smith et al. (2016) write, "Indigenous vulnerability and resilience to climate change cannot be detached from the context of colonialism, which created both the economic conditions for anthropogenic climate change and *the social conditions that limit indigenous resistance and resilience capacity*" (3, emphasis added).

When it comes to fires on the Klamath, many of the actions taken by the U.S. Forest Service and CALFIRE when fires occur do as much damage to cultural resources and tribal sovereignty as do the fires themselves. As Chapter 2 details, past management actions such as logging, road building, or fire suppression interact with fire events to influence the level of ecocultural vulnerability Karuk people face, as do management actions taken during a fire and those that may follow in the long term. For example, during fires, actions from retardant drops to the bulldozing of fire lines along ridge tops directly damage and often destroy culturally significant habitats and species assemblages that have been created over generations. These ecological conditions are themselves living archives of past management actions; they represent a repository of traditional knowledge in the land.

In the short- and long-term aftermath of such fires, Karuk traditional foods, cultural use species, political sovereignty, and management authority may be further affected. Long after fires have ceased to burn, further, long-term damage is caused by management actions such as reseeding, sediment control, road building, and salvage logging. Such activities create lasting impacts on the landscape by bringing in new species (i.e., invasives) that come into direct competition with culturally important Karuk species, increasing the future likelihood of high-severity fires (Brooks et al. 2004), increasing sedimentation, and causing vegetation assemblage shifts. But if all this is what "climate change" looks like in Karuk country, there is nothing new about it.

Colonialism has created the high fossil fuel, alienated, extractive world, and until we address its cosmologies and material practices, we will not respond to climate change in a way that gets us anywhere else, in part because colonial structures and mind-sets continue to erode the adaptive capacities of Indigenous peoples.

The flip side of the "this is not new" message relates to the discourse that "the world is about to end." This too is a remarkably privileged standpoint. Growing up in Berkeley, California, in the 1970s, this is a message I heard quite frequently. Then the specter of nuclear war and nuclear winter were the immediate causes; for my parents, it was Cuban missile crisis and, for my grandparents, World Wars I and II. Indeed, Rebecca Solnit has pointed out that every generation has believed it could be the last. There is an archetypal trope here the likes of Joseph Conrad have surely analyzed, and the deep and very real threats to life on earth posed by the changing climate lend themselves well to this theme. Yet on another level, climate change provides a window beyond these mythic framings, powerful though they may be. For certainly, the world as it was known has already ended for the many communities facing famine, civil war, and widespread ecological alteration. Many although hardly all of these communities are in the global South, where they are comfortably separated by boundaries of distance and border walls from wealthier communities in the global North. Within the United States, Indigenous communities in particular are living in what Kyle Whyte calls "their ancestors' dystopia." As I noted in the closing of my first book *Living in Denial*, it is only because of privilege that physical barriers and the elaborate boundaries created by norms of attention and other cognitive structures exist. Together, these make possible the social organization of denial such that some people can imagine "the apocalypse" will be happening in some future time.

Notions of apocalypse and crises work against Indigenous sovereignty at other levels too. Chapter 2 described multiple cases where the emergency management protocols that go into effect during wildfires erode Karuk tribal management authority, especially with respect to water quality standards but also concerning road access and practices of collaborative input. April Anson (2018) gives context to this situation in her challenge to the politics of using the apocalypse genre to narrate environmental crisis. Anson writes that "apocalyptic appeals provide a paradigm case of settler environmentalism," noting how "this state-of-emergency rhetoric, so pivotal to American apocalyptic appeals, easily erases, emboldens, and acquits the white violence of settler systems throughout the spectrum of U.S. political rhetoric."

SECOND IMPLICATION: THE ANTHROPOCENE AND HUMAN-NATURE CONNECTIONS

The term *Anthropocene* refers to the notion developed by natural scientists Crutzen and Stoermer (2000) to denote a new geological period or epoch in earth's history in which human activity is a significant force in the shaping of the planet. The term is meant to draw attention to the fact that human activity is fundamentally transforming the natural world, modifying even the planet's basic geochemical and atmospheric cycles. While climate change is the most obvious example, the term references chemical contamination and genetic engineering as well. In this new era, "natural forces and human forces became intertwined, so that the fate of one determines the fate of the other" (Zalasiewicz et al. 2010, 2231), and all environmental problems should be understood as having both natural science and human dimensions (Steffen et al. 2007; Crutzan 2002). Regardless of the debate within geology, at the level of popular discourse, the notion of the Anthropocene refers to widespread destabilization of the earth's systems in the face of climate change and other forms of environmental degradation.

The concept of the Anthropocene is about recognizing "the human role" in shaping the natural world. But is it a "human" role or is it the role of a particular group of people acting under particular social conditions that is causing this degree of negative impact? Anson (2018) emphasizes that "using the genre of environmental apocalypse to imagine an event yet to come may erase the ways that specific peoples have experienced, survived, adapted, resisted, and frustrated the ecological and humanitarian collapses associated with a settler anthropocene" (8). An understanding of this relationship between humans and the natural world is only new to some. As this view from the Klamath Basin has illustrated, humans have been intimate parts of the natural world, understanding their ability to wreak havoc on it, as well as how to work with other species in order to thrive. Just as the idea of a pristine untouched wilderness is an absurd concept, the "discovery" of the Anthropocene must equally be understood to be an event occurring only for a segment of the world's population. The universalizing language of "humanity's impact" enacts another round of erasure (Todd 2015; Collard, Dempsey, and Sundberg 2015).

Because Native people have been associated with the side of nature within the nature-society dualism, and the people whose experiences animate this book clearly have particularly strong connections with the natural world, it may be easy to dismiss my claims that the natural environment matters for social action more generally. While it may be true that the ways that the natural world shapes gender constructions, human health, and other sociological topics in this book *may be more apparent* for Karuk people quoted herein than for other communities, the notion that humans are now separate from the natural world is part of the myth of modernism and colonialism. Instead, to quote my colleague Steward Lockie: "What happens if we accept that despite the technological advances of the industrial age, human society has never transcended its ecological roots? If we accept that social change today is as much about ecosystem and climate processes as it is about institutions and power" (2015, 2) In *This Changes Everything Capitalism vs. the Climate*, Naomi Klein emphasizes that while some people have been able to temporarily shield ourselves from "the elements" or to the consequences of our impacts on the earth across time and space through the use of technologies, seeing beyond the illusion that humans are separate from or can control the natural world is one of the core opportunities presented by climate change. Through this lens, climate change is indeed an opportunity to understand the nature of our connection to one another and our connection to the earth.

One aim of this book has been to add my voice to the many others who have long been making a case that what we think of as "the environment" matters for a wide variety of social processes across the discipline of sociology—indeed, that the myth of a human-nature divide has always been just that. Instead, social phenomena from the strength of social networks to the factors that influence gender constructions may also be influenced by the resources available for "doing gender," and the tools in one's cultural toolkit may be found in the natural environment, whether or not sociologists have analyzed them that way. In the same way that sociologists have recently worked to bring the body and emotions into social theorizing, many sociological analyses on topics as diverse as the mechanisms of social movement mobilization, the collapse of labor markets, or the formation of physical health inequalities are misspecified if they do not include the natural environment.

So the strategic opportunity of climate change at the epistemic level is that this form of ecological decline is making interconnections between the natural and the social increasingly visible. The complex interplay between "natural" and "social" is evident through the presence of toxins in human bodies, crises in economic systems, and the deep emotional responses people experience to changing physical ecosystems. In the face of rapid environmental degradation, the importance of the natural world to the dynamics of power and inequality has become both more visible and more important to understand.

The nature-society dualism has left sociology with an inadequate understanding of the project that is most important to us—our understanding of the social world. Just as feminists Jagger (1989) and Tuana (1989) articulated how the presence of dualisms between mind-body, emotions-reason, and public-private works ideologically to subjugate women, the nature-society dualism serves to perpetuate racism against Indigenous people and advance the process of colonialism. Just as the notion of manifest destiny legitimated the actions of white settlers as inevitable, just as the myth of the vanished Indian is a settler logic that makes invisible the ongoing presence of Native people in the United States, the discourse that the environment is not part of the social world legitimates an understanding of society in which the experiences and injustices described here are invisible because they are in fact beyond the scope of social theorizing. Neglecting the natural world as a component of social action within sociological tradition is a remnant of colonialism. Overcoming this divide is essential for unsettling sociology and beyond.

THIRD IMPLICATION: THE NATURE OF POWER

> Power springs up between men when they act together.
> —Hannah Arendt

> Our power comes from the earth.
> —Winona LaDuke

A third idea Indigenous people have been articulating concerns the nature of power. Indigenous conceptions of power and proficiency regarding social change are highly relevant for scholars of social movements and political sociology, and they hold specific relevance for the discipline at large in light of the now rapidly changing climate. On one hand, for political sociologists, attending to Indigenous cosmologies makes visible important social movements that were hiding in plain sight. For example, in the first piece of scholarship to engage settler-colonial theory published by the discipline's flagship journal *American Journal of Sociology*, Erich Steinman (2012) describes the failure of social movements scholars to understand the achievement and significance of the Red Power movement of the 1970s: "To date, predominant sociological scholarship in the areas of social movements, political sociology, and race and ethnicity, as well as in the interdisciplinary field of postcolonial studies, has failed to acknowledge the ongoing settler colonial power dominating American Indians and thus has failed to categorize and analyze Indigenous movements in the United States as decolonizing in nature" (1074). As Steinman notes, this absence is particularly glaring when set alongside the attention given to the civil rights movement and other significant social movements of the past half century.[2] Steinman notes that in the absence of attention to sovereignty or a theory of North American colonialism, Indigenous forms of political action were simply invisible. Steinman writes, "Because settler colonialism is not identified in the social movements literature as a dimension of sociopolitical relations in the United States, it is difficult for scholars employing the structurally oriented contentious politics approach to perceive the existence of a movement reflecting colonial relations" (1076). Similarly, Cantzler and Hunyh (2016) emphasize that scholars of inequity have much to learn from the sophistication of Indigenous movement strategies for institutional change. The authors emphasize the sophistication of Indigenous political accomplishments with respect to salmon fishing and recovery, noting how tribes deployed a keen balance of legal structural and grassroots direct action tactics to leverage major political gains.

Yet if Indigenous political achievements are hidden by sociological blind spots, the means through which people account for having achieved such victories are further obscured. Take, for example, the notion that "power comes from the earth." Over 500 years of Indigenous resistance to colonialism has quite literally been achieved with the aid of the natural world. In chapter 1, I described how the natural environment mattered for Karuk resistance to genocide and racial categorization—people were able to hide in the mountains to avoid direct genocide. Indeed, even today the rough terrain of the Klamath area provides a buffer against incursions of extractive industry and tourism alike. Landslides take out roads each winter, extended power outages leave people dependent on woodstoves, and there is little or no cellular service. As described in chapter 2, Risling Baldy (2013) places such power at the center of the notion of biocultural sovereignty: "how could the indigenous outlast the European military invasion, the massive biological warfare, the systematic ecological imperialism and the meticulous destructuring of their institutions, and still initiate almost immediately a process of cultural and sociopolitical recuperation that allowed for their continuous and increasing presence in the social and biological history of the continent?" (Varese, quoted in Baldy 2013, 5). Indeed, she illustrates how this power to survive outright hostility, on one hand, and daily survival, on the other, comes from the earth. While the subfield of political ecology was developed in geography to theorize the dynamics of power relationships woven through human and ecological systems, sociological approaches to social movements have yet to theorize in this direction.

Karuk power is rooted not only in the earth but also in community ties. Indigenous notions of self and world stand in stark contrast to the neoliberal subjectivity (Pfister 2004), itself much critiqued in light of climate change. One of the central driving logics of capitalism and neoliberalism is the notion that we exist as individuals who lack connections to one another. Especially in the era of the Trump presidency, the architects of the present structure are doing everything possible to convince people that we are disconnected from one another. They seem to seek to create conditions of isolation and disconnection, be it through the use of racism and anti-Semitism to divide and reduce the power of the working class. They are also attempting to alter labor laws to reduce the organizing potential of unions. The rhetoric of disdain, contempt, and derision that characterizes Trump's persona is a message to people: "don't think you or your voice matters." So much of academic theory too has been centrally influenced by the tenets of neoliberalism from the rise of rational choice theory in social sciences and the direction of economics. And academic norms and structures that set up competition between faculty over ideas and resources are equally complicit.

Just as the logic of modernism rests upon the notion that humans are separate from the earth, so too does neoliberalism rely on severing and obscuring relationships between people. Ironically, however, the myth of individualism has gained traction at the same time as our lives have become increasingly *interconnected* through both material and symbolic flows. These connections are less visible in our complex modern societies where they are mystified by breaks in space and time, but as C. W. Mills emphasized in the concept of the sociological imagination, it is imperative that we learn to see them. It is because we are not actually individuals that we must engage in collective social action. These teachings relate to those of another of my favorite political theorists, Hannah Arendt (1958), who tells us, "Power springs up between men when they act together" and that powerlessness comes from being "inattentively caught in the web of human relationships" (151).

FOURTH IMPLICATION: BEYOND ENVIRONMENTAL PRIVILEGE AND NEOLIBERAL SUBJECTIVITY: WE HAVE A RESPONSIBILITY

> Perhaps the most difficult challenge is for more of the world to embrace and understand concepts like "mother earth" and spiritual-social relationship with land, air, sea, and all life that walks, flies, swims, or crawls upon it. Rather than view these alternative philosophies as "primitive" we need to see that they are holistic approaches better suited to the powerful and potentially lethal changes we will see as a result of climate change and global warming.
> —James Fenlon (2015a)

So where exactly is the strategic opportunity in climate change? According to Ron Reed, the opportunity is twofold: at least some non-Native people finally appear to be waking up to the seriousness of our situation. Furthermore, there is a chance they may listen to what Native people have been trying to say. A friend gave me a poster that hangs on my office wall commemorating the late Guatemalan Indigenous activist Berta Carceras with the following quote: "Wake up humanity: Time has run out." Perhaps the most important thing that so many Indigenous people have been saying regarding climate change is that now is the time to act and what it is that needs to be done.

My first book described the challenges faced by privileged communities in coming to terms with climate change. That project concerned the reasons for widespread public silence or "apathy" in the face of this most profound threat to humanity and life on earth. And although things have changed some in recent years, there is still a profound misfit between the level of public engagement on climate change and the seriousness of the problem. This gap has been a central concern to climate scientists, educators, social scientists, and others who believe in the necessity for democratic participation and hence seek to mobilize the public. I described how, in contrast to "information-deficit" models that presume we need better science education or more information to mobilize public response, or more cynical portrayals of privileged Westerners as inherently greedy or self-interested, the public silence with respect to climate change is a function of collective denial wrought from a sense of overwhelming powerlessness, guilt, and fear. Not only did thinking about climate change raise a number of disturbing emotions for the people with whom I spoke—from guilt to fear and helplessness—but I also described how people collectively employed a wide variety of cultural strategies to normalize the disturbing information and "keep it at a distance" where it was "no more than background noise." Socially organized climate denial is different from the outright rejection of climate science by so-called climate skeptics. Rather, this is the more pervasive and everyday denial in which people who report to be concerned about climate change manage to ignore it.

The cultural and emotional landscape concerning climate change is quite different in Karuk country. Not only is there virtually no traction for the "climate skeptic" frame, but there is also none of the handwringing or paralysis that pervades the more privileged progressive environmental community. Instead, the people I work with talk about the changes they see in the land on a regular basis and take seriously the responsibilities they feel to act on behalf of particular places, species, and people. These are sensibilities that lead people to pursue collective engagement on a wide variety of fronts from continuing to carry out the traditional use of fire, ceremonies, and language to legal challenges, public education, building ties with the non-Native community, and direct action.

What can be learned from this? If the key questions facing people around the world are "what will it take to mobilize our collective power to reduce climate emissions" and "what are the effective ways to engage," then there are a number of important take-home observations to be made from the Karuk experience regarding the nature of effective social action. There are also more personal lessons regarding what enacting an ethics of survival and living a good life might look like in the face of uncertainty.

First, setting the experiences of these communities side by side highlights how denial and apathy are at least in part functions of environmental privilege. But when compared against the Karuk case, climate apathy may also be understood as a function of a kind of disempowerment. In *Living in Denial*, it was my goal to take a "sympathetic but critical" stance to the problem of climate apathy. The people I listened to were not unkind or inherently greedy. They wanted good things for the world, for their neighbors, and for their families and their communities. These are people who represent the neoliberal subjectivity. They had been raised to believe in the possibility of democracy and modernism for providing "the good life," and to various degrees, they had experienced this outcome. Indeed, in contrast to the notion that people were not responding due to greed or self-interest, I described socially organized denial as a form of social cognitive dissonance in which people's ideas of what it meant to be a good and empowered person were in conflict with their understanding of the driving forces of climate change. Information about the changing climate is so disturbing precisely because people *do* care about one another and the world. Yet, socially organized denial is also about preserving privilege. And when the apathetic response is put alongside Indigenous activism, the implications of such paralysis become all the more clear. As Kyle Whyte (2018c) put it, "It would have been an act of imagining dystopia for our ancestors to consider the erasures we live through today, in which some Anishinaabek are finding it harder to obtain supplies of birch bark, or seeing algal blooms add to factors threatening whitefish populations, or fighting to ensure the legality in the eyes of the industrial settler state of protecting wild rice for harvest. Yet we do not give up by dwelling in a nostalgic past even though we live in our ancestors' dystopia" (3). I believe a critical takeaway from Karuk and other Indigenous communities' climate activism points to the many possible forms of engagement. And if climate change is not new, then there are examples for insight, strategy, and inspiration. Indigenous peoples in North America have endured over 500 years of invasion by multiple powerful military forces. Europeans believed Native people were destined to be eliminated, but Indigenous communities not only exist today but also continue to flourish in new ways. Indigenous people are on the forefront of climate adaption efforts, as well as climate resistance in every form imaginable across the board from innovative scientific approaches to community revitalization. If we observe the actions being taken and listen to the spokespeople from these communities, we can see that the paths being taken are deeper than shifting one's transportation choices or even changing from coal to solar fuels—but toward a reorganizing of the social, political, and economic systems that are destabilizing life on earth.

Tribal people and communities have excelled in a wide range of tactics for restoration and social change: from winning innovative legal cases to building relationships with non-Indigenous communities, developing environmental policies and public education campaigns, and engaging in a range of "direct action" protest tactics. Cantzler and Huynh highlight the sophistication of a multistrategy approach to salmon restoration efforts taken by treaty tribes in the Pacific Northwest, noting, "by simultaneously emphasizing legal transformation, diplomacy and relationship building, and the broader recognition of tribal cultural perspectives, tribal fishing rights advocates in the post-Boldt era demonstrate a nuanced understanding of the political and cultural dynamics of institutional power and the tactical methods that are most effective in contesting it" (218–219). The authors underscore the importance of Indigenous strategies for theoretical understanding "through which social inequality is produced and resisted" (220) in social movements literature. In particular, the authors highlight the Tribes' sophistication regarding the cultural dimensions of power:

> We present a model of decolonization that is successful specifically because it is sensitive to both the political and cultural foundations of inequality.... Many systems of inequality are perpetuated through legal and political mechanisms, and confronting these mechanisms through lawsuits, political lobbying, and collective action are essential strategies for marginalized groups. Often overlooked, however, are the cultural and ideological mechanisms that facilitate and reinforce hegemonic systems. We believe that the social inequality literature would be enhanced through greater interrogation of the dynamics through which law, politics, and culture interact in the production and dismantling of systems of inequality. (220–221)

These tactics are in stark contrast to the individualized response to climate change of the dominant culture that proposes recycling, riding bicycles, or taking shorter showers as the means to climate activism. Indeed, Karuk leaders have excelled in advancing their agendas at national and local levels simultaneously. Exciting recent achievements include the formation of the Western Klamath Restoration Partnership. This collaboration of federal, tribal, and nongovernmental organizations is working to restore traditional fire. The Karuk Tribe works actively with the local Mid-Klamath Watershed Council on a myriad of projects, including hosting annual fire training exchanges where people come together to learn techniques of prescribed burning, as well as with the Salmon River Restoration Council to organize and implement annual fish count and other fish and river enhancement projects.

These diverse approaches taken by my Karuk colleagues share similarities with other Indigenous restoration efforts. For example, Kyle Whyte (2018c) emphasizes how Anishinaabek/Neshnabék communities enact species restoration through the mutual restoration of human and ecological relationships: "Anishinaabek/Neshnabék throughout the Great Lakes region are at the forefront of native species conservation and ecological restoration projects that seek to learn from, adapt, and put into practice local human and nonhuman relationships and stories at the convergence of deep Anishinaabe history and the disruptiveness of industrial settler campaigns. These projects also seek to find ways to reconcile—as much as makes sense—with settler societies so that indigenous and settler conservation can share responsibilities and hold each other accountable" (4–5). Not only have Indigenous people engaged a wide range of tactics for survivance (Vizenor 2008) and social change, but they are also clear leaders in each of these areas. In 2016, the No Dakota Access Pipeline movement was a phenomenal illustration of the ability of Indigenous people to mobilize their own communities and inspire committed solidarity actions. While the pipeline itself was built, a generation of youth was deeply politicized from the experience of standing in prayerful witness to their truths. The events at Standing Rock are only the latest on a continuum of tens of thousands of powerful Indigenous actions from the resistance at Wounded Knee to the occupation of Alcatraz or the Idle No More movement. Yet as Steinman (2012, 2018) argues, this activism has not been adequately theorized within the dominant sociological movement theory frames.

Tribal achievements in the courts have been similarly profound. Even in the face of the highly problematic structure of U.S. law, Tribes have nonetheless managed to secure powerful legal victories establishing self-determination, upholding treaty rights, and reasserting tribal jurisdiction over gender relations, material culture and human remains (Wilkinson 2005; Tsosie 1996, 2007; Deer 2004, 2015). In the Klamath Basin, the Klamath, Karuk, Hoopa Valley, and Yurok Tribes have prevailed in numerous legal disputes focusing on water quality and quantity, as well as in the protection of endangered species. Tribal individuals and communities have achieved all these gains in the face of the ongoing targeting of state-sponsored and federally sponsored structures against their very existence and survival, brutal interpersonal racism, and severe economic disadvantage. Sociologists and other social movement scholars would do well to take note.

UNSETTLING SOCIOLOGY IN THE FACE OF CLIMATE CHANGE

Elsewhere, I have recently argued for the unique importance of sociological perspective for the broader public and interdisciplinary conversation on climate change (Norgaard 2016, 2018). With its attention to the interactive dimensions of social order between individuals, social norms, cultural systems, and political economy, sociology is uniquely positioned to be an important leader in this conversation. Among the pieces of sociological theory that I believe is most critical for colleagues in other disciplines to understand is Mill's ([1959] 2000) concept of the sociological imagination. This concept allows for sorely needed understanding of the deeper causes and consequences of climate change, as well as how we might meaningfully respond. I have argued that in the absence of a sociological imagination, most people in both the science community and the public at large lack the ability to see the social structure that surrounds us. Deprived of this view, the approaches that are put forward emphasize the role of individual consumption and decision making in the absence of social context. Yet when individuals are detached from their social context, we cannot account for where values or beliefs come from and thus how they might actually change. Instead, the potential role of individuals, especially individual consumption, in social change is drastically overemphasized, and there is little to no discussion of whether or how institutional, political, or economic transformation might be achieved. An individual can take shorter hot showers, but the U.S. military remains the biggest consumer of oil in the world. Imagination is power, especially in a time of crisis. If the broader public cannot imagine the reality of what is going on or imagine the level of change that is needed to change our course, then no forward movement will occur.

I have also argued here and elsewhere that sociologists ourselves need to develop another form of imagination: the ability to see the relationships between human actions and our impacts on earth's biophysical system or what I call an "ecological imagination." Climate change is a disciplinary opportunity for sociology, a field that emerged at the height of the modernist myth that humans had overcome natural "limits." It was presumed that the natural world was no longer a relevant influence on social outcomes. The central concern of our new discipline was in understanding the novel forms of social order that were emerging with modern capitalism—especially those in the rapidly growing urban areas. But as the social dimensions of climate change have become evident—thanks in large part to a still-marginalized set of environmental sociologists—the lack of attention paid to this urgent situation by mainstream sociologists is appalling. Just as those in the scientific community struggle to see social structure, it is time now for sociologists to develop an ecological imagination. Here too Indigenous people have much to teach.

For it turns out that Indigenous peoples have done much better than the general population in engaging the most urgent concerns raised by climate change. Not only have Indigenous people around the world identified changes in the earth's system early on through traditional ecological knowledge, but my Karuk colleagues and other Native leaders also have articulated clear, sophisticated, and coherent analyses of why climate change is happening—analyses that incorporate economic, political, and cultural dynamics. Most important, as much of the Western world continues to debate appropriate responses to climate change, Indigenous people offer a multitude of sophisticated, time-tested, and pragmatic solutions. Perhaps it is time we all listen, take their voices seriously, and figure out how to follow their lead.

METHODOLOGICAL APPENDIX

To say that there is an uneasy relationship between tribal communities and academics would be an understatement. If one looks at the arc of colonialism in North America, the 1700s and 1800s settlers took the lives and land from Native people through direct genocide, the 1800s and 1900s were about the usurpation of minerals and additional lands, and alongside and central to all of it has been the extraction of Native knowledge. Native knowledge extraction, appropriation, and stereotyping are an ongoing aspect of cultural genocide today, and academics are at the forefront—even "well-meaning" ones. So as a white, non-Native sociologist, it is important for me to be clear about not only my intent but also the process of research and writing. This project draws upon fifteen years of collaborative research with the Karuk Tribe to invite the discipline of sociology to more effectively theorize the importance of the so-called natural environment on the social, as well as the many ways that settler-colonialism shapes sociological theory, and to engage Indigenous experiences and perspectives on their own terms, that is, to engage Indigenous peoples as knowledge holders rather than merely as research "subjects."

My original work in Karuk country emerged in the context of the proposed relicensing of the Klamath River hydroelectric project. Beginning in 2003 Karuk cultural biologist and traditional dipnet fisherman Ron Reed and I began working together to assess impacts of the dams on the health, culture, and economy of the Karuk Tribe. I have done similar consulting work with the Yurok Tribe, and I continue to work closely with the Karuk Tribe on policy-relevant research regarding access to traditional foods, climate change, and fire policy. The data and examples that fill these pages come from the sixty-plus in-depth interviews that Ron Reed and I conducted with Karuk tribal members on at least five different projects over the past ten years. Information is also compiled from archived testimonies from Federal Energy Regulatory Commission (FERC) dam relicensing hearings as well as from direct observation, informal conversations, and participation in numerous meetings.

Although I have worked with and learned from many individuals in the course of this time, my closest research collaborator has been Ron Reed. Together, Ron and I have supervised nearly two dozen student theses, given a dozen talks at universities and colleges across the West, and coauthored journal articles and book chapters. In what methodology jargon terms an "insider," Ron used his knowledge of the situation to frame relevant research questions, identify existing data sources, and "gain access" to participants via social networks. Interviews were recorded, transcribed, and coded in relation to specific research questions. In the course of that work, research questions and framing always came from Ron and later or other key Tribal leaders, especially Bill Tripp and Leaf and Lisa Hillman. I have provided a sociological or policy framework. The impetus for the work has always been to effect positive outcomes for Karuk people. Research questions and reports were designed with specific policy interventions in mind—to improve water quality so that ceremonial uses of the river are protected, to increase awareness and understanding of understudied cultural use species, and to document the relationships between natural resource policies and human health. To remove dams. To restore fire to the land.

Over time, I became aware that much of what we were writing about and working on was not represented in academic circles and most glaringly within my own discipline of sociology. Not surprisingly, the papers I was writing up for natural resource policy audiences had contributions to theory in sociology and environmental studies. As this awareness developed, I began conversations with people about the idea of a book. The Karuk Tribe also has an internal review process, *Practicing Pikyav: A Guiding Policy for Research Collaborations with the Karuk Tribe*, that stipulates additional responsibilities of the researcher and that the work be overseen by an advisory team. I made a proposal to the Karuk Tribal Council, requesting permission to reanalyze and publish these materials in book form. It was important to all of us that the book be copyrighted to the Karuk Tribe. I am grateful to Peter Mickulas and Rutgers University Press for their willingness to accommodate this request. Ron Reed and I had already coauthored several articles and book chapters, but the book would be a more concentrated academic output. While our work to date has been clearly collaborative, as the conversation becomes more academic, the balance of our contributions has shifted. Furthermore, while he and I conducted the original research together, information in this book comes from multiple sources. Unfortunately, in the end, although I wanted Ron Reed to be a full coauthor on this book, it was not possible to do this and also have the copyright listed with the Karuk Tribe. I have therefore done my best to describe the nature of our collaboration in detail, emphasizing his role in the manuscript (see especially discussion in chapter 3). That said, real and unsatisfying internal tensions exist between tribal intellectual property and copyright, and the recognition of individual contributions and intellectual property.

It is also worth noting that while confidentiality and anonymity are the norm throughout sociological research, best research practices in Native communities generally identify speakers, especially where traditional knowledge is shared to give credit for intellectual property (Smith 1999). In this manuscript, people have been given the choice of being identified or not. Note that when quotes are drawn from public statements, actual names of individuals are given. In line with the Karuk policy on research (*Practicing Pikyav: A Guiding Policy for Research Collaborations with the Karuk Tribe*) and to "decolonize" the research process (Smith 2013; Denzin, Lincoln, and Smith 2008), copies of interviews were provided to people for confirmation of appropriate interpretation, and the full manuscript was shared with individuals and permission asked to use quotes in specific contexts.

Last, this book went through multiple stages of in-depth review. My Karuk colleagues, especially Leaf and Lisa Hillman and Dr. Frank Lake, took extensive time to review, comment upon, and clarify statements. Lisa and Leaf Hillman reviewed this manuscript word by word, making corrections and providing phenomenal direct quotes, historical material, and references. Lisa and Leaf inserted further material both as quoted "data" and directly into the text as authors. Lisa Hillman's attention to better engaging Karuk femininity in chapter 4 resulted in an entire overhaul of that chapter and significantly increased my own understanding and awareness. Academic readers, including especially Kyle Whyte, David Pellow, Frank Lake, and Erich Steinman, also gave wonderfully detailed, in-depth examples and commentary in their reviews. These conversations were another rich layer of learning for me. While I thank these individuals in the acknowledgments, I also acknowledge their specific intellectual contributions in a variety of ways in the text itself, such as adding their name and "personal communication" after insertions of direct quotes of their words into the text and/or noting their comment in the footnotes.

ACKNOWLEDGMENTS

I am forever changed by the many people whose teachings, visions, outrage, love, time, and voices are manifest in this book. May this work be in the service of your lives, of life on earth.

Not so long ago my closest friend told me I was delusional. With this project, I make a sweeping call for action across sociology and the social sciences. Perhaps no one would tackle such an undertaking without some measure of blind optimism, sheer outrage, and delusion. It may be that I will offend many people who I respect. It may be that I will be ignored. Time will tell. Regardless, if I have managed anything here, it is because of the generosity of a great many others who I am blessed to have in my life.

This book and all that it represents would not exist were it not for my long-term research collaborator and friend Ron Reed. In 2003, Ron was working as the Karuk cultural biologist when he reached out to me in the context of his work on the proposed relicensing of the Klamath River hydroelectric project. Without the hundreds of hours that Ron has invested explaining how people, culture, rivers, fish, and fire are connected—what he calls Karuk social management—not to mention schooling me on what I should and should not do, say, or write as a non-Indian person living and working on the river, none of this book or any of the other policy work I have done for the Karuk Tribe would exist. The original "data" in these pages come from what became known as the Altered Diet report. Ron was a collaborator in that work in the fullest sense—the research questions and design came from him, we did the interviews and other data collection together, and I wrote up the report. The deep friendship that has developed not only between Ron and me but also between our families over the past decade and a half are among my most cherished relationships. It was my original hope that Ron could be a coauthor on this manuscript, as he has been on several of the earlier articles. For technical reasons relating to the Karuk tribal copyright, this was not possible.

Over the years many, many Karuk people have spent time speaking to me, sharing their views, and investing in my understanding of what is happening on the river. Many of you are quoted here. Thank you for trusting me with your words. I hope you find what I have done here useful. My understanding of Karuk perspectives, traditional management practices, and how colonialism operates via land management practices has been significantly shaped by the countless hours that many individuals have invested in me in my capacity as a consultant for the Karuk Tribe. In particular, working with Karuk Department of Natural Resources Director and Founder Leaf Hillman, Deputy Director of Ecocultural Management

William Tripp, and U.S. Forest Service Research Ecologist Dr. Frank Lake have had a formative impact on my life. In each and every of our interactions, I learn from you not only about ecology or fire policy but also ways of being, teaching, and doing. Within the Karuk Department of Natural Resources, I have also been particularly privileged to learn from Lisa Hillman, Earl Crosby, Toz Soto, Aja Conrad, Vikki Preston, Sinead Talley, and Analisa Tripp. All of your dedication, vision, and accomplishments inspire me on a daily basis.

My academic colleagues on the Klamath, especially Frank Lake, Sibyl Diver, Daniel Sarna, Jennifer Sowerwine, and Tom Carlson continue to be wonderful sounding boards and sources of support as I seek to apply my academic training in the service of Indigenous survivance.

As this project made its way into a book, I began to receive other forms of assistance. I thank the Karuk Tribal Council for their support of this work, their willingness to trust a settler academic, and their time in reviewing my proposals. Peter Mickulas of Rutgers University Press did a great deal of legwork from start to finish to make this project possible, not the least of which was figuring out how to have it copyrighted by the Karuk Tribe when other presses could not. Thank you Peter! HSU Humboldt Room library staff assisted in tracking down and scanning historical materials. Jenny Stormy Staats of the Klamath Salmon Media Collaborative and Geoffrey Marcus provided invaluable technical assistance with photos. I received fabulous editorial assistance from Gina Sbrilli and Alexa Foor. Thanks to Casey Thoreson of Twin Oaks Indexing. Lisa Hillman provided much appreciated consultation on the cover art. Tanya Golosh-Boza led a phenomenal faculty writing workshop on the Oregon coast, and somehow Dan HoSang convinced the University of Oregon to fund it. My thanks to all involved—especially to Raahi Redy and Dan HoSang, who read the very first draft of chapter 1. At the University of Oregon, I sat down in many a writing group with various configurations of Jessica Vasquez-Tokos, Eileen Otis, Theresa May, Michelle Jacob, and Analisa Taylor. Thank you for keeping me going!

There are hazards of writing across multiple fields, and in my attempts to learn as much as I can about how other disciplines are engaging Indigenous voices and theories of settler-colonialism, I have leaned on the generosity many people. Article versions of this material, chapters, and in some cases the entire manuscript were read by Jules Bacon, Phil Brown, Clare Evans, Linda Fuller, Julian Go, Dan HoSang, Jocelyn Hollander, Michelle Jacob, Frank Lake, Raoul Lievanos, Lisa Park, C. J. Pascoe, David Pellow, Laura Pulido, Erich Steinman, Noel Sturgeon, Barbara Sutton, Julie Sze, Malcolm Terence, David Vasquez, Jessica Vasquez-Tokos, Jerome Viles, Nicholas Viles, Kirsten Vinyeta, Sarah Wald, and last, but far from least, Kyle Powys Whyte!

I thank also the anonymous reviewers who gave input at various stages. These kind souls have taken time from their own families and intellectual endeavors in attempt to save me from my most flagrant delusional tendencies. Despite their collective best efforts, there are surely errors and omissions. Any forthcoming hate mail should be addressed to me only. Thank you. Tusen takk. Yôotva. Wow, look at this list! I am wealthy!

In the later phases of this project, I received what must surely be the most remarkable gift any author can receive. Lisa and Leaf Hillman reviewed the entire manuscript word by word, making corrections, providing references, and inserting additional material (and surely reframing many an outraged remark made to one another into tactful suggestions before communicating them to me!). Since both of them are among the busiest and most strategic people I have ever met, these tasks were taking place at odd hours of the night, while on the road and at other inconvenient times. While I know your engagement was part of the tribal review process, the personal interest, knowledge, and thoughtfulness you provided brought my own understanding and this book to another level. I owe a particular debt to Lisa for her insistence that chapter 4 include much more on Karuk femininity even when the manuscript was months overdue and I was ready to make that part to the next project! In what universe does a person receive such a blessing? Yôotva. The two of you are a constant source of inspiration in my life in more ways than you know.

Much gratitude as well to my graduate students, especially Jules Bacon, Aja Conrad, Kirsten Vinyeta, Allison Ford, Mirranda Willette, Sara Worl, Bruno Seraphin and the students in my Race, Gender, and Environment seminars. I learn from each of you. I appreciate how you question me, introduce me to new ideas, and connect me to the flow of generations of mentors and students across time. Working with you is hands down the best part of my time spent at the university. I am buoyed and supported on a daily basis by my incredible University of Oregon colleagues in sociology and environmental studies, the Native strategies community, and our expanding cross-campus community of environmental justice scholars. Conversations, friendships, and encouragement from my colleagues within the national and international community of environmental sociologists, especially Brett Clark, Erich Steinman, Robert Brullle, Andrew Jorgensen, David Pellow, Tammy Lewis, Jill Lindsey Harrison, and Richard York, have made all the difference to me in the past year alone. Thank you to Barbara Sutton for listening, making me laugh, sharing so many ideas about this and other projects, reading drafts, and never failing to pick up the phone when I call in seemingly dire need of some last-minute council. You may not have been able to save me from my delusions, but your friendship is my lifeline. Debora Loft, Carl Jorgenson, Rhea Cramer, Jane Keating, and Jennifer Raymond listened. A lot. For years.

An earlier version of chapter 4 appears as K. M. Norgaard, R. Reed, and J. M. Bacon, "How Environmental Decline Restructures Indigenous Gender Practices: What Happens to Karuk Masculinity When There Are No Fish?," *Sociology of Race and Ethnicity* 4, no. 1 (2018): 98–113. An earlier version of chapter 5 appeared as K. M. Norgaard and R. Reed, "Emotional Impacts of Environmental Decline:

What Can Native Cosmologies Teach Sociology about Emotions and Environmental Justice?," *Theory and Society* 46, no. 6 (2017): 463–495.

 Of course, the contributions of our parents and grandparents to everything we do remain the most incomprehensible—and my awareness of this has only become all the more so as I move through my life. I thank you for all that I comprehend of what you have given me and more—in particular for supporting me in retaining a love of learning, a connection to nature, a sense of justice, and a sense that my voice could matter. Sam "Salmon" Norgaard-Stroich, my life partner of going on three decades, has also been a partner in my work on the Klamath and my personal process of coming to terms with settler-colonialism and its resistance. He has shared resources and ideas as well as love, wonder, and outrage. Sam has been a partner in activism, parenting, and keeping our household going. He has spent much time listening, providing encouragement, cooking, and getting me out of the house to "do something fun." "Is this the same book you've been working on since I was born, Mom?" With thanks to our son, Cody, for the joy and perspective you bring into my life. For sure you are one of my biggest teachers.

 It is hot in summer on the Klamath, and only some of the spaces where I spend time are air conditioned. On innumerable hundred-degree days after work, I have stepped into the Salmon and Klamath rivers and experienced physical, mental, and spiritual renewal. I have known much joy and wonder in thousands of hours spent snorkeling underwater watching salmon, playing in your currents. My thanks to these rivers, this earth, and all the beings who are part of you. May you flourish. From such generosity comes much responsibility. It is with great humility and love that I offer this book to the world.

NOTES

INTRODUCTION

1. Karuk Department of Natural Resources Founder and Director Leaf Hillman provides this note on the term *management*: "our work in the forest is land management. However, this is not a style of management which violates and attempts to overwhelm the nature of Nature, but an approach which is drawn from human need and responsibility based on close observation of the processes of nature" (Hillman and Salter 1997, 1).
2. Most Karuk people I have spent time with use the terms *Karuk*, *Indian*, or *Native* to refer to themselves and *Tribe* to refer to the political entity. Sociologists tend to use the term *American Indian*. I will most frequently use the terms most common "on the river" but will otherwise use the terms *Indigenous peoples*, *Native Americans*, *tribal peoples*, and *American Indians* interchangeably. See Moreton-Robinson (2014) for further discussion of the politics of this etymology.
3. While the absence of Indigenous voices and theoretical perspectives is particularly acute in U.S. sociology, the erasure is ongoing globally (see Bhambra 2007 and 2014 for valuable discussions of the disciplinary tendency in this regard globally).
4. I use this term very specifically to denote the promotion of Indigenous self-determination. In their landmark essay, Eve Tuck and K. Wayne Yang (2012) write that "decolonization brings about the repatriation of Indigenous land and life; it is not a metaphor for other things we want to do to improve our societies and schools." My use of the term here is with the intention of turning the discipline of sociology from an instrument of the settler state to a vehicle of Indigenous revitalization.
5. See Methodological Appendix for discussion of the longstanding collaboration from which this book emerges.
6. See online Karuk Dictionary: http://linguistics.berkeley.edu/~karuk/.
7. Other actions such as academic research are also part of this assault. Food Security Project Coordinator and Pikyav Field Institute Manager Lisa Hillman notes the process of "taking culture away, such as through academic research (and the resulting transfer of knowledge ownership) and material cultural collection (much of which was just flat-out stolen from under our feet, and all of which we are trying to repatriate under terms the U.S. has set)."
8. Central to settler-colonialism has been erasure of Indigenous genocide in North America. Academics and Native peoples who use this word have been challenged as to their terminology. Longstanding work by Jack Norton (1979), James Fenelon (2014, 2015), and many others is now being recognized. Benjamin Madley's (2016) work, *An American Genocide*, gives specific justification for this term and detailed events in California.
9. Much of what sociology takes as common sense with respect to Native peoples ranges on a scale from patently absurd to deeply offensive.
10. See Whyte (2015 and 2016d), Grey and Patel (2015), Bradley and Herrera (2016), and Alkon and Mares (2012) for further elaboration.
11. See Bacon (2018) for a detailed elaboration of this concept.

CHAPTER 1 MUTUAL CONSTRUCTIONS OF RACE AND NATURE ON THE KLAMATH

1. 1868, cited in Anderson, *Tending the Wild* (2005, 15).
2. Frank Lake notes that as the currency system and ideas of wealth changed from Indigenous notions of wealth in the form of food, knowledge, and wisdom to gold and other commodities, so did the mental and philosophical connection from good land stewardship and shifts to the exploitive and extractive development and degraded management under colonization.
3. This is also true for theories of empire that neglect to engage settler-colonialism.
4. See, for example, work on race, climate change, and the Anthropocene, including Sze et al. (2009), Sze 2018, London et al. (2008), Haraway (2015), Pulido (2018), Davis and Todd (2017), and Mitman, Armiero, and Emmett (2018).
5. Chinese were also present but will not be discussed in detail; few blacks were living in the region.
6. See, for example, Pulido (2000), Pulido et al. (2016), Shilling et al. (2009), London et al. (2008), and Pellow (2018).
7. See, for example, Claborn (2014). See Du Bois's (1935) work in *Black Reconstruction*.
8. In *The Scholar Denied*, Morris (2015) argues that much of sociology and social theory broadly would look very different had Du Bois not faced stiff racist marginalization.
9. While the focus of this chapter is to systematically detail how the environment matters for racial formation theory, a settler-colonial framework usefully reorients a number of the debates explored in the chapter. I will periodically reference the usefulness of Marxist and settler-colonial framework in particular places.
10. See Wald (2016) for further rich discussion of the importance of the California environment for Japanese American racial formations.
11. Again, see important scholarship by Dorceta Taylor (2014, 2016), David Pellow (2018), Laura Pulido (2016, 2017c), Sarah Ray (2013), Carolyn Finney (2014), Carl Zimring (2017), and others elaborating this theme.
12. Again, while I emphasize race here in order to underscore the importance of the natural environment to the theory of racial formation, these racial formation processes are in turn taking place within the context of settler-colonialism and global capitalist expansion, hence my periodic use of the term *racial-colonial formation*.
13. Note that the Spanish were *not* players this far north in what would become California.
14. Quote taken from the title of a presentation uploaded to the California Forest Pest Council website by Rick A. Sweitzer, Department of Environmental Science, Policy, and Management, Center for Forestry, University of California, Berkeley.
15. The archaeologist and ethnographer A. L. Kroeber noted in 1916, "In general, Spanish occupation has been more favorable than American settlement to preservation of native designations of localities." And while the etymology of several California place names can be traced to their Indigenous inhabitants, the "Karok . . . have furnished no terms to modern California geography" (Kroeber 1916).
16. From the 1964 Wilderness Act: Definition of Wilderness "A wilderness, in contrast with those areas where man and his own works dominate the landscape, is hereby recognized as an area where the earth and its community of life are untrammeled by man, where man himself is a visitor who does not remain. An area of wilderness is further defined to mean in this Act an area of undeveloped Federal land retaining its primeval character and influence, without permanent improvements or human habitation, which is protected and managed so as to preserve its natural conditions and which (1) generally appears to have been affected primarily by

the forces of nature, with the imprint of man's work substantially unnoticeable; (2) has outstanding opportunities for solitude or a primitive and unconfined type of recreation; (3) has at least five thousand acres of land or is of sufficient size as to make practicable its preservation and use in an unimpaired condition; and (4) may also contain ecological, geological, or other features of scientific, educational, scenic, or historical value."

17. As Glenn (2015) underscores, colonialism is not just about Indigenous people but fundamentally shapes the racial formation processes of all groups in North America.

18. These dynamics lead to complicated circumstances for Native people with respect to the navigation of racism and identity (see Robertson 2013a, 2013b).

19. Note that this is a very high percentage of the total California population participating—by one estimate in 1851, the entire non-Native population of California was 314,000 (http://cprr.org/Museum/California_1851.html).

20. Frank Lake notes there is a report/document that states that the reason for the creation of the Klamath Reservation at Ft. Tewer (near Klamath Glenn) circa 1850s was because it was far enough down river to be away from the gold-rich placer flats (village/terraces) found farther upriver near Weitchpec.

21. Karuk people from the northern territory were supposed to go to Fort Jones in Scott Valley. That was what the "Treaty R" called for, although this, too, was never congressionally ratified.

22. See Spence (1999) for a discussion of this in relation to national parks.

23. Quinn, personal communication.

24. Section 31 of the act notes "that the Secretary of the Interior is hereby authorized, in his discretion, to make allotments within the national forests in conformity with the general allotment laws as amended by section of this act, to any Indian occupying, living on, or having improvements on land included within any such national forest who is not entitled to an allotment on any existing Indian reservation, or for whose tribe no reservation has been provided, or whose reservation was not sufficient to afford an allotment to each member thereof."

25. Hoxie (1984), Fixico (1986), and Child (1998).

26. From Happy Camp, Karuk children were sent via the mail carrier to the nearest Indian boarding school in Hoopa Valley—a distance of nearly 100 miles. There were no roads at this time, and so this journey was completed through a combination of horseback and foot travel. One can assume that these children did not return home often—or at all—once they were taken away from their family. Once children "aged out" of the school in Hoopa, they were sent away to other—more distant—Indian boarding schools.

27. Letter to Fred Baker, superintendent of the Sacramento Indian Agency, dated December 12, 1931, from Rose Temple Sutherland. This letter accompanied an appeal to the denial she received regarding her application for enrollment.

28. Numbers reflect the costs set in 2016 by the California Fish and Wildlife Commission for California residents.

29. Cutting mushrooms in half was a technique developed to indicate noncommercial subsistence harvesting.

CHAPTER 2   ECOLOGICAL DYNAMICS OF SETTLER-COLONIALISM

1. Marxist scholars, including John Foster (1999) and Foster, Clark, and York (2011), have produced a cannon of crucial work in theorizing the role of the environment for the simultaneous production of wealth and inequality. I hope this discussion will extend their efforts to center the environment on additional important themes in our discipline, from the construction of

race and gender to the operation of emotions and social power. Sociology needs to better theorize the centrality of settler-colonialism. Critiques of capitalism are crucial for settler-colonialism, although Indigenous and Marxist critiques do vary. I will weave discussion on these topics from Coulthard (2014), Dunbar-Ortiz (2014), and Grande (2004) throughout this chapter. See Fenelon (2016), Tuck and Yang (2012), Klopotek (2011), Bruyneel (2007), Goldstein (2014), Byrd (2011), and others for further useful discussion of the complex relationships between the operation of genocide, racism, capitalism, and colonialism.

2. Note that fire suppression was mandated as well by the first Spanish governors—an event that was especially important in the southern part of the state. See Timbrook et al. (1993, 129) and Lake (2007).

3. The term *fire exclusion* is essentially synonymous with *fire suppression*.

4. See Merchant (1980), Cronon (1983), Anderson (2005), Ortiz (2014), and Turner (2008) for discussion and a range of illustration of these phenomena.

5. There is a range in the degree to which this is true. U.S. sociologists have been particularly slow in engaging settler-colonial theory, whereas Canadian and Australia sociologists are on the forefront.

6. For further discussion of limitations of postcolonial theory, see Veracini (2011), Magubane (2013), Ray (2013), and Tuck and Yang (2012).

7. See Gowlett (2016) and Pyne (2016).

8. See Hoover (2018) for a discussion of these dynamics in Akwesasne context and the notion of "environmental reproductive justice."

9. See Grinde and Johansen (1995).

10. Letter dated December 5, 1922, from Edger K. Miller, superintendent of the Greenville Indian Industrial School in California to the commissioner of Indian affairs in Washington, D.C.

11. Lake, personal communication.

12. A 1916 letter to the California Fish and Game Commission by Klamath River Jack, published in 1916 in a Requa, California, newspaper as "An Indian's View of Burning and a Reply." Response to Klamath River Jack's letter of May 27, 1916, addressed to the Fish & Game Commission in San Francisco, California. Both this letter and the forest ranger's response were published because, as the editor noted, "they entertainingly express, in informal English, two opposing views of a mooted question." Page 194 of California Fish and Game: Conservation of Wildlife through Education, San Francisco, CA January 1916 Volume 2, Number 1.

13. See Wilson (2001), Cornford (1983), and Raphael and Freeman House (2007).

14. Note Karuk territory lies in both Humboldt and Siskiyou counties.

15. For example, seeing animals as diminutive and cute.

16. Modern attempts to "control" and "protect" the natural environment are often undertaken by agencies with no regard or a blatant disregard for the Indigenous management of these land spaces. In the past ten years, the state of California has set up regulations for gathering on coastal lands through the Coastal Commission and the Marine Life Protection Act (MLPA) (Middleton 2011). Many of these policies were written without tribal consultation and seem designed to degrade biocultural sovereignty, not only for California Indian tribes but also without acknowledgment or understanding of the sovereignty and rights of the land itself. Baldy (2013).

17. These included ceanothus, poison oak, tanoak, madrone, and fir that had moved up from the slope below.

18. As Dr. Frank Lake explained to me in 2017, backburning is a technique to contain fires or protect structures by lighting a new fire that spreads and connects to the existing fire. To get a fire line that is adequate to prevent the approaching fire from burning over/through the back-burned area along the fire line, you need "good black," meaning significant torching and burning off of so-called surface and ladder fuels along a control feature.

19. See Wofford et al. (2003), Matthewson (2007), and Bill (2006).
20. See Champagne (2005, 2008) and Goldtooth (1995).
21. This is in part a function of usufruct rights, the ability of tribal members to steward and claim rights based on the active management of resources. See Silvern (1999): "The court declared that the Indians understood that settlement would affect their treaty rights and, therefore, they "did not possess absolute usufructuary rights" but rights that "were subject to extinction if the lands were needed for settlement" (658).
22. Other principles, statutes, and actions that underscore Karuk management authority include the Tribal Trust doctrine (Warner 2015; Tsosie 2003; Wood 2000), Executive Order 13175 regarding consultation with tribes, the 2009 Presidential Memorandum on Tribal Consultation, and Secretarial Order 3206 (American Indian Tribal Rights, Federal-Tribal Trust Responsibilities, and the Endangered Species Act). Karuk tribal management authority is fundamental to self-determination and is thus further supported by the Indian Self-Determination and Educational Assistance Act and multiple articles in the 2007 United Nations Declaration of the Rights of Indigenous Peoples (UNDRIP), which the United States has endorsed as aspirational.
23. A new study is just out regarding the unknown human health effects of pesticides and other chemicals on the health of firefighting personnel (Carratt et al. 2017).

CHAPTER 3   RESEARCH AS RESISTANCE

1. As of this writing, the process is moving forward with an expected removal date of January 2021.
2. Internationally, the food sovereignty movement began from Indigenous movement Via Campesina, but in the United States, this radical vision has not been primary. I will discuss environmental health and environmental justice together here, but these are in fact distinct fields. Both are interdisciplinary fields that emphasize environmental dimensions of human health; both are fields in which sociologists have played leading roles. Environmental justice has a central emphasis on inequality, as well as a history of strong community-academic partnerships, and has recently taken off as an area within the environmental humanities. The field of environmental health tends to have more contributions from the natural sciences. It is mainly from life sciences, not natural sciences.
3. Luis Neuner, Karuk Tribe Peekaavíchvaan (Tribal Youth Summer Environmental Work Program) (2016, personal communication).
4. For important work in this area, see, for example, Mary Arquette, Margaret Peters, Katsi Cook, LM LeBruyn, Nina Wallerstein, James Ransom, Elizabeth Hoover, and Vanessa Watts Simmons, among others.
5. Indigenous perspectives are underrepresented in academia and very necessary to advance accurate understanding of Indigenous experiences and the processes of colonialism. At the same time, it is critical for non-Native scholars to educate themselves about the very significant ways that academic research has been complicit in colonialism. Anyone considering doing work in this area should also become fluent with the issues and writing on appropriate protocols by Indigenous peoples. Denzin, Lincoln, and Smith (2008), Smith (2013), and Wilson (2004, 2008) are great starting places. Online, see Guidelines for Considering Traditional Knowledge in Climate Change Initiatives (https://climatetkw.wordpress.com) and NEIHS Resources on TEK (https://www.niehs.nih.gov/research/supported/translational/peph/webinars/tribal/index.cfm).
6. Phoebe Maddux 1932 in Harrington (1932).

7. Karuk origin story, where the Orleans Maiden refuses to listen to Elders and harvest "twin" corms rather than leave them in the ground to grow.
8. The 2016 Klamath Basin Food Assessment reported more than 151 diabetic tribal members.
9. See, for example, Roht-Arriaza (1995), Bannister, Solomon, and Brunk (2009), Harding et al. (2012), and Hardison and Bannister (2011).
10. The Karuk word *Ikxaréeyav*, or First/Spirit People, refers to divine entities that are plural and rarely gender specified (but appear to be male or female in certain circumstances).
11. Pulido (2017b) raises the issue of colonialism, noting, "A focus on racial capitalism requires greater attention to the essential processes that shaped the modern world, such as colonization, primitive accumulation, slavery, and imperialism." And "By insisting that we are still living with the legacy of these processes, racial capitalism requires that we place contemporary forms of racial inequality in a materialist, ideological and historical framework" (526–527). David Pellow's wonderful new book, *Critical Environmental Justice*, discusses settler-colonialism in the context of Palestine.

CHAPTER 4 ENVIRONMENTAL DECLINE AND CHANGING GENDER PRACTICES

1. Not all Karuk men fish. Particular families hold "rights" to fish at certain places. This chapter centers the experience of men from fishing families. Men who are not from fishing families have other responsibilities (e.g., to hold specific ceremonies and dances, to hunt, and more).
2. See Fausto-Sterling (2000, 2005) and Joan Roughgarden (2013) for discussion of this point from a biological perspective.
3. Copyright 2016; this excerpt is taken from one of the lessons called "Oral Traditions." Karuk Tribe (2016b), Karuk Oral Traditions, Grade 4, Lesson 4.
4. See also Trask (1991).
5. Nazaryan (2018).
6. Karuk word for the World Renewal priest's female assistant.
7. Karuk word for the World Renewal priest.
8. Karuk word for deer.
9. Quoted from the K–12 Environmental Education Coordinator for the Karuk Department of Natural Resources' Píkyav Field Institute. She (the speaker) is referring to the "Native Food and Culture" Grade 7, Lesson 3, published by the Karuk Tribe (2016b). The speaker teaches K–12 level classes at schools located within the Karuk Aboriginal Territory, where the student population is reported to range from 55% to 100% Native.

CHAPTER 5 EMOTIONS OF ENVIRONMENTAL DECLINE

1. For example, the issue of mental health has yet to be included in the many anthologies, college courses, or journal review articles devoted to the field of environmental justice.
2. See Willox et al. (2013) and Gill and Picou (1998) for important exceptions.
3. With appreciation for Stewart Lockie's (2016) call for more work in this area. See Jacobson (2016) for work on emotions and gender in contested illnesses, as well as Davidson (2018a, 2018b) for recent work that links this terrain through the application of theories of trauma in relation to natural gas fracking and the importance of emotions in the process of reflexivity. While sociology is only now beginning to incorporate theorizing on emotions and the natural world, other social sciences have made important advances. Within geography, there is a new focus on "emotional geographies of space." Similarly, the discipline of psychology has

the subfield of "ecopsychology" and has explored the emotional dimensions of environmental problems of climate change in significant detail. Others describe the notion of biophyllia or "the innately emotional affiliation of human beings to other living organisms" and outline a number of corresponding emotional consequences of environmental degradation, including grief, anger, and hopelessness. While powerfully important, such work is primarily descriptive and does not interrogate the meaning of these emotions in structuring power or normative experiences.

4. The final line is Auntie's addition to the original statement.

5. There are notable exceptions to this trend: at the forefront of early literature are Kai Erikson's (1976) texts *Everything in Its Path: Destruction of Community in the Buffalo Creek Flood* and *A New Species of Trouble*, as well as Michael Edelstein's (1988) landmark study *Contaminated Communities: The Social and Psychological Effects of Residential Toxic Exposure*, in which he describes how residential toxic exposure negatively affects multiple aspects of individual psychological experience, as well as family, social, and community relations (1988, 2004). More recently, Auyero and Swistun's (2009, 2007) work on "environmental suffering" builds on the concept of social suffering from Bourdieu (1999) to articulate long-term environmental decline as an interactive component of state and corporate power. Similarly, Shriver and Webb (2009) use a "ecological-symbolic" perspective (Kroll-Smith and Couch 1993) to describe the experiences of Ponca tribal members with air contamination. Alongside material impacts of contamination and degradation are "symbolic" ones, and here too through emotional experiences, "harms" are manifested (Auyero and Switsun 2009).

CONCLUSION

1. See also Todd (2015) and Davis and Todd (2017), among others.
2. See Joane Nagel's (1997) important work for a clear exception.

WORKS CITED

Adams, J., and G. Steinmetz. 2015. "Sovereignty and Sociology: From State Theory to Theories of Empire." In *Patrimonial Capitalism and Empire*, 269–285. Emerald Group Publishing Limited.

Adas, Michael. 1989. *Machines as the Measure of Men: Science, Technology, and Ideologies of Western Dominance*. Ithaca, NY: Cornell University Press.

Agyeman, J., D. Schlosberg, L. Craven, and C. Matthews. 2016. "Trends and Directions in Environmental Justice: From Inequity to Everyday Life, Community, and Just Sustainabilities." *Annual Review of Environment and Resources* 41:321–340.

Akaba, Azibuike. 2004. "Science as a Double-Edged Sword: Research Has Often Rewarded Polluters, but EJ Activists Are Taking It Back." *Race, Poverty & the Environment* 11 (2): 9–11.

Alaimo, Stacy. 2010. *Bodily Natures: Science, Environment, and the Material Self*. Bloomington: Indiana University Press.

Alfred, T. 1999. *Peace, Power and Righteousness*. Ontario: Oxford University Press.

———. 2005. *Wasase: Indigenous Pathways of Action and Freedom*. Toronto: University of Toronto Press.

Alkon, Alison Hope, and Christie Grace McCullen. 2011. "Whiteness and Farmers Markets: Performances, Perpetuations ... Contestations?" *Antipode* 43 (4): 937–959.

Alkon, A. H., and T. M. Mares. 2012. "Food Sovereignty in US Food Movements: Radical Visions and Neoliberal Constraints." *Agriculture and Human Values* 29:347–359.

Almaguer, Tomas. 2008. *Racial Fault Lines: The Historical Origins of White Supremacy in California*. Berkeley: University of California Press.

Alves, Rômulo R. N., and Ierecê M. L. Rosa. 2007. "Biodiversity, Traditional Medicine and Public Health: Where Do They Meet?" *Journal of Ethnobiology and Ethnomedicine* 3 (1): 14.

Anderson, Kat. 2005. *Tending the Wild: Native American Knowledge and the Management of California's Natural Resources*. Berkeley: University of California Press.

Anderson, K., and J. Ball. 2011. "Foundations: First Nation and Metis Families." In *Visions of the Heart: Canadian Aboriginal Issues*, edited by D. Long and O. Dickason, 55–89. Ontario: Oxford University Press.

Anderson, M. K., and F. Lake. 2013. "California Indian Ethnomycology and Associated Forest Management." *Journal of Ethnobiology* 33 1 (1): 33–85.

———. 2016. "Beauty, Bounty, and Biodiversity: The Story of California Indian's Relationship with Edible Native Geophytes." *Fremontia* 44 (3): 44–51.

Anson, April. 2018. "'The President Just Stole Your Land': Naturalizing Settler Environmentalism." Presented at the Race, Environmental Justice and Public Lands Symposium. May 10, 2018, University of Oregon.

Anthony, R. Michael, Jambs Evans, and Gerald D. Lindsey. 1986. "Strychine-Salt Blocks for Controlling Porcupines in Pine Forests: Efficacy and Hazards." In *Proceedings of the Twelfth Vertebrate Pest Conference*, 4. http://digitalcommons.unl.edu/vpc12/4.

Arendt, Hannah. 1958. *The Human Condition*. Chicago: University of Chicago Press.

Armitage, Derek, Fikret Berkes, and Nancy Doubleday, eds. 2010. *Adaptive Co-Management: Collaboration, Learning, and Multi-Level Governance*. Vancouver: University of British Columbia Press.

Arvin, Maile, Eve Tuck, and Angie Morrill. 2013. "Decolonizing Feminism: Challenging Connections between Settler Colonialism and Heteropatriarchy." *Feminist Formations* 25 (1): 8–34.
Auyero, J., and D. Swistun. 2007. "Confused Because Exposed: Towards an Ethnography of Environmental Suffering." *Ethnography* 8 (2): 123–144.
Auyero, Javier, and Débora Alejandra Swistun. 2009. *Flammable: Environmental Suffering in an Argentine Shantytown.* Oxford: Oxford University Press.
Bacon, J. M. 2017. "'A Lot of Catching Up,' Knowledge Gaps and Emotions in the Development of a Tactical Collective Identity among Students Participating in Solidarity with the Winnemem Wintu." *Settler Colonial Studies* 7 (4): 441–455.
———. 2018. "Settler Colonialism as Eco-Social Structure and the Production of Colonial Ecological Violence." *Environmental Sociology* 5 (1): 1–11.
Bacon, J. M., and Matthew Norton. 2019. "Colonial America Today: U.S. Empire and the Political Status of Native American Nations." *Comparative Studies in Society and History* 61 (2).
Baldy, C. R. 2013. "Why We Gather: Traditional Gathering in Native Northwest California and the Future of Bio-cultural Sovereignty." *Ecological Processes* 2:17.
———. 2017. "'mini-k'iwh'e: n (For That Purpose—I Consider Things) (Re) writing and (Re) righting Indigenous Menstrual Practices to Intervene on Contemporary Menstrual Discourse and the Politics of Taboo." *Cultural Studies↔ Critical Methodologies* 17 (1): 21–29.
———. 2018. *We Are Dancing for You: Native Feminisms and the Revitalization of Women's Coming-of-Age Ceremonies.* Seattle: University of Washington Press.
Bang, M., L. Curley, A. Kessel, A. Marin, E. S. Suzukovich III, and G. Strack. 2014. "Muskrat Theories, Tobacco in the Streets, and Living Chicago as Indigenous Land." *Environmental Education Research* 20 (1): 37–55.
Bang, M., D. Medin, K. Washinawatok, and S. Chapman. 2010. "Innovations in Culturally Based Science Education through Partnerships and Community." In *New Science of Learning.* 569–592. New York: Springer.
Bannister, Kelly, Maui Solomon, and Conrad G. Brunk. 2009. "Appropriation of Traditional Knowledge: Ethics in the Context of Ethnobiology." In *The Ethics of Cultural Appropriation,* 140–172.
Barker, Joanne, ed. 2017. *Critically Sovereign: Indigenous Gender, Sexuality, and Feminist Studies.* Durham, NC: Duke University Press.
Beauvais, F. 2000. "Indian Adolescence: Opportunity and Challenge." In *Adolescent Diversity in Ethnic, Economic, and Cultural Contexts,* edited by R. Montmeyer, G. Adams, and T. Gullotta, 110–140. Thousand Oaks, CA: Sage.
Bhambra, Gurminder K. 2007. *Rethinking Modernity: Postcolonialism and the Sociological Imagination.* New York: Springer, 2007.
———. 2014. *Connected Sociologies.* London: Bloomsbury Publishing.
Bill, Amber. 2006. "California Indian Women Basketweavers as Grassroots Political Activists." *McNair Scholars Journal* 7:18–32.
Biswell, Harold. 1999. *Prescribed Burning in California Wildlands Vegetation Management.* Berkeley: University of California Press.
Blackburn, Thomas C., and Kat Anderson. 1993. *Before the Wilderness: Environmental Management by Native Californians.* No. 40. Menlo Park, CA: Ballena Press.
Blauner, R. 1969. "Internal Colonialism and Ghetto Revolt." *Social Problems* 16 (4): 393–408.
———. 1972. *Racial Oppression in America.* New York: HarperCollins College Division.
Bobrow-Strain, Aaron. 2012. *White Bread: A Social History of the Store-Bought Loaf.* Boston: Beacon Press.

Bonilla-Silva, E. 2012. "The Invisible Weight of Whiteness: The Racial Grammar of Everyday Life in Contemporary America." *Ethnic and Racial Studies* 35 (2): 173–194.

———. 2018. "'Racists,' 'Class Anxieties,' Hegemonic Racism, and Democracy in Trump's America." *Social Currents* 6 (1): 14–31.

Bourdieu, Pierre. 1999. *The Weight of the World: Social Suffering in Contemporary Society*. Stanford, CA: Stanford University Press.

Bowcutt, F. 2011. "Tanoak Target: The Rise and Fall of Herbicide Use on a Common Native Tree." *Environmental History* 16 (2): 197–225.

———. *The Tanoak Tree: An Environmental History of a Pacific Coast Hard*. Seattle: University of Washington Press, 2015.

Boyce, Vicky L., and Boyd A. Swinburn. 1993. "The Traditional Pima Indian Diet: Composition and Adaptation for Use in a Dietary Intervention Study." *Diabetes Care* 16 (1): 369–371.

Bradley, K., and H. Herrera. 2016. "Decolonizing Food Justice: Naming, Resisting, and Researching Colonizing Forces in the Movement." *Antipode* 48 (1): 97–114.

Brave Heart, M. Y., and Lemyra DeBruyn. 1998. "The American Indian Holocaust: Healing Historical Unresolved Grief." *American Indian and Alaska Native Mental Health Research* 8 (2): 56–78.

Brooks, Matthew L., Carla M. D'antonio, David M. Richardson, James B. Grace, Jon E. Keeley, Joseph M. DiTomaso, Richard J. Hobbs, Mike Pellant, and David Pyke. 2004. "Effects of Invasive Alien Plants on Fire Regimes." *BioScience* 54 (7): 677–688.

Brown, Kirby. 2018. *Stoking the Fire: Nationhood in Cherokee Writing, 1907–1970*. Norman: University of Oklahoma Press.

Brown, Phil. 2013. "Integrating Medical and Environmental Sociology with Environmental Health: Crossing Boundaries and Building Connections through Advocacy." *Journal of Health and Social Behavior* 54 (2): 145–164.

Brown, Phil, and Edwin J. Mikkelsen. 1997. *No Safe Place: Toxic Waste, Leukemia, and Community Action*. Berkeley: University of California Press.

Brown, Tony. 2003. "Critical Race Theory Speaks to the Sociology of Mental Health: Mental Health Problems Produced by Racial Stratification." *Journal of Health and Social Behavior* 44 (3): 292–301.

Brulle, R. J. 2015. "Sociological Theory after the Death of Nature." In *Emerging Trends in the Social and Behavioral Sciences*, edited by Robert Scott and Stephen Kosslyn. Hoboken, NJ: John Wiley.

Brulle, R. J., and D. N. Pellow. 2006. "Environmental Justice: Human Health and Environmental Inequalities." *Annual Review of Public Health* 27:103–124.

Bruyneel, Kevin. 2007. *The Third Space of Sovereignty: The Postcolonial Politics of US Indigenous Relations*. Minneapolis: University of Minnesota Press.

Bryant, Bunyan I., and Paul Mohai, eds. 1992. *Race and the Incidence of Environmental Hazards: A Time for Discourse*. Boulder, CO: Westview Press.

Bryson, Lois, Kathleen McPhillips, and Kathyrn Robinson. 2001. "Turning Public Issues into Private Troubles: Lead Contamination, Domestic Labor and the of Women's Unpaid Labor in Australia." *Gender & Society* 15 (5): 754–772.

Bullard, Robert D. 2008. *Dumping in Dixie: Race, Class, and Environmental Quality*. Boulder, CO: Westview Press.

Busam, Heather M. 2006. "Characteristics and Implications of Traditional Native American Fire Management on the Orleans Ranger District, Six Rivers National Forest." Master's thesis, California State University, Sacramento.

Byrd, Jodi A. 2011. *The Transit of Empire: Indigenous Critiques of Colonialism*. Minneapolis: University of Minnesota Press.

Cantrell, B. G. 2001. "Access and Barriers to Food Items and Food Preparation among Plains Indians." *Wicazo Sa Review* 16 (1): 65–74.

Cantzler, Julia Miller, and Megan Huynh. 2016. "Native American Environmental Justice as Decolonization." *American Behavioral Scientist* 60 (2): 203–223.

Carbado, Devon W., and Cheryl I. Harris. 2012. "The New Racial Preferences." In *Racial Formation in the Twenty-First Century*, edited by Daniel HoSang, Oneka LaBennett, and Laura Pulido, 183. Berkeley: University of California Press.

Carmichael, Stokely, and Charles Hamilton. 1967. *Black Power: The Politics of Liberation in America*. New York: Vintage Books.

Carratt, Sarah, H. Flayer, M. E. Kossack, and J. A. Last. 2017. "Pesticides, Wildfire Suppression Chemicals and California Wildfires: A Human Health Perspective." *Current Topics in Toxicology* 13:1–12.

Casas, T. 2014. "Transcending the Coloniality of Development: Moving beyond Human/Nature Hierarchies." *American Behavioral Scientist* 58 (1): 30–52.

Champagne, Duane. 2005. "From Sovereignty to Minority: As American as Apple Pie." *Wicazo Sa Review* 20: 21–36.

———. 2006. *Social Change and Cultural Continuity among Native Nations*. Walnut Creek, CA: Rowman Altamira.

———. 2008. "From First Nations to Self-Government: A Political Legacy of Indigenous Nations in the United States." *American Behavioral Scientist* 51 (12): 1672–1693.

Chan, H. M., K. Fediuk, S. Hamilton, L. Rostas, A. Caughey, H. Kuhnlein, G. Egeland, and E. Loring, E. 2006. "Food Security in Nunavut, Canada: Barriers and Recommendations." *International Journal of Circumpolar Health* 65 (5): 416–431.

Chang, D. A. 2010. *The Color of the Land: Race, Nation, and the Politics of Landownership in Oklahoma, 1832–1929*. Chapel Hill: University of North Carolina Press.

Charnley, Susan, and M. R. Poe. 2007. "Community Forestry in Theory and Practice: Where Are We Now? *Annual Review of Anthropology* 36:301–303.

Child, Brenda J. 1998. *Boarding School Seasons: American Indian families, 1900–1940*. Lincoln: University of Nebraska Press.

Claborn, John. 2014. "W.E.B. Du Bois at the Grand Canyon." In *The Oxford Handbook of Ecocriticism*, edited by G. Garand, 118–131. New York: Oxford University Press.

Cochran, P. A., C. A. Marshall, C. Garcia-Downing, E. Kendall, D. Cook, L. McCubbin, and R.M.S. Gover. 2008. "Indigenous Ways of Knowing: Implications for Participatory Research and Community." *American Journal of Public Health* 98 (1): 22–27.

Cocking, Matthew I., J. Morgan Varner, and Rosemary L. Sherriff. 2012. "California Black Oak Responses to Fire Severity and Native Conifer Encroachment in the Klamath Mountains." *Forest Ecology and Management* 270:25–34.

Collard, Rosemary-Claire, Jessica Dempsey, and Juanita Sundberg. 2015. "A Manifesto for Abundant Futures." *Annals of the Association of American Geographers* 105 (2): 322–330.

Collins, Patricia Hill. 1994. "Shifting the Center: Race, Class, and Feminist Theorizing about Motherhood." In *Mothering: Ideology, Experience, and Agency*, edited by Evelyn Nakano Glenn, Grace Chang, and Linda Rennie Forcey, 45–65. Abington on Thames, UK: Routledge.

Collins, Patricia Hill, and Sirma Bilge. 2016. *Intersectionality*. New York: John Wiley.

Collins, Randal. 2004. *Interaction Ritual Chains*. Princeton, NJ: Princeton University Press.

Colomeda, L. A., and E. R. Wenzel. "Medicine Keepers: Issues in Indigenous Health." *Critical Public Health* 10 (2): 243–256.

Colorado, P., and D. Collins. 1987. "Western Scientific Colonialism and the Reemergence of Native Science." *Practice: Journal of Politics, Economics, Psychology, Sociology and Culture* 1:50–65.

Connell, R. W. 1990. "A Whole New World: Remaking Masculinity in the Context of the Environmental Movement." *Gender and Society* 4 (4): 452–478.
Conners, Pamela. 1998. *A History of the Six Rivers National Forest: Commemorating the First Fifty Years.* Manuscript on file at the Heritage Resources Program, Six Rivers National Forest, Eureka.
Coombes, B., J. T. Johnson, and R. Howitt. 2012. "Indigenous Geographies I Mere Resource Conflicts? The Complexities in Indigenous Land and Environmental Claims." *Progress in Human Geography* 36 (6): 810–821.
Cordner, A., D. Ciplet, P. Brown, and R. Morello-Frosch. 2012. "Reflexive Research Ethics for Environmental Health and Justice: Academics and Movement Building." *Social Movement Studies* 11 (2): 161–176.
Cornford, Daniel Allardyce. 1983. "Lumber, Labor, and Community in the Progressive Era in Humboldt County, California, 1900–1920." Diss., University of California, Santa Barbara.
Coulthard, G. S. 2014. *Red Skin, White Masks: Rejecting the Colonial Politics of Recognition.* Minneapolis: University of Minnesota Press.
Crawford, J. N., S. A. Mensing, F. K. Lake, and S. R. Zimmerman. 2015. "Late Holocene Fire and Vegetation Reconstruction from the Western Klamath Mountains, California, USA: A Multi-Disciplinary Approach for Examining Potential Human Land-Use Impacts." *The Holocene* 25 (8): 1351–1357.
Cronon, William. 1983. *Changes in the Land: Indians, Colonists, and the Ecology of New England.* New York: Hill and Wang.
Crossley, Nick. 2001. *The Social Body: Habit, Identity and Desire.* Thousand Oaks, CA: Sage.
Crowder, Kyle, and Liam Downey. 2010. "Inter-Neighborhood Migration, Race, and Environmental Hazards: Modeling Micro-Level Processes of Environmental Inequality." *American Journal of Sociology* 115 (4): 1110.
Crutzen, Paul J., and Eugene F. Stoermer. 2000. "Global Change Newsletter." *The Anthropocene* 41:17–18.
Cunsolo, Ashlee, and Karen Landman, ed. 2017. *Mourning Nature: Hope at the Heart of Ecological Loss and Grief.* Montreal: McGill-Queen's Press-MQUP.
David, A. T., J. E. Asarian, and F. K. Lake. 2018. "Wildfire Smoke Cools Summer River and Stream Water Temperatures." *Water Resources Research* 54 (10): 7273–7290.
Davidson, Debra J. 2018a. "Emotion, Reflexivity and Social Change in the Era of Extreme Fossil Fuels." *British Journal of Sociology.* https://doi.org/10.1111/1468-4446.
———. 2018b. "Evaluating the Effects of Living with Contamination from the Lens of Trauma: A Case Study of Fracking Development in Alberta, Canada." *Environmental Sociology* 4 (2): 196–209.
Davis, H., and Z. Todd. 2017. "On the Importance of a Date, or Decolonizing the Anthropocene." *ACME: An International E-Journal for Critical Geographies* 16 (4).
DeBano, Leonard F., Leonard F. DeBano, Daniel G. Neary, and Peter F. Ffolliott. 1998. *Fire Effects on Ecosystems.* New York: John Wiley.
Deer, Sarah. 2004. "Toward an Indigenous Jurisprudence of Rape." *Kansas Journal of Law and Public Policy* 14: 121.
———. 2015. *The Beginning and End of Rape: Confronting Sexual Violence in Native America.* Minneapolis: University of Minnesota Press.
Delaney, Lisa. 1981. *Klamath National Forest Karuk Allotment Situation Assessment Past, Present and Future.* Unpublished U.S. Forest Service document.
———. 1998. "Intellectual Self-Determination and Sovereignty: Looking at the Windmills in Our Minds." *Wicazo Sa Review* 13 (1): 25–31.

Denzin, Norman K., Yvonna S. Lincoln, and Linda Tuhiwai Smith, ed. 2008. *Handbook of Critical and Indigenous Methodologies*. Thousand Oaks, CA: Sage.

Dillon, L., and J. Sze. 2018. "Equality in the Air We Breathe: Police Violence, Pollution and the Politics of Sustainability." In *Sustainability: Approaches to Environmental Justice and Social Power*, edited by Julie Sze, 246–270. New York: New York University Press.

Diver, S. 2016. "Co-management as a Catalyst: Pathways to Post-Colonial Forestry in the Klamath Basin, California." *Human Ecology* 44 (5): 533–546.

Diver, Sibyl, Lisa Liu, Naomi Canchela, Sara Rose Tannenbaum, and Raphael Siberblatt. 2010. *Karuk Lands Management Historical Timeline*. https://karuktimeline.wordpress.com/.

Doka, Kenneth. 1989. *Disenfranchised Grief: Recognizing Hidden Sorrow*. Lexington, KY: Lexington Books.

Downey, Liam, and Marieke Van Willigen. 2005. "Environmental Stressors: The Mental Health Impacts of Living Near Industrial Activity." *Journal of Health and Social Behavior* 46 (3): 289–305.

Du Bois, W.E.B. 1903. *The Souls of Black Folk*. Chicago: Chicago University Press.

———. 1935. *Black Reconstruction in America 1860–1880*. New York, NY: Free Press.

Ducre, Kishi Animashaun. 2018. "The Black Feminist Spatial Imagination and an Intersectional Environmental Justice." *Environmental Sociology* 4 (1): 22–35.

Dunbar-Ortiz, Roxanne. 2014. *An Indigenous People's History of the United States*. Boston: Beacon Press.

Dunlap, Riley. 2002. "Paradigms, Theories and Environmental Sociology." In *Sociological Theory and the Environment: Classical Foundations, Contemporary Insights*, edited by Riley H. Dunlap, Frederick H. Buttel, Peter Dickens, and August Gijswijt, 329–351. Lanham, MD: Rowman and Littlefield.

———. 2010. "The Maturation and Diversification of Environmental Sociology: From Constructivism and Realism to Agnosticism and Pragmatism." In *The International Handbook of Environmental Sociology*, edited by M. Redclift and G. Woodgate, 15–32. 2nd ed. Cheltenham, UK: Edward Elgar.

Edelstein, Michael. 2004. *Contaminated Communities: The Social and Psychological Effects of Residential Toxic Exposure*. Boulder, CO: Westview Press.

Erikson, Kai T. 1976. *Everything in Its Path: Destruction of Community in the Buffalo Creek Flood*. New York: Simon and Schuster.

———. 1995. *A New Species of Trouble: The Human Experience of Modern Disasters*. New York: W.W. Norton & Company.

Escobar, Arturo. 1999. "After Nature: Steps to an Antiessentialist Political Ecology." *Current Anthropology* 40 (1): 1–30.

Faber, Daniel. 2008. *Capitalizing on Environmental Injustice: The Polluter-Industrial Complex in the Age of Globalization*. Lantham MD: Rowman and Littlefield.

Fausto-Sterling, A. 2000. "The Five Sexes, Revisited." *The Sciences* 40 (4): 18–23.

———. 2005. "The Problem with Sex/Gender and Nature/Nurture." In *Debating Biology*, 133–142. London: Routledge.

Federici, Silvia. 2012. *Revolution at Point Zero: Housework, Reproduction, and Feminist Struggle*. Oakland, CA: PM Press.

Fenelon, James V. 2014. *Culturicide, Resistance, and Survival of the Lakota (Sioux Nation)*. New York: Routledge.

———. 2015a. "Colonial Genocide in Indigenous North America." *American Indian Culture and Research Journal* 39:3131–133.

———. 2015b. "The Haunting Question of Genocide in the Americas." *Great Plains Quarterly* 35 (2): 203–213.

———. 2015c. "Indigenous Alternatives to the Global Crises of the Modern World-System." In *Overcoming Global Inequalities*, edited by Immanuel Wallerstein, Christopher Chase Dunn, and Christian Suter. New York: Routledge.

———. 2016. "Genocide, Race, Capitalism: Synopsis of Formation within the Modern World-System." *Journal of World-Systems Research* 22 (1): 23–30.

———. 2017. "Standing Rock, Epicenter of Resistance to American Empire." Paper presented in Section on Comparative-Historical Sociology, Empires, Colonies, Indigenous Peoples, American Sociological Association, Montreal, August 15, 2017.

Fenelon, J. V., and C. E. Trafzer. 2014. "From Colonialism to Denial of California Genocide to Misrepresentations: Special Issue on Indigenous Struggles in the Americas." *American Behavioral Scientist* 58 (1): 3–29.

Ferreira, Mariana Leal, and Gretchen Chesley Lang. 2005. *Indigenous Peoples and Diabetes*. Durham, NC: Carolina Academic Press.

Fiege, M. 2012. *The Republic of Nature: An Environmental History of the United States*. Seattle: University of Washington Press.

Finney, Carolyn. 2014. *Black Faces, White Spaces: Reimagining the Relationship of African Americans to the Great Outdoors*. Chapel Hill: University of North Carolina Press.

Fites-Kaufman, J. A., A. F. Bradley, and A. G. Merrill. 2006. "Fire and Plant Interactions." In *Fire in California's Ecosystems*, edited by N. G. Sugihara, J. W. van Wagtendonk, K. E. Shaffer, J. Fites-Kaufman, and A. E. Thode, 94–117. Berkeley: University of California Press.

Fixico, Donald L. 1986. *Termination and Relocation: Federal Indian Policy, 1945–1960*. Albuquerque: University of New Mexico Press.

Foster, John Bellamy. 1999. "Marx's Theory of Metabolic Rift: Classical Foundations for Environmental Sociology." *American Journal of Sociology* 105 (2): 366–405.

Foster, John Bellamy, Brett Clark, and Richard York. 2011. *The Ecological Rift: Capitalism's War on the Earth*. New York: New York University Press.

Garcia, Matthew. 2012. "The Importance of Being Asian." In *Racial Formation in the Twenty-First Century*, edited by Daniel HoSang, Oneka LaBennett, and Laura Pulido, 95. Berkeley: University of California Press.

Garro, L. C., and G. C. Lang. 1994. "Explanations of Diabetes: Anishinaabe and Dakota Deliberate upon a New Illness." In *Diabetes as a Disease of Civilization: The Impact of Culture Change on Indigenous Peoples*, 293–328. Berlin: Mouton de Gruyter.

Garroutte, Eva Marie. 2001. "The Racial Formation of American Indians: Negotiating Legitimate Identities within Tribal and Federal Law." *American Indian Quarterly* 25 (2): 224–239.

———. 2003. *Real Indians: Identity and the Survival of Native America*. Berkeley: University of California Press.

Giddens, A. 1991. "Structuration Theory: Past, Present and Future." In *Giddens' Theory of Structuration: A Critical Appreciation*, edited by C. Bryant and D. Jary. London: Routledge.

Gilio-Whitaker, D. 2015. "Idle No More and Fourth World Social Movements in the New Millennium." *South Atlantic Quarterly* 114 (4): 866–877.

———. 2017. "Indigenizing Environmental Justice." Public Lecture, University of Oregon Law School, October 2017.

Gill, Duane, and J. Steven Picou. 1998. "Technological Disaster and Chronic Community Stress." *Society & Natural Resources* 11 (8): 795–815.

Glenn, E. N. 1985. "Racial Ethnic Women's Labor: The Intersection of Race, Gender and Class Oppression." *Review of Radical Political Economics* 17 (3): 86–108.

———. 2015. "Settler Colonialism as Structure: A Framework for Comparative Studies of US Race and Gender Formation." *Sociology of Race and Ethnicity* 1 (1): 52–72.

Go, Julian. 2013. "Sociology's Imperial Unconscious: The Emergence of American Sociology in the Context of Empire." In *Sociology and Empire*, edited by George Steinmetz, 83–105. Durham, NC: Duke University Press.

———. 2015. "Colonialism and Neocolonialism." In *The Wiley Blackwell Encyclopedia of Race, Ethnicity, and Nationalism*, edited by Polly Rizova and Anthony Smith, 1–3. Hoboken, NJ: John Wiley and Sons.

———. 2016. *Postcolonial Thought and Social Theory*. Oxford: Oxford University Press.

———. 2017. "Decolonizing Sociology: Epistemic Inequality and Sociological Thought." *Social Problems* 64 (2): 194–199.

Goeman, Mishuana. 2009. "Notes toward a Native Feminism's Spatial Practice." *Wicazo Sa Review* (2): 169–187.

———. 2013. *Mark My Words: Native Women Mapping Our Nations*. Minneapolis: University of Minnesota Press.

Goeman, Mishuana R., and Jennifer Nez Denetdale. 2009. "Native Feminisms: Legacies, Interventions, and Indigenous Sovereignties." *Wicazo Sa Review* 24 (2): 9–13.

Goldberg, David Theo. 2002. *The Racial State*. Malden, MA: Blackwell.

Goldstein, Alyosha, ed. 2014. *Formations of United States Colonialism*. Durham, NC: Duke University Press.

Goldtooth, Tom. 1995. "Indigenous Nations: Summary of Sovereignty and Its Implications for Environmental Protection." In *Environmental Justice: Issues, Policies, and Solutions*, edited by Bunyan Bryant, 138–148. Washington DC: Island Press.

Goodwin, Jeff, James M. Jasper, and Francesca Polletta. 2004. *Emotional Dimensions of Social Movements*. London: Blackwell.

Gordon, Anne, and Vanessa Oddo. 2012. *Addressing Child Hunger and Obesity in Indian Country: Report to Congress*. Princeton, NJ: Mathematica Policy Research.

Gosnell, Hannah, and Erin Clover Kelly. 2010. "Peace on the River? Social-Ecological Restoration and Large Dam Removal in the Klamath Basin, USA." *Water Alternatives* 3 (2): 362.

Gowlett, J. A. J. 2016. "The Discovery of Fire by Humans: A Long and Convoluted Process." *Philosophical Transactions B* 371:1–12.

Grande, Sandy. 2004. "Whitestream Feminism and the Colonialist Project: Toward a Theory of Indigenista." *Red Pedagogy: Native American Social and Political Thought*, 123–157. Lanham, MD: Rowman and Littlefield.

———. 2015. *Red Pedagogy: Native American Social and Political Thought*. 10th anniversary edition. Lanham, MD: Rowman and Littlefield.

Grey, Sam, and Raj Patel. 2015. "Food Sovereignty as Decolonization: Some Contributions from Indigenous Movements to Food System and Development Politics." *Agriculture and Human Values* 32 (3): 1–14.

Grinde, Donald A., and Bruce E. Johansen. 1995. *Ecocide of Native America: Environmental Destruction of Indian Lands and Peoples*. Santa Fe, NM: Clear Light.

Grossman, Zoltan. 2017. *Unlikely Alliances: Native Nations and White Communities Join to Defend Rural Lands*. Seattle: University of Washington Press.

Grosz, E. 2010. "The Untimeliness of Feminist Theory." *NORA—Nordic Journal of Feminist and Gender Research* 18 (1): 48–51.

Guthman, Julie. 2008. "'If They Only Knew': Color Blindness and Universalism in California Alternative Food Institutions." *The Professional Geographer* 60 (3): 387–397.

———. 2011. *Weighing in: Obesity, Food Justice, and the Limits of Capitalism*. Berkeley: University of California Press.

Hankins, D. L. 2005. "Pyrogeography: Spatial and Temporal Relationships of Fire, Nature, and Culture." PhD diss., University of California, Davis.

Haraway, Donna. 1989. *Primate Visions: Gender, Race, and Nature in the World of Modern Science.* New York: Routledge.
———. 1991. *Simians, Cyborgs, and Women: The Reinvention of Nature.* New York: Routledge.
———. 2015. "Anthropocene, Capitalocene, Plantationocene, Chthulucene: Making Kin." *Environmental Humanities* 6 (1): 159–165.
Harding, Anna, et al. 2012. "Conducting Research with Tribal Communities: Sovereignty, Ethics, and Data-Sharing Issues." *Environmental Health Perspectives* 120 (1): 6.
Hardison, P. D., and Kelly Bannister. 2011. "Ethics in Ethnobiology: History, International Law and Policy, and Contemporary Issues." In *Ethnobiology*, edited by E. Anderson, Deborah Pearsall, Eugene Hunn, and Nancy Turner, 27–49. Hoboken, NJ: John Wiley.
Harrington, John P. 1932. "Karuk Indian Myths." Smithsonian Institution, Bureau of American Ethnology, Bulletin 107.
Harrison, J. L. 2011. *Pesticide Drift and the Pursuit of Environmental Justice.* Cambridge, MA: MIT Press.
———. 2014. "Neoliberal Environmental Justice: Mainstream Ideas of Justice in Political Conflict over Agricultural Pesticides in the United States." *Environmental Politics* 23 (4): 650–669.
———. 2015. "Coopted Environmental Justice? Activists' Roles in Shaping EJ Policy Implementation." *Environmental Sociology* 1 (4): 241–255.
Heizer, R. F. 1972. *The Eighteen Unratified Treaties of 1851–1852 between the California Indians and the United States Government.* Archaeological Research Facility, Dept. of Anthropology, University of California.
Hewes, Gordon W. 1973. "Indian Fisheries Productivity in Pre-Contact Times in the Pacific Salmon Area." *Northwest Anthropological Research Notes* 7 (2): 133–155.
Hillman, L., and J. F. Salter. 1997. "Environmental Management: American Indian Knowledge and the Problem of Sustainability." *Forests, Trees and People Newsletter* (FAO) (Sweden).
Hinton, Alexander Laban, Andrew Woolford, and Jeff Benvenuto, eds. 2014. *Colonial Genocide in Indigenous North America.* Durham, NC: Duke University Press.
Hitchcock, Robert K., and Samuel Totten. 2011. *Genocide of Indigenous Peoples.* Piscataway, NJ: Transaction Publishers.
Hochschild, Arlie Russell. 1983. *The Managed Heart: Commercialization of Human Feeling.* Berkeley: University of California Press.
Hoover, Elizabeth. 2017. *The River Is in Us: Fighting Toxics in a Mohawk Community.* Minneapolis: University of Minnesota Press.
———. 2018. "Environmental Reproductive Justice: Intersections in an American Indian Community Impacted by Environmental Contamination." *Environmental Sociology* 4 (1): 8–21.
Hoover, E., K. Cook, R. Plain, K. Sanchez, V. Waghiyi, P. Miller, R. Dufault, C. Sislin, and D. O. Carpenter. 2012. "Indigenous Peoples of North America: Environmental Exposures and Reproductive Justice." *Environmental Health Perspectives* 120 (12): 1645.
Hoover, Elizabeth, Mia Renauld, Michael R. Edelstein, and Phil Brown. 2015. "Social Science Collaboration with Environmental Health." *Environmental Health Perspectives* 123 (11): 1100.
HoSang, Daniel, Oneka LaBennett, and Laura Pulido, eds. 2012. *Racial Formation in the Twenty-First Century.* Berkeley: University of California Press.
Hoxie, Frederick E. 1984. *A Final Promise: The Campaign to Assimilate the Indians, 1880–1920.* Lincoln: University of Nebraska Press.
Hubbard, Tasha. 2014. "'Kill, Skin, and Sell': Buffalo Genocide." In *Colonial Genocide in Indigenous North America*, edited by Alexander Laban Hinton, Andrew Woolford, and Jeff Benvenuto, 292–305. Durham, NC: Duke University Press.

Huffman, M. R. 2014. "Making a World of Difference in Fire and Climate Change." *Fire Ecology* 10 (3): 90–101.

Huhndorf, Shari M. 2001. *Going Native: Indians in the American Cultural Imagination.* Ithaca, NY: Cornell University Press.

Hummel, S., Frank Lake, and A. Watts. 2015. "Using Forest Knowledge: How Silviculture Can Benefit from Ecological Knowledge Systems about Beargrass Harvesting Sites." *General Technical Report.* PNW-GTR-912. Portland, OR: US Department of Agriculture, Forest Service, Pacific Northwest Research Station 9:912.

Huntsinger, Lynn, and Sarah McCaffrey. 1995. "A Forest for the Trees: Forest Management and the Yurok Environment, 1850 to 1994." *American Indian Culture and Research Journal* 19 (4): 155–192.

Hurtado, Albert L. 1990. *Indian Survival on the California Frontier.* New Haven, CT: Yale University Press.

Huyser, Kimberly. 2017. "Understanding the Construction of American Indian and Alaska Native Diabetes Using Critical Race Theory." Presentation at the American Sociological Association Meetings, August 2017, Montreal, CA.

Jacob, Michelle M. 2013. *Yakama Rising: Indigenous Cultural Revitalization, Activism, and Healing.* Tucson: University of Arizona Press.

———. 2016. *Indian Pilgrims: Indigenous Journeys of Activism and Healing with Saint Kateri Tekakwitha.* Tucson: University of Arizona Press.

———. 2017. "Indigenous Studies Speaks to American Sociology: The Need for Individual and Social Transformations of Indigenous Education in the USA." *Social Sciences* 7 (1): 1–10.

Jacob, Michelle M., and H. Blackhorn. 2018. "Building an Indigenous Traditional Ecological Knowledge Initiative at a Research University: Decolonization Notes from the Field. *Journal of Sustainability Education* 18. http://www.susted.com/wordpress/content/building-an-indigenous-traditional-ecological-knowledge-initiativeat-a-research-university-decolonization-notes-from-the-field_2018_04.

Jacobson, G. 2016. "The Sociology of Emotions in a Contested Environmental Illness Case: How Gender and the Sense of Community Contribute to Conflict." *Environmental Sociology* 2 (3): 238–253.

Jaggar, Alison. 1989. *Gender-Body-Knowledge: Feminist Reconstruction of Being and Knowing.* New Brunswick, NJ: Rutgers University Press.

Jaimes, M. Annette, and Theresa Halsey. 1997. "American Indian Women: At the Center of Indigenous Resistance in Contemporary North America." In *Early California Laws and Policies Related to California Indians,* edited by Kimberly Johnston-Dodds, 298. California Research Bureau (CRB 02-014). Sacramento CA: California State Library.

Jasper, James. 2011. "Emotions and Social Movements: Twenty Years of Theory and Research." *Annual Review of Sociology* 37:285–303.

Joe, Jennie Rose, and Francine Gauchpin, ed. 2012. *Health and Social Issues of Native American Women.* Santa Barbara, CA: Prager.

Joe, Jennie Rose, and Robert S. Young, ed. 1994. *Diabetes as a Disease of Civilization: The Impact of Culture Change on Indigenous Peoples.* Berlin: Walter de Gruyter.

Johnson, J. T., and S. C. Larsen. 2013. *A Deeper Sense of Place: Stories and Journeys of Collaboration in Indigenous Research.* Corvallis: Oregon State University Press.

Johnston, Fay H., Shannon Melody, and David M.J.S. Bowman. 2016. "The Pyrohealth Transition: How Combustion Emissions Have Shaped Health through Human History." *Philosophical Transactions of the Royal Society B* 371 (1696): 20150173.

Johnston-Dodds, Kimberly. 2002. *Early Laws and Policies Related to California Indians.* CRB 02-014. California Research Bureau.

Kann, J. 2008. *Technical Memorandum: Microcystin Bioaccumulation in Klamath River Fish and Mussel Tissue: Preliminary 2007 Results*. Ashland, OR: Aquatic Ecosystem Sciences.

Kann, J., and S. Corum. 2009. "Toxigenic Microcystis Aeruginosa Bloom Dynamics and Cell Density/Chlorophyll a Relationships with Microcystin Toxin in the Klamath River, 2005–2008." 46. Prepared for Karuk Tribe Department of Natural Resources, Orleans, CA.

Kann, J., S. Corum, and K. Fetcho. 2010. "Microcystin Bioaccumulation in Klamath River Freshwater Mussel Tissue: 2009 Results." Prepared by Aquatic Ecosystem Sciences, LLC, the Karuk Tribe Natural Resources Department, and the Yurok Tribe Environmental Program.

Karuk Tribe. 2010. "Karuk Tribe Department of Natural Resources Eco Cultural Resource Management Plan: An Integrated Approach to Adaptive Problem Solving, in the Interest of Managing the Restoration of Balanced Ecological Processes Utilizing Traditional Ecological Knowledge Supported by Western Science." Orleans, CA: Karuk Tribe. http://www.karuk.us/karuk2/images/docs/dnr/ECRMP.

———. 2016a. "Karuk K-12 Needs Assessment." Karuk Department of Natural Resources internal document.

———. 2016b. "Karuk Oral Traditions." Karuk Department of Natural Resources internal document.

Karuk Tribe and Partners. 2013. "Practicing *Pikyav*: A Guiding Policy for Research Collaborations with the Karuk Tribe." Available from the Karuk Department of Natural Resources by request.

Kauanui, J. Kēhaulani. 2008. *Hawaiian Blood: Colonialism and the Politics of Sovereignty and Indigeneity*. Durham, NC: Duke University Press.

KBSRA, Karuk Department of Natural Resources, Megan Mucioki, and Jennifer Sowerwine. 2016. *Klamath Basin Food System Assessment: Karuk Tribe Data*. Orleans, CA: Karuk Tribe and University of California at Berkeley.

Kemple, Thomas M., and Renisa Mawani. 2009. "The Sociological Imagination and Its Imperial Shadows." *Theory, Culture & Society* 26 (7–8): 228–249.

Key, J. 2000. "Effects of Clearcuts and Site Preparation on Fire Severity, Dillon Creek Fire 1994." Master's thesis, Department of Forestry, Humboldt State University.

Kimmerer, R. W., and F. K. Lake. 2001. "The Role of Indigenous Burning in Land Management." *Journal of Forestry* 99 (11): 36–41.

Klein, Naomi. 2015. *This Changes Everything: Capitalism vs. the Climate*. New York: Simon and Schuster.

Klopotek, Brian. 2011. *Recognition Odysseys: Indigeneity, Race, and Federal Tribal Recognition Policy in Three Louisiana Indian Communities*. Durham, NC: Duke University Press.

Kosek, Jake. 2006. *Understories: The Political Life of Forests in Northern New Mexico*. Durham, NC: Duke University Press.

Krieger, Nancy. 1994. "Epidemiology and the Web of Causation: Has Anyone Seen the Spider?" *Social Science & Medicine* 39 (7): 887–903.

———. 2001. "The Ostrich, the Albatross, and Public Health: An Ecosocial Perspective—or Why an Explicit Focus on Health Consequences of Discrimination and Deprivation Is Vital for Good Science and Public Health Practice." *Public Health Reports* 116 (5): 419.

———. 2011. *Epidemiology and the People's Health: Theory and Context*. Oxford: Oxford University Press.

Kroeber, A. L. 1916. "California Place Names of Indian Origin." *University of California Publications in American Archaeology and Ethnology* 12 (2): 33.

Kroll-Smith, J. S., and S. R. Couch. 1993. "Technological Hazards." In *International Handbook of Traumatic Stress Syndromes*, edited J. P. Wilson and B. Raphael, 79–91. Boston: Springer.

Kuhnlein, H. V., and H. M. Chan. 2000. "Environment and Contaminants in Traditional Food Systems of Northern Indigenous Peoples." *Annual Review of Nutrition* 20 (1): 595–626.

Kuhnlein, H. V., and M. M. Humphries. 2017. *Traditional Animal Foods of Indigenous Peoples of Northern North America*. Centre for Indigenous Peoples' Nutrition and Environment, Montreal, Quebec: McGill University.

Kuhnlein, H. V., and O. Receveur. 1996. "Dietary Change and Traditional Food Systems of Indigenous Peoples." *Annual Review of Nutrition* 16 (1): 417–442.

Kuhnlein, Harriet V., Bill Erasmus, and Dina Spigelski. 2009. *Indigenous Peoples' Food Systems: The Many Dimensions of Culture, Diversity and Environment for Nutrition and Health*. Food and Agriculture Organization of the United Nations Rome, Italy (FAO).

Kurtz, H. E. 2009. "Acknowledging the Racial State: An Agenda for Environmental Justice Research." *Antipode* 41 (4): 684–704.

LaDuke, Winona. 1999. *All Our Relations: Native Struggles for Land and Life*. Boston MA: South End Press.

Laird, Brian D., Alexey B. Goncharov, Grace M. Egeland, and Hing Man Chan. 2013. "Dietary Advice on Inuit Traditional Food Use Needs to Balance Benefits and Risks of Mercury, Selenium, and n3 Fatty Acids." *Journal of Nutrition* 143 (6): 923–930.

Lake, F. K. 2007. *Traditional Ecological Knowledge to Develop and Maintain Fire Regimes in Northwestern California, Klamath-Siskiyou Bioregion: Management and Restoration of Culturally Significant Habitats*. Corvalis, OR: Oregon State University.

———. 2013. "Trails, Fires, and Tribulations: Tribal Resource Management and Research Issues in Northern California." *Occasion* 5:1–22.

Lake, F. K., and J. W. Long. 2014. "Fire and Tribal Cultural Resources." In *Science Synthesis to Support Socio-ecological Resilience in the Sierra Nevada and Southern Cascade Range*, edited by J. W. Long, L. Quinn-Davidson, and C.N. Skinner, 173–186. Arcata CA: USDA Forest Service, Pacific Southwest Research Station.

Lake, F. K., V. Wright, P. Morgan, M. McFadzen, D. McWethy, and C. Stevens-Rumann. 2017. "Returning Fire to the Land: Celebrating Traditional Knowledge and Fire." *Journal of Forestry* 115 (5): 343–353.

LaLande, Jeffrey M. 1981. "Sojourners in the Oregon Siskiyous: Adaptation and Acculturation of the Chinese Miners in the Applegate Valley, CA. 1855–1900." Diss., Oregon State University.

———.1985. "Sojourners in Search of Gold: Hydraulic Mining Techniques of the Chinese on the Oregon Frontier." *IA: The Journal of the Society for Industrial Archeology* 11 (1): 29–52.

Large, Judith. 1997. "Disintegration Conflicts and the Restructuring of Masculinity." *Gender & Development* 5 (2): 23–30.

Latour, B. 2004. *Politics of Nature*. Cambridge, MA: Harvard University Press.

Lawrence, B. 2003. "Gender, Race, and the Regulation of Native Identity in Canada and the United States: An Overview." *Hypatia* 18 (2): 3–31.

LeBeau, M. L.1998. "Federal Land Management Agencies and California Indians: A Proposal to Protect Native Plant Species." *Environs: Environmental Law & Policy Journal* 21:27.

Lefevre, Tate A. 2015. *Settler Colonialism*. Oxford: Oxford University Press, 2015.

Leimbach, J. 2009. *Preparation for FERC Hydropower Relicensing: An Activist's Guide for the Six Months to Two Years before a Relicensing*. Washington, DC: Hydropower Reform Net.

Leonetti, C. 2010. *Indigenous Stewardship Methods and NRCS Conservation Practices*. U.S. Department of Agriculture Natural Resources Conservation Service. http://www.fws.gov/nativeamerican/traditional-knowledge.html.

Leung, Peter. 2001. *One Hundred and Fifty Years of the Chinese Presence in California (1848–2001): Honor the Past, Engage the Present, Build the Future*. Sacramento, CA: Chinese Culture Foundation.

Lewis, Diane. 1973. "Anthropology and Colonialism." *Current Anthropology* 14 (5): 581–602.
Lewis, Henry T. 1973. *Patterns of Indian Burning in California: Ecology and Ethnohistory*. No. 1. Menlo Park, CA: Ballena Press.
Lindsay, B. C. 2012. *Murder State: California's Native American Genocide, 1846–1873*. Lincoln: University of Nebraska Press.
———. 2014. "Humor and Dissonance in California's Native American Genocide." *American Behavioral Scientist* 58 (1): 97–123.
Lockie, Stewart. 2015. "What Is Environmental Sociology?" *Environmental Sociology* 1 (3): 139–142.
———. 2016. "The Emotional Enterprise of Environmental Sociology." *Environmental Sociology* 2 (3): 233–237.
———. 2018. "Privilege and Responsibility in Environmental Justice Research." *Environmental Sociology* 4 (2): 175–180.
Lomawaima, K. T., and T. L. McCarty. 2006. *"To Remain an Indian": Lessons in Democracy from a Century of Native American Education*. New York: Teachers College Press.
London, Jonathan K., Julie Sze, and Raoul S. Lievanos. 2008. "Problems, Promise, Progress, and Perils: Critical Reflections on Environmental Justice Policy Implementation in California." *UCLA Journal of Environmental Law and Policy* 26:255.
Long, Jonathan W., Leland W. Tarnay, and Malcolm P. North. 2017. "Aligning Smoke Management with Ecological and Public Health Goals." *Journal of Forestry* 116 (1): 76–86.
Lowry, C. 1999. *Northwest Indigenous Gold Rush History: The Indian Survivors of California's Holocaust*. Arcata, CA: Indian Teacher and Educational Personnel Program.
Lynn, Frances M. 2000. "Community-Scientist Collaboration in Environmental Research." *American Behavioral Scientist* 44 (4): 649–663.
Madley, Benjamin. 2016. *An American Genocide: The United States and the California Indian Catastrophe, 1846–1873*. New Haven, CT: Yale University Press.
Magubane, Z. 2013. "Common Skies and Divided Horizons? Sociology, Race, and Postcolonial Studies." *Political Power and Social Theory* 24:81–116.
———. 2014. "Science, Reform, and the 'Science of Reform': Booker T Washington, Robert Park, and the Making of a 'Science of Society.'" *Current Sociology* 62 (4): 568–583.
Malacrida, Claudia, and Tiffany Boulton.2012. "Women's Perceptions of Childbirth "Choices" Competing Discourses of Motherhood, Sexuality, and Selflessness." *Gender & Society* 26 (5): 748–772.
Manning, Beth Rose Middleton. 2018. *Upstream: Trust Lands and Power on the Feather River*. Tucson: University of Arizona Press.
Marable, Manning. 1983. *How Capitalism Underdeveloped Black America*. London: Pluto.
Maracle, Lee. 1996. *I Am Woman: A Native Perspective on Sociology and Feminism*. London: Global Professional Publishing.
Marquez, John. 2014. *Black-Brown Solidarity: Racial Politics in the New Gulf South*. Austin: University of Texas.
Martinez, D. 2011. "Indigenous Ecosystem-Based Adaptation and Community-Based Ecocultural Restoration during Rapid Climate Disruption: Lessons for Western Restorationists." http://www.scribd.com/doc/76322289/Dennis-Martinez-2011.
Matthewson, M. 2007. "California Indian Basketweavers and the Landscape." In *To Harvest, to Hunt: Stories of Resource Use in the American West*, edited by Judy Li. Corvallis: Oregon State Press.
McCovey, Mavis, and John Salter. 2009. *Medicine Trails: A Life in Many Worlds*. Berkeley, CA: Heyday Books.
McEvoy, Arthur. 1986. *The Fisherman's Problem: Ecology and Law in the California Fisheries, 1850–1980*. Cambridge: Cambridge University Press.

McGregor, D. 2005. "Traditional Ecological Knowledge: An Anishnabe Woman's Perspective." *Atlantis: Critical Studies in Gender, Culture & Social Justice* 29 (2): 103–109.

———. 2008. "Linking Traditional Ecological Knowledge and Western Science: Aboriginal Perspectives from the 2000 State of the Lakes Ecosystem Conference." *Canadian Journal of Native Studies* 28 (1): 139.

———. 2009. "Honouring Our Relations: An Anishnaabe Perspective on Environmental Justice." In *Speaking for Ourselves: Environmental Justice in Canada*, edited by J. Agyeman, P. Cole, R. Haluza DeLay, and P. O'Riley, 27–41. Vancouver: University of British Columbia Press.

Meissner, S. N., and K. P. Whyte. 2017. "Theorizing Indigeneity, Gender, and Settler Colonialism." In *The Routledge Companion to the Philosophy of Race*, edited by P. C. Taylor, L. Alcoff, and L. Anderson, 1–22. London: Routledge.

Merchant, Carolyn. 1980. *The Death of Nature: Women, Ecology, and the Scientific Revolution*. New York: Harper & Row.

———. 1998. *Green versus Gold: Sources in California's Environmental History*. Washington, DC: Island Press.

Messerschmidt, J. W. 2009. "'Doing Gender': The Impact and Future of a Salient Sociological Concept." *Gender & Society* 23 (1): 85–88.

Middleton, Beth Rose. 2011. *Trust in the Land: New Directions in Tribal Conservation*. Tucson: University of Arizona Press.

———. 2015. "40. Jahát Jatítotòdom: Toward an Indigenous Political Ecology." In *The International Handbook of Political Ecology*, edited by Raymond Bryant, 561–576. Kings College, London: Edward Elger Publishing.

Middleton, Elisabeth. 2010. "A Political Ecology of Healing." *Journal of Political Ecology* 17 (1): 1–28.

Miller, R. J. 2005. "The Doctrine of Discovery in American Indian Law." *Idaho Law Review* 42:1–96.

Million, D. 2013. *Therapeutic Nations: Healing in an Age of Indigenous Human Rights*. Tucson: University of Arizona Press.

Mills, Charles W. (1959). 2000. *The Sociological Imagination*. Oxford: Oxford University Press.

———. 1997. *The Racial Contract*. Ithaca, NY: Cornell University Press.

———. 2001. "Black Trash." In *Faces of Environmental Racism*, edited by L. Westra and B. Lawson. Boulder, CO: Rowman & Littlefield.

———. 2007. "White Ignorance." In *Agnotology: The Making and Unmaking of Ignorance*, edited by Robert N. Proctor and Londa Schiebinger. Stanford, CA: Stanford University Press.

Miner, D. 2014. *Creating Aztlán: Chicano Art, Indigenous Sovereignty, and Lowriding across Turtle Island*. Tucson: University of Arizona Press.

Miranda, Deborah A. 2013. *Bad Indians: A Tribal Memoir*. Berkeley, CA: Heyday.

Mirchandani, Kiran. 2003. "Challenging Racial Silences in Studies of Emotion Work: Contributions from Anti-Racist Feminist Theory." *Organization Studies* 24 (5): 721–742.

Mirowsky, J., and C. Ross. 1989. *Social Causes of Psychological Distress*. New York: Aldine de Gruyter.

Mitman, Gregg, Marco Armiero, and Robert Emmett, eds. 2018. *Future Remains: A Cabinet of Curiosities for the Anthropocene*. Chicago: University of Chicago Press.

Mohr, J. A., C. Whitlock, and C. N. Skinner. 2000. "Postglacial Vegetation and Fire History, Eastern Klamath Mountains, California, USA." *The Holocene* 10:587–601.

Mojola, Sanyu. 2011. "Fishing in Dangerous Waters: Ecology, Gender, Economy and HIV Risk." *Social Sciences and Medicine* 72 (2): 149–156.

―――. 2014. *Love, Money and HIV: Becoming a Modern African Woman in the Age of AIDS*. Berkeley: University of California Press.

Moore, Donald S., Jake Kosek, and Anand Pandian, eds. 2003. *Race, Nature, and the Politics of Difference*. Durham, NC: Duke University Press.

Moreton-Robinson, Aileen. 2014. "Race Matters: The 'Aborigine' as a White Possession." In *The Indigenous World of North America*, 467–486. London: Routledge.

―――. 2015. *The White Possessive: Property, Power, and Indigenous Sovereignty*. Minneapolis: University of Minnesota Press.

Morrill, Angie. 2017. "Time Traveling Dogs (and Other Native Feminist Ways to Defy Dislocations)." *Cultural Studies↔Critical Methodologies* 17 (1): 14–20.

Morris, Aldon. 2015. *The Scholar Denied: WEB Du Bois and the Birth of Modern Sociology*. Berkeley: University of California Press.

Mortimer-Sandilands, Catriona, and Bruce Erickson. 2010. *Queer Ecologies: Sex, Nature, Politics, Desire*. Bloomington: Indiana University Press.

Mott, J. A., P. Meyer, D. Mannino, S. C. Redd, E. M. Smith, C. Gotway-Crawford, and E. Chase. 2002. "Wildland Forest Fire Smoke: Health Effects and Intervention Evaluation, Hoopa, California, 1999." *Western Journal of Medicine* 176:157–162.

Nagel, Joane. 1994. "Constructing Ethnicity: Creating and Recreating Ethnic Identity and Culture." *Social Problems* 41 (1): 152–176.

―――. 1997. *American Indian Ethnic Renewal: Red Power and the Resurgence of Identity and Culture*. Oxford: Oxford University Press.

Nazaryan, A. 2018. "California Slaughter: The State-Sanctioned Genocide of Native Americans." *Newsweek*. http://www.newsweek.com/2016/08/26/california-native-americans-genocide-490824.html.

Newcomb, Steven T. 2008. *Pagans in the Promised Land: Decoding the Doctrine of Christian Discovery*. Golden, CO: Fulcrum Publishing.

Ngo, A. D., C. Brolan, L. Fitzgerald, V. Pham, and H. Phan. 2013. "Voices from Vietnam: Experiences of Children and Youth with Disabilities, and Their Families, from an Agent Orange Affected Rural Region." *Disability & Society* 28 (7): 955–969.

Ngo, A. D., R. Taylor, C. L. Roberts, and T. V. Nguyen. 2006. "Association between Agent Orange and Birth Defects: Systematic Review and Meta-Analysis." *International Journal of Epidemiology* 35 (5): 1220–1230.

Nixon, Rob. 2011. *Slow Violence and the Environmentalism of the Poor*. Cambridge, MA: Harvard University Press.

Norgaard, Kari Marie. 2004. *The Effects of Altered Diet on the Health of the Karuk People: A Preliminary Report*. Orleans, CA: Karuk Tribe Department of Natural Resource.

―――. 2005. *The Effects of Altered Diet on the Health of the Karuk People*. Orleans, CA: Karuk Tribe Department of Natural Resource.

―――. 2007. "The Politics of Invasive Weed Management: Gender, Race, and Risk Perception in Rural California." *Rural Sociology* 72 (3): 450–477.

―――. 2011. *Living in Denial: Climate Change, Emotions, and Everyday Life*. Cambridge, MA: MIT Press.

―――. 2012. "Climate Denial and the Construction of Innocence: Reproducing Transnational Environmental Privilege in the Face of Climate Change." *Race, Gender & Class* 19 (1–2): 80–103.

―――. 2014a. "The Politics of Fire and the Social Impacts of Fire Exclusion on the Klamath." *Humboldt Journal of Social Relations* 39:73–97.

―――. 2014b. "Karuk Traditional Ecological Knowledge and the Need for Knowledge Sovereignty: Social, Cultural and Economic Impacts of Denied Access to Traditional

Management." Karuk Tribe, https://karuktribeclimatechangeprojects.wordpress.com/about/karuk-tek-knowledge-sovereignty/.

———. 2016. "Climate Change is a Social Issue." *Chronicle of Higher Education* (available online).

———. 2018. "The Sociological Imagination in a Time of Climate Change." *Global and Planetary Change* 163:171–176.

Norgaard, Kari Marie, Spenser Meeks, B. Crayne, and Frank Dunnivant. 2013. "Trace Metal Analysis of Karuk Traditional Foods in the Klamath River." *Journal of Environmental Protection* 4 (4): 319–328.

Norgaard, Kari Marie, and Ron Reed. 2017. "Emotional Impacts of Environmental Decline: What Can Native Cosmologies Teach Sociology about Emotions and Environmental Justice?" *Theory and Society* 46 (6): 463–495.

Norgaard, Kari Marie, Ron Reed, and J. M. Bacon. 2018. "How Environmental Decline Restructures Indigenous Gender Practices: What Happens to Karuk Masculinity When There Are No Fish?" *Sociology of Race and Ethnicity* 4 (1): 98–113.

Norgaard, Kari Marie, Ron Reed, and Carolina Van Horn. 2011. "A Continuing Legacy: Institutional Racism, Hunger and Nutritional Justice on the Klamath." In *Cultivating Food Justice: Race, Class and Sustainability*, edited by A. Alkon and J. Agyeman, 23–46. Cambridge, MA: MIT Press.

Norgaard, Richard B. 1994. *Development Betrayed: The End of Progress and a Co-evolutionary Revisioning of the Future*. New York: Routledge.

Norton, Jack. 1979. *When Our Worlds Cried: Genocide in Northwestern California*. San Francisco: Indian Historian Press.

———. 2014. "If the Truth Be Told: Revising California History as a Moral Objective." *American Behavioral Scientist* 58 (1): 83–96.

Norton-Smith, Kathryn, Kathy Lynn, Karletta Chief, Karen Cozzetto, Jamie Donatuto, Margaret Hiza Redsteer, Linda E. Kruger, Julie Maldonado, Carson Viles, and Kyle P. Whyte. 2016. *Climate Change and Indigenous Peoples: A synthesis of Current Impacts and Experiences*. United States Department of Agriculture, Forest Service, Pacific Northwest Research Station.

O'Brien, Jean. 2010. "Firsting and Lasting."In *Writing Indians Out of Existence in New England*. Minneapolis: University of Minnesota Press.

Odion, Dennis C., Evan J. Frost, James R. Strittholt, Hong Jiang, Dominick A. Dellasala, and Max A. Moritz. 2004. "Patterns of Fire Severity and Forest Conditions in the Western Klamath Mountains, California." *Conservation Biology* 18 (4): 927–936.

Omi, Michael, and Howard Winant. 1994. *Racial Formation in the United States* New York: Routledge.

———. 2014. *Racial Formation in the United States*. 3rd ed. New York: Routledge.

Ommer, Rosemary E. 2007. *Coasts under Stress: Restructuring and Social-Ecological Health*. Montreal: McGill-Queen's Press.

Outka, Paul. 2016. *Race and Nature from Transcendentalism to the Harlem Renaissance*. New York: Springer.

Owens, Patricia. 2015. *Economy of Force: Counterinsurgency and the Historical Rise of the Social*. Vol. 139. Cambridge: Cambridge University Press.

Park, Lisa Sun-Hee, and David Pellow. 2004. "Racial Formation, Environmental Racism, and the Emergence of Silicon Valley." *Ethnicities* 4 (3): 403–424.

Pascoe, Cheri J., and Tristan Bridges. 2015. *Exploring Masculinities: Identity, Inequality, Continuity, and Change*. Oxford: Oxford University Press.

Peek, Lori, and Alice Fothergil. 2006. "Displacement, Gender and the Challenge of Parenting after Hurricane Katrina." *National Women's Studies Association Journal* 20 (3): 69–105.

Pellow, David Naguib. 2002. *The Garbage Wars: The Struggle for Environmental Justice in Chicago*. Cambridge, MA: MIT Press.

———. 2009. "The State and Policy: Imperialism, Exclusion and Ecological Violence as State Policy." In *Twenty Lessons in Environmental Sociology*, edited by Kenneth Alan Gould and Tammy L. Lewis, 47–58. Oxford: Oxford University Press.

———. 2016. "Toward a Critical Environmental Justice Studies: Black Lives Matter as an Environmental Justice Challenge." *Du Bois Review: Social Science Research on Race* 13 (2): 221–236.

———. 2017. *What Is Critical Environmental Justice?* New York: John Wiley.

Pellow, David Naguib, and Robert J. Brulle. 2005. "Power, Justice, and the Environment: Toward Environmental Justice Studies." In *Power, Justice, and the Environment: A Critical Appraisal of the Environmental Justice Movement*, edited by Pellow D. and R Brulle, 1–19. Cambridge MA: MIT Press.

Perry, D. A., P. F. Hessburg, C. N. Skinner, T. A. Spies, S. L. Stephens, A. H. Taylor, J. F. Franklin, B. McComb, and G. Riegel. 2011. "The Ecology of Mixed Severity Fire Regimes in Washington, Oregon, and Northern California." *Forest Ecology and Management* 262 (5): 703–717.

Peters, J. G., and B. Ortiz. 2016. *After the First Full Moon in April: A Sourcebook of Herbal Medicine from a California Indian Elder*. London: Routledge.

Pfister, J. 1993. "The Politics of Reason: Towards a Feminist Logic." *Australasian Journal of Philosophy* 71 (4): 436–462.

———. 2004. *Individuality Incorporated: Indians and the Multicultural Modern*. Durham, NC: Duke University Press.

Pollan, Michael. 2001. *The Botany of Desire: A Plant's-Eye View of the World*. New York: Penguin.

———. 2006. *The Omnivore's Dilemma: A Natural History of Four Meals*. New York: Penguin.

Prus, Robert. 1987. "Generic Social Processes Maximizing Conceptual Development in Ethnographic Research." *Journal of Contemporary Ethnography* 16 (3): 250–293.

Pulido, Laura. 2000. "Rethinking Environmental Racism: White Privilege and Urban Development in Southern California." *Annals of the Association of American Geographers* 90 (1): 12–40.

———. 2016. "Flint, Environmental Racism, and Racial Capitalism." *Capitalism, Nature, Socialism* 27 (3): 1–16.

———. 2017a. "Conversations in Environmental Justice: An Interview with David Pellow." *Capitalism Nature Socialism* 28 (2): 43–53.

———. 2017b. "Geographies of Race and Ethnicity II: Environmental Racism, Racial Capitalism and State-Sanctioned Violence." *Progress in Human Geography* 41 (4): 524–533.

———. 2017c. "Historicizing the Personal and the Political: Evolving Racial Formations and the Environmental Justice Movement." In *The Routledge Handbook of Environmental Justice*, 15–24. New York: Routledge.

———. 2018. "Racism and the Anthropocene." In *Future Remains: A Cabinet of Curiosities for the Anthropocene*, edited by Gregg Mitman, Marco Armiero, and Robert Emmett. Chicago: University of Chicago Press.

Pulido, L., and J. De Lara. 2018. "Reimagining 'Justice' in Environmental Justice: Radical Ecologies, Decolonial Thought, and the Black Radical Tradition." *Environment and Planning E: Nature and Space* 1 (1–2): 76–98.

Pulido, L., E. Kohl, and N. M. Cotton. 2016. "State Regulation and Environmental Justice: The Need for Strategy Reassessment." *Capitalism Nature Socialism* 27 (2): 12–31.

Pyne, S. J. 2016. "Fire in the Mind: Changing Understandings of Fire in Western Civilization." *Philosophical Tranactions of the Royal Society B* 371.

Quijano, A. 2000. "Coloniality of Power and Eurocentrism in Latin America." *International Sociology* 15 (2): 215–232.
Quinn, Scott. 2007. *Karuk Tribe of California Aboriginal Territory Acreage Assessment.* Orleans, CA: Karuk Department of Natural Resources.
Rahman, M., and A. Witz. 2003. "What Really Matters? The Elusive Quality of the Material in Feminist Thought." *Feminist Theory* 4 (3): 243–261.
Ranco, Darren, and Dean Suagee. 2007. "Tribal Sovereignty and the Problem of Difference in Environmental Regulation: Observations on 'Measured Separatism' in Indian country." *Antipode* 39 (4): 691–707.
Raphael, Ray, and Freeman House. 2007. *Two Peoples, One Place.* Eureka CA: Humboldt County Historical Society for the Writing Humboldt History Project.
Ray, Sarah Jaquette. 2013. *The Ecological Other: Environmental Exclusion in American Culture.* Tucson: University of Arizona Press.
Reed, R., and K. Norgaard. 2010. "Salmon Feeds Our People: Challenging Dams on the Klamath River." *Indigenous People and Conservation: From Rights to Resource Management*, 7–17. Arlington, VA: Conservation International.
Reo, N. J., and K. P. Whyte. 2012. "Hunting and Morality as Elements of Traditional Ecological Knowledge." *Human Ecology* 40 (1): 15–27.
Reynolds, Kristin. 2015. "Disparity Despite Diversity: Social Injustice in New York City's Urban Agriculture System." *Antipode* 47 (1): 240–259.
Richards, R. T., and M. Creasy. 1996. "Ethnic Diversity, Resource Values, and Ecosystem Management: Matsutake Mushroom Harvesting in the Klamath Bioregion." *Society & Natural Resources* 9 (4): 359–374.
Richardson, T. 2011. "Navigating the Problem of Inclusion as Enclosure in Native Culture-Based Education: Theorizing Shadow Curriculum." *Curriculum Inquiry* 41 (3): 332–349.
Robbins, P. 2004. *Political Ecology: A Critical Introduction.* New York: Blackwell.
Robertson, Dwanna L. 2013a. "Navigating Indigenous Identity." PhD diss., University of Massachusetts Amherst.
———. 2013b. "A Necessary Evil: Framing an American Indian Legal Identity." *American Indian Culture and Research Journal* 37 (4): 115–140.
———. 2016. "Decolonizing the Academy with Subversive Acts of Indigenous Research: A Review of Yakama Rising and Bad Indians." *Sociology of Race and Ethnicity*, 248–252.
Robinson, C. J. 2000. *Black Marxism: The Making of the Black Radical Tradition.* Chapel Hill: University of North Carolina Press.
Roht-Arriaza, N. 1995. "Of Seeds and Shamans: The Appropriation of the Scientific and Technical Knowledge of Indigenous and Local Communities." *Michigan Journal of International Law* 17: 919.
Ross, L. 2009. "From the 'F' Word to Indigenous/Feminisms." *Wicazo Sa Review* 24 (2): 39–52.
Roughgarden, Joan. 2013. *Evolution's Rainbow: Diversity, Gender, and Sexuality in Nature and People.* Berkeley: University of California Press.
Ruffin, Kimberly N. 2010. *Black on Earth: African American Ecoliterary Traditions.* Athens: University of Georgia Press.
Sabzalian, L. 2018. "Curricular Standpoints and Native Feminist Theories: Why Native Feminist Theories Should Matter to Curriculum Studies." *Curriculum Inquiry* 48 (3): 359–382.
Salleh, Ariel Kay. 1984. "Deeper Than Deep Ecology: The Eco-Feminist Connection." *Environmental Ethics* 6 (4): 339–345.
Salmón, Enrique. 2000. "Kincentric Ecology: Indigenous Perceptions of the Human–Nature Relationship." *Ecological Applications* 10 (5): 1327–1332.

Salter, John F. 2003. *A Context Statement concerning the Effect of the Klamath Hydroelectric Project on Traditional Resource Uses and Cultural Patterns of the Karuk People within the Klamath River Corridor.* White Paper on Behalf of the Karuk Tribe of California, 1–82. https://sipnuuk.mukurtu.net/system/files/atoms/file/AFRIFoodSecurity_UCB_Jennifer Sowerwine_001_009.pdf.

Saperstein, A., A. M. Penner, and R. Light. 2013. "Racial Formation in Perspective: Connecting Individuals, Institutions, and Power Relations." *Annual Review of Sociology* 39:19–39.

Saunders, P. 1998. "The Quiet Rebellion: Chinese Miners Accepted in Orleans Despite 1885 Expulsion." *Humboldt Historian* (Summer): 11–19.

Scheff, Thomas. 1994. *Microsociology: Discourse, Emotion, and Social Structure.* Chicago: University of Chicago Press.

———. 2000. "Shame and the Social Bond: A Sociological Theory." *Sociological Theory* 18 (1): 84–99.

———. 2014. "The Ubiquity of Hidden Shame in Modernity." *Cultural Sociology* 8 (2): 129–141.

Schlosberg, D. 2013. "Theorising Environmental Justice: The Expanding Sphere of a Discourse." *Environmental Politics* 22 (1): 37–55.

Schmidt, R. W. 2011. "American Indian Identity and Blood Quantum in the 21st Century: A Critical Review." *Journal of Anthropology* 17:1–9.

Schrock, D., and M. Schwalbe. 2009. "Men, Masculinity, and Manhood Acts." *Annual Review of Sociology* 35:277–295.

Schwalbe, Michael, Daphne Holden, Douglas Schrock, Sandra Godwin, Shealy Thompson, and Michele Wolkomir. 2000. "Generic Processes in the Reproduction of Inequality: An Interactionist Analysis." *Social Forces* 79 (2): 419–452.

Scott, James C. 1998. *Seeing Like a State: How Certain Schemes to Improve the Human Condition Have Failed.* New Haven, CT: Yale University Press.

———. 2008. *Weapons of the Weak: Everyday Forms of Peasant Resistance.* New Haven, CT: Yale University Press.

Secrest, W. B. 2003. *When the Great Spirit Died: The Destruction of the California Indians, 1850–1860.* Quill Driver Books.

Segal, Lynne. 1990. *Slow Motion: Changing Masculinities, Changing Men.* New Brunswick, NJ: Rutgers University Press.

Senos, R., F. K. Lake, N. Turner, and D. Martinez. 2006. "Traditional Ecological Knowledge and Restoration Practice." In *Restoring the Pacific Northwest: The Art and Science of Ecological Restoration in Cascadia*, edited by D. Apostol and M. Sinclair, 393–426. Washington, DC: Island Press.

Sherman, Jennifer. 2009. "Bend to Avoid Breaking: Job Loss, Gender Norms, and Family Stability in Rural America." *Social Problems* 56: 599–620.

Shilling, F. M., J. K. London, and R. S. Liévanos. 2009. "Marginalization by Collaboration: Environmental Justice as a Third Party in and beyond CALFED." *Environmental Science & Policy* 12 (6): 694–709.

Show, S. B., and E. I. Kotok. 1923. *Forest Fires in California, 1911–1920: An Analytical Study.* Department Circular #243. Washington, DC: United States Department of Agriculture, U.S. Government Printing Office.

Shriver, Thomas, and Gary Webb. 2009. "Rethinking the Scope of Environmental Injustice: Perceptions of Health Hazards in a Rural Native American Community Exposed to Carbon Black." *Rural Sociology* 74 (2): 270–292.

Silvern, S. E. 1999. "Scales of Justice: Law, American Indian Treaty Rights and the Political Construction of Scale." *Political Geography* 18 (6): 639–668.

Simonds, V. W., and S. Christopher. 2013. "Adapting Western Research Methods to Indigenous Ways of Knowing." *American Journal of Public Health* 103 (12): 2185–2192.

Simonds, V. W., N. Wallerstein, B. Duran, and M. Villegas. 2013. "Peer Reviewed: Community-Based Participatory Research: Its Role in Future Cancer Research and Public Health Practice." *Preventing Chronic Disease* 10. DOI: http://dx.doi.org/10.5888/pcd10.120205.

Simpson, Audra, and Andrea Smith. 2014. *Theorizing Native Studies.* Durham, NC: Duke University Press.

Simpson, Leanne. 2001. "Aboriginal Peoples and Knowledge: Decolonizing Our Processes." *Canadian Journal of Native Studies* 21 (1): 137–148.

———. 2003. "Toxic Contamination Undermining Indigenous Food Systems and Indigenous Sovereignty." *Pimatiziwin: A Journal of Aboriginal and Indigenous Community Health* 1 (2): 130–134.

Simpson, L. R. 2004. "Anticolonial Strategies for the Recovery and Maintenance of Indigenous Knowledge." *American Indian Quarterly* 28 (3): 373–384.

Simpson, Leanne Betasamosake. 2013. *Islands of Decolonial Love.* Winnipeg: Arbeiter Ring Publishing.

———. 2017. *As We Have Always Done: Indigenous Freedom through Radical Resistance.* Minneapolis: University of Minnesota Press.

Simpson, L. & Coulthard, G. 2014. "Leanne Simpson and Glen Coulthard on Dechinta Bush University, Indigenous Land-Based Education and Embodied Resurgence" [Web blog post]. https://decolonization.wordpress.com/2014/11/26/leanne-simpson-and-glen-coulthard-on-dechinta-bush-university-indigenous-land-based-education-and-embodied-resurgence/.

Skinner, Carl N., Alan H. Taylor, and James K. Agee. 2006. "Klamath Mountains Bioregion." *Fire in California's Ecosystems,* edited by N. G. Sugihara, J. W. van Wagtendonk, J. Fites-Kaufmann, K. E. Shaffer, and A. E. Thode, 170–194. Berkeley: University of California Press.

Smith, Andrea. 2005. "Native American Feminism, Sovereignty, and Social Change." *Feminist Studies* 31 (1): 116–132.

———. 2012. "Indigeneity, Settler Colonialism, White Supremacy." In *Racial Formation in the Twenty-First Century,* edited by Daniel Martinez HoSang, Oneka LaBennett, and Laura Pulido, 66–90. Berkeley: University of California Press.

Smith, Andrea, and J. Kēhaulani Kauanui. 2008. "Native Feminisms Engage American Studies." *American Quarterly* 60 (2): 241–249.

Smith, H. A., and K. Sharp. 2012. "Indigenous Climate Knowledges." *Wiley Interdisciplinary Reviews: Climate Change* 3:467–476.

Smith, Linda Tuhiwai. 2013. *Decolonizing Methodologies: Research and Indigenous Peoples.* London: Zed Books Ltd.

Smith, S., S. Jacob, M. Jepson, and G. Israel. 2003. "After the Florida Net Ban: The Impacts on Commercial Fishing Families." *Society & Natural Resources* 16 (1): 39–59.

Smith-Lovin, Lynn. 2007. "The Strength of Weak Identities: Social Structural Sources of Self, Situation and Emotional Experience" *Social Psychological Quarterly* 70 (2): 106–124.

Snipp, C. Matthew. 1986. "The Changing Political and Economic Status of the American Indians: From Captive Nations to Internal Colonies." *American Journal of Economics and Sociology* 45 (2): 145–158.

———. 1992. "Sociological Perspectives on American Indians." *Annual Review of Sociology* 18 (1): 351–371.

Snyder, John O. 1931. *Fish Bulletin No. 34. Salmon of the Klamath River California. I. The Salmon and the Fishery of Klamath River. II. A Report on the 1930 Catch of King Salmon in Klamath*

River. UC San Diego: Library—Scripps Collection. https://escholarship.org/uc/item/6bx937pf.

Spence, M. D. 1999. *Dispossessing the Wilderness: Indian Removal and the Making of the National Parks*. Oxford: Oxford University Press.

Stark, H. 2010. "Respect, Responsibility, and Renewal: The Foundations of Anishinaabe Treaty Making with the United States and Canada." *American Indian Culture and Research Journal* 34 (2): 145–164.

Steffen, Will, Paul J. Crutzen, and John R. McNeill. 2007. "The Anthropocene: Are Humans Now Overwhelming the Great Forces of Nature." *Ambio: A Journal of the Human Environment* 36 (8): 614–621.

Steinman, Erich W. 2012. "Settler Colonial Power and the American Indian Sovereignty Movement: Forms of Domination, Strategies of Transformation." *American Journal of Sociology* 117 (4): 1073–1130.

———. 2016. "Decolonization Not Inclusion: Indigenous Resistance to American Settler Colonialism." *Sociology of Race and Ethnicity* 2 (2): 219–236.

———. 2018. "Why Was Standing Rock and the #NoDAPL Campaign So Historic? Factors Affecting American Indian Participation in Social Movement Collaborations and Coalitions." *Ethnic and Racial Studies*, 1–21. DOI: https://doi.org/10.1080/01419870.2018.1471215.

Steinmetz, George. 2013a. "A Child of the Empire: British Sociology and Colonialism, 1940s–1960s." *Journal of the History of the Behavioral Sciences* 49 (4): 353–378.

———, ed. 2013b. *Sociology and Empire: The Imperial Entanglements of a Discipline*. Durham, NC: Duke University Press.

———. 2014. "The Sociology of Empires, Colonies, and Postcolonialism." *Annual Review of Sociology* 40:77–103.

Steinmetz, K. F., B. P. Schaefer, and H. Henderson. 2017. "Wicked Overseers: American Policing and Colonialism." *Sociology of Race and Ethnicity* 3 (1): 68–81.

Stephens, Scott L., and Lawrence W. Ruth. 2005. "Federal Forest-Fire Policy in the United States." *Ecological Applications* 15 (2): 532–542.

Stephens, Scott L., and Neil G. Sugihara. 2006. "Fire Management and Policy since European Settlement." In *Fire in California's Ecosystems*, edited by N. G. Sugihara, 431–443. Berkeley: University of California Press.

Stoll, Shannan. 2016. "Transforming Regulatory Processes: Karuk Participation in the Klamath River Total Maximum Daily Load (TMDL) Process." Master's thesis, University of Oregon.

Stremlau, R. 2005. "To Domesticate and Civilize Wild Indians." *Journal of Family History* 30 (3): 265–286.

Stretesky, P., and M. J. Hogan. 1998. "Environmental Justice: An Analysis of Superfund Sites in Florida." *Social Problems* 45 (2): 268–287.

Sturgeon, N. 1997. *Ecofeminist Natures*. New York: Routledge.

———. 2009. *Environmentalism in Popular Culture: Gender, Race, Sexuality, and the Politics of the Natural*. Tucson: University of Arizona Press.

Suman, Seth. 2009. "Putting Knowledge in Its Place: Science, Colonialism, and the Postcolonial." *Postcolonial Studies* 12 (4): 373–388.

Swyngedouw, Erik, and Nik Heynen. 2003. "Urban Political Ecology, Justice and the Politics of Scale." *Antipode* 35 (5): 898–918.

Sze, J. 2007. *Noxious New York: The Racial Politics of Urban Health and Environmental Politics*. New York: New York University Press.

———, ed. 2018. *Sustainability: Approaches to Environmental Justice and Social Power*. New York: New York University Press.

Sze, J., J. London, F. Shilling, G. Gambirazzio, T. Filan, and M. Cadenasso. 2009. "Defining and Contesting Environmental Justice: Socio-Natures and the Politics of Scale in the Delta." *Antipode* 41 (4): 807–843.

TallBear, K. 2013. "Genomic Articulations of Indigeneity." *Social Studies of Science* 43 (4): 509–533.

———. 2014. "Standing with and Speaking as Faith: A Feminist-Indigenous Approach to Inquiry." *Journal of Research Practice* 10 (2): 17.

———. 2015. "An Indigenous Reflection on Working beyond the Human/Not Human." *GLQ: A Journal of Lesbian and Gay Studies* 21 (2): 230–235.

Tamez, M. 2016. "Indigenous Women's Rivered Refusals in El Calaboz." *Diálogo* 19 (1): 7–21.

Taylor, Dorceta E. 2009. *The Environment and the People in American Cities, 1600s–1900s: Disorder, Inequality, and Social Change*. Durham, NC: Duke University Press.

———. 2014. *Toxic Communities: Environmental Racism, Industrial Pollution, and Residential Mobility*. New York: New York University Press.

———. 2016. *The Rise of the American Conservation Movement: Power, Privilege, and Environmental Protection*. Durham, NC: Duke University Press.

Tengan, Ty P. Kāwika. 2008. *Native Men Remade: Gender and Nation in Contemporary Hawai'i*. Durham, NC: Duke University Press.

Teves, Stephanie. 2011. "'Bloodline Is All I Need': Defiant Indigeneity and Hawaiian Hip-Hop." *American Indian Culture and Research Journal* 35 (4): 73–101.

———. 2015. "Tradition and Performance." In *Native Studies Keywords*, edited by Stephanie Nohelani Teves, Andrea Smith, and Michelle Raheja. Tucson: University of Arizona Press.

Teves, Stephanie Nohelani, Andrea Smith, and Michelle Raheja, eds. 2015. *Native Studies Keywords*. Tucson: University of Arizona Press.

Thoits, Peggy. 2010. "Stress and Health: Major Findings and Policy Implications." *Journal of Health and Social Behavior* 51:S41–S53.

———.2012. "Emotional Deviance and Mental Disorder." In *Emotions Matter: A Relational Approach to Emotions*, edited by Alan Hunt, Kevin Walby, and Dale Spencer. Toronto: University of Toronto Press.

Timbrook, Jan, John Johnson, and David Earle. 1993. "Veg. Burning by the Chumash, Arrillaga's Proclamation May 31, 1793." In *Before the Wilderness: Environmental Management by Native Californians*, edited by Thomas Blackburn and Kat Anderson, 117–150. Menlo Park, CA: Ballena Press.

Todd, Z. 2014. "Fish Pluralities: Human-Animal Relations and Sites of Engagement in Paulatuuq, Arctic." *Études/Inuit/Studies* 38 (1–2): 217–238.

———. 2015. "Indigenizing the Anthropocene." In *Art in the Anthropocene: Encounters among Aesthetics, Politics, Environments and Epistemologies*, edited by H. Davis and E. Turpin, 241–254. London: Open Humanities Press.

Trafzer, Clifford E., and Diane Weiner, eds. 2001. *Medicine Ways: Disease, Health, and Survival among Native Americans*. Walnut Creek CA: AltaMira Press.

Trask, H. K. 1991. "Lovely Hula Lands: Corporate Tourism and the Prostitution of Hawaiian Culture." *Border/Lines* 23:22–29.

Trosper, R. L. 1995. "Traditional American Indian Economic Policy." *American Indian Culture and Research Journal* 19 (1): 65–95.

———. 2003. "Resilience in Pre-Contact Pacific Northwest Social Ecological Systems." *Conservation Ecology* 7 (3). http://www.consecol.org/vol7/iss3/art6.

Tsosie, Rebecca. 1996. "Tribal Environmental Policy in an Era of Self-Determination: The Role of Ethics, Economics, and Traditional Ecological Knowledge." *Vermont Law Review* 21:225.

———. 2003. "The Conflict between the 'Public Trust' and the 'Indian Trust' Doctrines: Federal Public Land Policy and Native Nations." *Tulsa Law Review* 39:271.
———. 2007. "Indigenous People and Environmental Justice: The Impact of Climate Change." *University of Colorado Law Review* 78:1625.
———. 2013. "Climate Change and Indigenous Peoples: Comparative Models of Sovereignty." *Tulane Environmental Law Journal* 26 (2): 239–257.
Tuana, Nancy. 1989. *Feminism and Science*. Bloomington: Indiana University Press.
Tuck, E. 2009. "Suspending Damage: A Letter to Communities." *Harvard Educational Review* 79 (3): 409–428.
Tuck, E., and R. A. Gaztambide-Fernández. 2013. "Curriculum, Replacement, and Settler Futurity." *Journal of Curriculum Theorizing* 29 (1): 72–89.
Tuck, E., and K. W. Yang. 2012. "Decolonization Is Not a Metaphor." *Decolonization: Indigeneity, Education & Society* 1 (1): 1–40.
Turner, Nancy J. 2008. *The Earth's Blanket: Traditional Teachings for Sustainable Living*. Vancouver: Douglas and McIntyre Publishers.
Udel, Lisa J. 2001. "Revision and Resistance: The Politics of Native Women's Motherwork." *Frontiers: A Journal of Women Studies* 22 (2): 43–62.
U.S. Department of Agriculture (USDA). 2012. "The National Cohesive Wildland Fire Management Strategy: Phase III Western Science-Based Risk Analysis Report. Final report of the Western Regional Strategy Committee." https://www.forestsandrangelands.gov/strategy/documents/reports/phase3/WesternRegionalRiskAnalysisReportNov2012.pdf.
Varese, S., and A. Chirif. 2006. *Witness to Sovereignty: Essays on the Indian Movement in Latin America*. Copenhagen: Iwgia.
Vasquez, J. M., and C. Wetzel. 2009. "Tradition and the Invention of Racial Selves: Symbolic Boundaries, Collective Authenticity, and Contemporary Struggles for Racial Equality." *Ethnic and Racial Studies* 32 (9): 1557–1575.
Veracini, L. 2011. "Introducing: Settler Colonial Studies." *Settler Colonial Studies* 1 (1): 1–12.
———. 2013. "'Settler Colonialism': Career of a Concept." *Journal of Imperial and Commonwealth History* 41 (2): 313–333.
Verma, P., K. Vaughan, K. Martin, E. Pulitano, J. Garrett, and D. D. Piirto. 2016. "Integrating Indigenous Knowledge and Western Science into Forestry, Natural Resources, and Environmental Programs." *Journal of Forestry* 114 (6): 648–655.
Vickery, J., and L. M. Hunter. 2016. "Native Americans: Where in Environmental Justice Research?" *Society & Natural Resources* 29 (1): 36–52.
Vinyeta, K., K. Whyte, and K. Lynn. 2016a. *Climate Change through an Intersectional Lens: Gendered Vulnerability and Resilience in Indigenous Communities in the United States*. Washington, DC: U.S. Forest Service.
———. 2016b. "Indigenous Masculinities in a Changing Climate." In *Men, Masculinities and Disaster*, edited by Elaine Enarson and Bob Pease, 140–151. New York: Routledge.
Viramontes, Helena María. 1996. *Under the Feet of Jesus*. New York: Penguin.
Vizenor, Gerald, ed. 2008. *Survivance: Narratives of Native Presence*. Lincoln: University of Nebraska Press.
Voyles, Traci Brynne. 2015a. "Environmentalism in the Interstices: California's Salton Sea and the Borderlands of Nature and Culture." *Resilience: A Journal of the Environmental Humanities* 3 (1): 211–241.
———. 2015b. *Wastelanding: Legacies of Uranium Mining in Navajo Country*. Minneapolis: University of Minnesota Press.

Wald, Sarah D. 2016. *The Nature of California: Race, Citizenship, and Farming since the Dust Bowl.* Seattle: University of Washington Press.

Warner, Elizabeth Ann Kronk. 2015. "Working to Protect the Seventh Generation: Indigenous Peoples as Agents of Change." *Santa Clara Journal of International Law* 13:273.

Watts, V. 2013. "Indigenous Place-Thought and Agency amongst Humans and Non-Humans (First Woman and Sky Woman Go on a European World Tour!)." *Decolonization: Indigeneity, Education & Society* 2 (1): 20–34.

Weatherspoon, C. Phillip, and Carl N. Skinner. 1995. "An Assessment of Factors Associated with Damage to Tree Crowns from the 1987 Wildfires in Northern California." *Forest Science* 41 (3): 430–451.

Weaver, Jace. 1996. *Defending Mother Earth: Native American Perspectives on Environmental Justice.* Maryknoll, NY: Orbis.

———. 1997. *That the People Might Live: Native American Literatures and Native American Community.* Oxford: Oxford University Press on Demand.

West, C., and D. H. Zimmerman. 1987. "Doing Gender." *Gender & Society* 1 (2): 25–151.

Whitbeck, Les, Xiaojin Chen, Dan Hoyt, and Gary Adams. 2004. "Discrimination, Historical Loss and Enculturation: Culturally Specific Risk and Resiliency Factors for Alcohol Abuse among American Indians." *Journal of Studies on Alcohol and Drugs* 65 (4): 409.

Whyte, K. 2013a. "Justice Forward: Tribes, Climate Adaptation and Responsibility." *Climatic Change* 120 (3): 517–530.

———. 2013b. "On the Role of Traditional Ecological Knowledge as a Collaborative Concept: A Philosophical Study." *Ecological Processes* 2:1–12.

———. 2015. "Indigenous Food Systems, Environmental Justice and Settler Industrial States." In *Global Food, Global Justice: Essays on Eating under Globalization*, edited by M. Rawlinson and C. Ward, 143–156. Cambridge: Cambridge Scholars Publishing.

———. 2016a. "Indigeneity." In *Keywords for Environmental Studies*, edited by J. Adamson, W. Gleason, and D. Pellow, 143–144. New York: New York University Press.

———. 2016b. "Indigeneity and U.S. Settler Colonialism." In *Oxford Handbook of Philosophy and Race*, edited by N. Zack, 91–101. Oxford: Oxford University Press.

———. 2016c. "Indigenous Experience, Environmental Justice and Settler Colonialism." In *Nature and Experience: Phenomenology and the Environment*, edited by B. Bannon, 157–174. New York: Rowman and Littlefield.

———. 2016d. "Indigenous Food Sovereignty, Renewal and U.S. Settler Colonialism." In *The Routledge Handbook of Food Ethics*, edited by M. Rawlinson and C. Ward, 354–365. New York: Routledge.

———. 2017a. "Is It Colonial Déjà vu? Indigenous Peoples and Climate Injustice." In *Humanities for the Environment: Integrating Knowledges, Forging New Constellations of Practice*, edited by J. Adamson, M. Davis, and H. Huang, 88–104. London: Earthscan Publications.

———. 2017b. "The Dakota Access Pipeline, Environmental Injustice and U.S. Colonialism." *Red Ink—An International Journal of Indigenous Literature, Arts and Humanities* 19 (1): 154–169.

———. 2018a. "Critical Investigations of Resilience: A Brief Introduction to Indigenous Environmental Studies & Sciences." *Daedalus* 147 (2): 136–147.

———. 2018b. "Indigenous Science (Fiction) for the Anthropocene: Ancestral Dystopias and Fantasies of the Climate Change Crisis." *Environment & Planning E: Nature and Space* 1 (1–2): 224–242.

———. 2018c. "Food Sovereignty, Justice and Indigenous Peoples: An Essay on Settler Colonialism and Collective Continuance." In *Oxford Handbook of Food Ethics*, edited by A. Barnhill, T. Doggett, and A. Egan, 345–366. Oxford: Oxford University Press.

———. 2018d. "What Do Indigenous Knowledges Do for Indigenous People." In *Traditional Ecological Knowledge: Learning from Indigenous Practices for Environmental Sustainability*, edited by Melissa K. Nelson and Daniel Shilling, 57–81. Cambridge: Cambridge University Press.

———. Forthcoming. "Way Beyond the Lifeboat: An Indigenous Allegory of Climate Justice." In *Climate Futures: Reimagining Global Climate Justice*, edited by D. Munshi, K. Bhavnani, J. Foran, and P. Kurian. Berkeley: University of California Press.

Whyte, K., C. Caldwell, and M. Schaefer. 2018. "Indigenous Lessons about Sustainability Are Not Just for 'All Humanity.'" In *Sustainability: Approaches to Environmental Justice and Social Power*, edited by J. Sze, 149–179. New York: New York University Press.

Widick, Richard. 2009. *Trouble in the Forest: California's Redwood Timber Wars*. Minneapolis: University of Minnesota Press.

Wildcat, Daniel. 2010. *Red Alert! Saving the Planet with Indigenous Knowledge*. Golden, CO: Fulcrum Publishing.

Wilkes, Rima. 2017. "Settler Colonialism and Indigenous Resurgence." Presentation at meeting of the American Sociological Association, Montreal Canada, August 12, 2017.

Wilkes, Rima, and Michelle M. Jacob. 2006. "Introduction to Indigenous Peoples: Canadian and US Perspectives." *American Behavioral Scientist* 50 (4): 423–427.

Wilkins, Amy. 2012. "'Not Out to Start a Revolution': Race, Gender, and Emotional Restraint among Black University Men." *Journal of Contemporary Ethnography* 41 (1): 34–65.

Wilkins, Amy C., and Jennifer A. Pace. 2014. "Class, Race, and Emotions." In *Handbook of the Sociology of Emotions: Volume II*, edited by J. Stets and J. T. Turner, 385–409. Dordrecht: Springer.

Wilkinson, Charles F. 2005. *Blood Struggle: The Rise of Modern Indian Nations*. New York: W. W. Norton.

Willette, Mirranda, Kari Norgaard, and Ron Reed. 2016. "You Got to Have Fish: Families, Environmental Decline and Cultural Reproduction." *Families, Relationships and Societies* 5 (3): 375–392.

Williams, D. R., and M. Sternthal. 2010. "Understanding Racial-Ethnic Disparities in Health: Sociological Contributions." *Journal of Health and Social Behavior* 51:S15–S27.

Williams, T., and P. Hardison. 2013. "Culture, Law, Risk and Governance: Contexts of Traditional Knowledge in Climate Change Adaptation." *Climatic Change* 120:531–544.

Willox, A. C. 2012. "Climate Change as the Work of Mourning." *Ethics & the Environment* 17 (2): 137–164.

Willox, A. C., S. L. Harper, J. D. Ford, V. L. Edge, K. Landman, K. Houle, S. Blake, and C. Wolfrey. 2013. "Climate Change and Mental Health: An Exploratory Case Study from Rigolet, Nunatsiavut, Canada." *Climatic Change* 121 (2): 255–270.

Willox Cunsolo, A., and N. R. Ellis. 2018. "Ecological Grief as a Mental Health Response to Climate Change-Related Loss." *Nature Climate Change* 8 (4): 275.

Wilson, Paul G. 2001. "The Legacy of the Log Boom: Humboldt County Logging from 1945 to 1955." http://humboldt-dspace.calstate.edu/bitstream/handle/10211.3/132765/Wilson_Paul_Barnum_otf.pdf.

Wilson, Shawn Stanley. 2004. "Research as Ceremony: Articulating an Indigenous Research Paradigm." PhD diss., Monash University, School of Humanities, Communications and Social Sciences.

Wilson, Shawn. 2008. *Research Is Ceremony: Indigenous Research Methods*. Halifax: Fernwood Publishing.

Winant, Howard. 2004. *The New Politics of Race: Globalism, Difference, Justice*. Minneapolis: University of Minnesota Press.

Wing, S. 1994. "Limits of Epidemiology." *Medicine and Global Survival* 1 (2): 74–86.
Wingfield, Adia Harvey. 2010. "Are Some Emotions Marked 'Whites Only'? Racialized Feeling Rules in Professional Workplaces." *Social Problems* 57 (2): 251–268.
Witz, Anne. 2000. "Whose Body Matters? Feminist Sociology and the Corporeal Turn in Sociology and Feminism." *Body & Society* 6 (2): 1–24.
Wofford, P., K. Goh, D. Jones, H. Casjens, H. Feng, J. Hsu, D. Tran, J. Medina, and J. White. 2003. *Forest Herbicide Residues in Surface Water and Plants in the Tribal Territory of the Lower Klamath River Watershed of California.* Sacramento, CA: California Environmental Protection Agency Department of Pesticide Regulation.
Wolfe, Patrick. 2006. "Settler Colonialism and the Elimination of the Native." *Journal of Genocide Research* 8 (4): 387–409.
Wood, Mary Christina. 1994. "Indian Land and the Promise of Native Sovereignty: The Trust Doctrine Revisited." *Utah Law Review*, 1471–1569.
———. 2000. "The Tribal Property Right to Wildlife Capital (Part I): Applying Principles of Sovereignty to Protect Imperiled Wildlife Populations." *Idaho Law Review* 37:1.
Wright, Harold A., Henry A. Wright, and Arthur W. Bailey. 1982. *Fire Ecology: United States and Southern Canada.* New York: John Wiley & Sons.
Ybarra, Priscilla Solis. 2016. *Writing the Goodlife: Mexican American Literature and the Environment.* Tucson: University of Arizona Press.
Zalasiewicz, Jan, Mark Williams, Will Steffen, and Paul Crutzen. 2010. "The New World of the Anthropocene." *Environmental Science and Technology* 44 (7): 2228–2231.
Zimring, Carl A. 2017. *Clean and White: A History of Environmental Racism in the United States.* New York: New York University Press.

INDEX

academic research, Indigenous collaboration in, 135–137, 154–156, 158–159
Achviivich. *See* Arwood, David
acorns: as actively managed, 10; in creation stories, 151; and cultural reproduction, 93, 109, 162, 185, 207; denial of access to, 12, 66, 114, 116, 145, 146, 153, 210; destruction of as genocide, 150; fire as beneficial for, 2, 12, 41, 89, 97, 109; and fire retardants, 122–123; and gender constructions, 166, 182, 186; health consequences from lack of, 149, 151; in Karuk diet, 42, 66, 143, 144, 153, 158, 185, 208; as relatives, 160, 171; and reorganization of natural world, 56, 76; and transfer of knowledge, 92; as tribal identity, 186, 203, 207; yields of, 100, 146
"Act for the Government and Protection of Indians" (1850), 48, 94
activism: for dam removal, 191, 220; masculine identity as, 191–193; and sociology, 238; and traditional responsibilities, 192–193. *See also* resistance
Adelson, Naomi, 163
affective emotions, 219
Agent Orange, 107
agriculture, 55, 57–59
Akaba, Azibuike, 134
alcohol use: and cosmology of connection, 149, 150; and environmental decline, 151; and gender identity, 184; and role strain, 187; settler acquisition of land through, 174; and social capital, 192, 205
Alfred, Taiaiake, 72, 108
algae, 12, 42, 203, 236
Alkon, Alison Hope, 163
allotment system, 58–60, 102. *See also* Dawes Act
Almaguer, Tomas, 36
American Diabetes Association, 153
American Journal of Sociology, 232
American Sociological Association (ASA): lack of Indigenous-centered theories in, 136; lack of section on colonialism in, 79, 128, 224; lack of section on Indigenous peoples, 6, 79, 128, 224
Anderson, Kat, 3, 4, 54, 55, 89, 90–91
anger, 187, 209–211, 210t, 219t
Anishinaabek people, 139, 236, 237–238
Anson, April, 230
Anthony, Carl, 30
"Anthropocene," 195, 226–227, 230–231
anthropology, 15, 29–30, 176, 194
anti-capitalism, Indigenous people and, 83–84
anxiety, 190
apathy to climate change, 235, 236
apocalypse, privilege and notions of, 229–230
Arendt, Hannah, 232, 234
Arwood, David, 61, 64, 65–66, 210–211, 215
assimilation. *See* forced assimilation
Association of Black Sociologists, 224
asthma, 69
atomism, 154–155
attendance policies, forced assimilation and, 115
attitude, food preparation and, 149–150, 182
Aubry, Earl "Scrub," 209

Bacon, Jules: "colonial ecological violence," 93, 110, 172, 190, 209; on pathologizing Indigenous people, 136, 214
Baldy, Cutcha Risling, 118, 233
barter economy, 205
basketweaving, 64–65, 100, 171, 187, 204, 214
Beale, Edward F., 52, 55
Beaucage, Glenna, 31, 84
"being Indian," 11, 93
Bigler, John, 51, 60
Biko, Steven, 200
Binx. *See* Brink, Kenneth "Binx"
biocultural sovereignty, 118, 233
birth defects, 107
bison, American, 38, 150
black settlers, 40
blood quantum, 36, 38, 44, 48
boarding schools, 6, 61, 173, 251n26

283

Bonilla-Silva, Eduardo, 26, 199–200
Boscana, Geronimo, 51–52
Botany of Desire (Pollan), 161
bounties on Indigenous people, 6, 173
Bowman, David M. J. S., 112
Brave Heart, M. Y., 113, 217, 218
Brink, Kenneth "Binx": on family responsibilities, 41; on fishermen as providers, 188, 204; on importance of salmon, 190, 209; on intergenerational knowledge transfer, 92, 162; on regulations, 9; on salmon restoration, 220; on use of fire for river health, 90
Brown, Phil, 157
Bryson, Lois, 194–195
Bureau of Indian Affairs (BIA), 59–60, 69, 97, 119
Burnett, Peter, 50

California: abundance of natural resources in, 25; bounty hunters in, 6, 173; emergence of capitalism in, 47–48, 49, 54, 101, 114, 228; forced relocation of Indigenous people, 52–53; genocide in, 6, 38, 39, 50–51, 57, 228; militias in, 50–51, 57; population of Karuk tribe in, 4–5, 50, 209, 251n19; racial formation in, 26, 30, 36, 39, 47, 51; refusal to recognize Karuk land title, 56; transfer of wealth to non-Indigenous settlers, 114; treaties with Indigenous tribes, 49–50; water quality standards in, 132
California Constitution, prohibition of slavery in, 40
California Constitutional Convention, 48
California Department of Fish and Wildlife, 62, 119
California Department of Forestry and Fire Protection (CALFIRE), 119, 121, 228–229
California Fish and Game Act (1852), 60
California Research Bureau, 48
California Rural Indian Health Board, 142
California State Legislature, 48, 75, 94
California Water Board, 119
cancers, 107
canneries, 42
Cantrell, Betty, 149–150
Cantzler, Julia Miller, 233, 237
capitalism, 26–29, 31, 48, 83–84, 233–234
Captain Jack's Stronghold, 57
Carceras, Berta, 234–235

Cartesian reductionism, 144
Casas, T., 169
Casey, Jim, 98, 208
census data, 174
ceremonies: and fishing, 10, 207; and forced assimilation, 115; to renew the world, 9, 40, 59, 181–182, 210, 213; as resistance, 220; use of fire in, 40, 213
Champagne, Duane, 20
China, 39
Chinese Exclusion Laws, 40
Chinese people in Klamath region, 39–40
Chinook salmon. *See* salmon
Civilian Conservation Corps, 96
civilization as inevitable, 57–58
civilized *vs.* savage dualism, 47
climate change: awareness of current management practices, 8, 223; collective denial of, 235; and colonialism, 126, 155, 224, 226, 229; and cross-species connections, 14; dominant society's response to, 228; and fire suppression, 125–126, 227–228; holistic approaches to, 234; as human caused, 126; importance of sociology in, 238–239; Indigenous perspectives on, 104–105, 106, 226–238; as nothing new, 227–229; as about race and racism, 70; as strategic opportunity, 234–238
climate injustice, settler-colonialism and, 47–48, 49, 54, 101, 114, 228–229
climate resistance, Indigenous people and, 220, 226–238, 236–238
Climate Vulnerability Assessment (Karuk Tribe), 228
Cochran, P. A., 135
cod fisheries, 205
coho salmon. *See* salmon
collaborations: of Indigenous in academic research, 132–135, 140–143, 145, 154–159; in Karuk Department of Natural Resources, 139
collective continuance, 180, 183, 185–186, 189, 190–191, 192
collective culture, emotional toll of loss of, 151
collective engagement, 235
Collins, Patricia Hill, 196, 200
colonial amnesia in sociology, 82
"colonial ecological violence," 93, 110, 127, 172, 190, 209

colonialism: climate change as outgrowth of, 126, 224, 226, 229; collective continuance as resistance to, 23, 177, 190–194, 197; and ecological erasure, 14–15, 17, 43, 46–48, 54–56, 73, 78–79, 87–88, 93–102, 109–110, 197, 233; and environmental health research, 17, 157–158; and gender constructions, 167–168; and grief, 113, 202–209, 204t, 217, 219t; Indigenous resistance to, 18–19, 52–53, 97, 110, 116–118, 127, 137, 156, 162, 175, 177, 190–196, 220, 233; lack of section in ASA, 79, 128, 224; and land management policies, 32, 72–128; and loss of traditional roles, 184, 187–188, 196, 214, 219; as metaphor, 80; and misrepresentation of Indigenous cosmologies, 129; and nature-society dualism, 232; as ongoing, 15, 17, 20, 73, 78–79, 82; police activity in, 95; and racial formation, 26–27, 34–38, 39–47, 56, 60, 66–67; and the state, 31–32; theories of, 8, 79–80, 82, 196
"colonial politics of recognition," 118
colonial violence: environmental decline as, 12, 103; fire suppression as, 94–102; as gendered, 169–170, 172–175, 177, 194, 195
commodity food programs, 69, 113, 143, 153
commodity production: extraction of, 41, 48; and fire suppression policies, 101–102, 114; natural resources as, 83–84; and regulations on hunting, 62–65. *See also* extractive management activities; "primitive accumulation"
community-based science, 140, 147, 158
Comte, Auguste, 176
Connell, Raewyn, 177, 186
Conners, Pamela, 212
Conrad, Joseph, 229
containment, 109
contamination. *See* toxins
cosmologies, colonial, 11–12, 15, 63t, 155, 229
cosmologies, Indigenous: and conceptions of power, 232–234; in environmental health research, 133, 135, 139, 154–156, 158–159; interconnections within, 154–156; misrepresentation of in colonialism, 129; responsibility to animate nature in, 91–93
cosmologies of connection. *See* "kincentricity"

Coulthard, Glen: on anti-capitalism and land, 28–29; dispossession and the state, 57, 108; on grounded normativity, 41, 111; on political sovereignty, 118; on settler-colonialism and dispossession, 88; on settler-colonialism and restructuring of power, 85, 105
court victories, 238
"covenant of reciprocity," 92–93
Creasy, M., 39
creation stories, 91, 93, 151, 171, 185
criminalization of traditional management practices, 25, 60, 61–67, 97, 115. *See also* harassment by law enforcement; imprisonment; regulations
crisis management mode, long-term impacts of, 120, 124
"critical environmental justice," 29
Cronise, Titus Fey, 25
cross-species responsibilities, 12–13, 14, 91–93, 137–138, 173, 179, 191. *See also* "kincentricity"
Crutzen, Paul J., 230
cultural artifacts, damage to, 96–97, 103–104, 120, 121, 228–229
cultural genocide: denial of use of fire as, 109–110; environmental degradation as, 13, 71, 189–190, 207, 209, 225
cultural practices, Karuk: criminalization of as forced assimilation, 61–67, 97, 115–116; replaced by extractive technologies, 55–56
cultural production, traditional knowledge and, 110–111, 177–178
cultural roles: colonialism and loss of, 184, 187–188, 196, 214, 219; importance of, 177–178; and management activities, 93
cultural sovereignty, 118
cultural survival, tradition and, 178. *See also* collective continuance
"cultural use species," 11, 13, 21, 116, 162, 229
cultural values and traditional knowledge, 92, 110–111
culture, loss of, 151–152, 205–206. *See also* collective continuance; forced assimilation; genocide
Curtis, Edward, 38
customs, Karuk, 57

286   Index

dams: activism against, 191, 220; and algal blooms, 42; and diabetes rates, 19, 132, 154; relicensing of, 129–131, 241; removal of, 5, 132, 133, 192; and salmon access, 12, 130, 209; and traditional foods, 12, 130, 132, 144
Davis, Vera Vena, 204
Dawes Act (Indian General Allotment Act, 1887), 6, 38, 47, 58–59
DeBruyn, Lemyra, 113, 217, 218
decolonization, 15, 82–88, 106, 133, 201, 232–233, 237, 249n4; of environmental justice, 19, 138–139, 156, 159–161
"Decolonization Is Not a Metaphor" (Tuck and Yang), 133, 138
deer: abundance of, 2, 25; commodity foods and decline of, 66; in creation stories, 151; denial of access to, 12; and extractive technologies, 55; fire suppression and decline of, 109; fire to make good habitat for, 2, 12, 41, 89–90, 97, 116; and herbicides, 107; hunting regulations on, 60, 62, 63t, 145, 210, 211; in Karuk diet, 143, 144, 158, 206; loss of as sign of the end, 209; providing of, 182, 183; and reorganization of natural world, 56; socialization while processing, 207. *See also* foods, traditional
Deer, Sarah, 172
Deloria, Philip J., 44
denial of access, 116, 132, 141, 210
diabetes: appearance of, 148f; economic costs of, 153; and environmental degradation, 157; loss of salmon and rise of, 19, 147–149, 148f; and loss of traditional foods, 69, 130, 132, 157; rates of, 147
diet-related diseases. *See* diabetes; heart disease
diets: historic Karuk, 143–144; traditional *vs.* store-bought foods in, 66
diet shifts: as forced assimilation, 12, 208; loss of culture and identity through, 151–152; and loss of traditional foods, 4, 69, 116, 130, 141, 143–144, 145; reasons for, 144–147. *See also* foods, traditional
"digger" as derogatory, 175
dioxin, 107
direct action, 220, 235, 236–238. *See also* activism; resistance
"disenfranchised grief," 202, 217–218
dispossession. *See* land dispossession
Doctrine of Discovery, 46–47, 56

Doka, Ken, 202, 217
"domestic dependent nations," 117–118
domestic violence, 184
dominance: and forced alteration of gender structures, 173; and masculinity, 189; as not a part of culture, 167
Douglas fir, 42, 76
Downey, Liam, 216
Du Bois, W.E.B., 78, 136
Dunbar-Ortiz, Roxanne, 1, 7, 16, 30, 39, 78
Dunlap, Riley, 169
Durkheim, Emile, 168

ecocide, 93
ecocultural vulnerability, 228
ecological alteration, political sovereignty and, 13–14, 46–47, 104–105, 117–118
ecological domination, racial domination and, 34
ecological erasure, colonialism and, 14–15, 17, 43, 46–48, 54–56, 73, 78–79, 87–88, 93–102, 109–110, 197, 233
"ecological imagination," 239
ecologies, 76, 87–88
educational system, 6, 217, 220
eels. *See* lamprey
"The Effects of Altered Diet on the Health of the Karuk People" (Norgaard and Reed), 19, 132, 157
elderberries, 198
elders, providing for, 11, 180f, 183, 189
elk: abundance of, 2, 25; commodity foods and decline of, 66; extractive technologies and decline of, 55; fire to make good habitat for, 2, 12, 41, 89, 97, 116; fire suppression and decline of, 100, 109; hunting regulations on, 60, 62, 145; in Karuk diet, 143, 144; socialization while processing, 207. *See also* foods, traditional
Elliott, Wallace, 53
emotional harm, 199, 200, 214
emotion norms, 218
emotions, 198–222; and decline of Klamath River, 189–190, 202, 205, 207, 208, 210, 212, 215, 215t, 220; of environmental decline and power structures, 18, 200–201, 204, 211, 214, 217, 218, 219–220, 219t, 222, 225; and identity, 200, 204, 204t, 205–206, 209, 210, 210t, 211, 212–214, 215t, 216–217, 216t, 218–219, 219t;

and intact landscapes, 201–202; and loss of salmon, 19, 150, 151, 202–203, 206, 210, 215, 217; and mental health, 200; and natural environment, 199–200, 201, 216, 225, 254n3 (chap. 5); and resistance, 219–220; as shared, 200; visibility of in sociology, 17–18, 217–218, 222. *See also* anger; disenfranchised grief; grief; hopelessness; shame
empire framework, 80–82, 86
Endangered Species Act (1973), 5
environment: and gender, 18; link to human health, 19–20; role of in wealth production, 28; use within sociology, 221
environmental degradation: alteration of social relationships by, 19, 30, 68–69, 92–93, 110–111, 193, 201–220, 216–217; as colonial violence, 12, 103, 110, 125, 177, 211; as cultural genocide, 13, 71, 109, 189–190, 207, 209, 225; and diabetes, 157; and diet shifts, 145; as disruption to Indigenous cultural management, 45, 55–56, 110–111, 121; elimination of racial categories through, 38; emotional experience of in sociology, 17–18, 199, 217–218, 222; emotions of and power structures, 200, 219–220, 219t, 225; and gendered violence, 18, 166–175, 187, 189–191, 193–197, 194; and gender identities, 182–183, 184, 193; and grief, 113, 202–209, 204t, 215, 217, 218, 219t; loss of traditional roles through, 188; and shame, 189, 202, 212–215, 215t, 218, 219t; as threat to collective continuance, 190–191; and wealth transfer, 190
environmental health, 253n2; collaboration with Indigenous people in, 158–159; and colonialism, 17, 157–158; interconnections in, 19–20, 139; and TEK, 137, 157–159
environmental justice, 253n2; and appeals to the state for protection, 160; community-based science in, 140; decolonization of, 19, 138–139, 156, 159–161; emotional dimension of, 199, 221–222, 225; erasure of Indigenous presence in, 225; expansion of, 198–199; Indigenous collaboration in, 135; Indigenous resistance to colonialism as, 19, 137; Indigenous visions for, 161; mental health impacts in, 19; nature-society dualism in, 221; and racial capitalism, 33; and racial formation, 30–31; responsibilities to animate nature in, 139, 160; shift of focus from wilderness in, 18
Environmental Protection Agency (EPA), 119, 124
environmental racism, 46, 52
environmental review policies, firefighting activities and, 124
"environmental sociology," 16, 29–30, 169
epidemiology, 144, 158
erasure of colonialism in sociology, 80–82
erasure of Indigenous presence: and alteration of gender structures, 168, 172–175, 197; and climate change, 226–227, 230–231; and fire suppression, 76, 101, 104–105; "firsts" lists as, 53, 101; from food sovereignty movement, 21, 137–138, 225; as ongoing, 43, 81–82, 103, 133; and settler-colonialism, 73, 133–134, 170, 221–222; in sociology, 6, 134, 136; through geographic names, 42–43, 250n15; and wilderness areas, 43, 250n16
Eureka, California, 40
extraction of Indigenous knowledge, 28, 241
extractive management activities, 28, 41–42, 55–56, 113–114, 250n2

Faber, Daniel, 70
families: forced reorganization of, 174; as resistance to racism, 196; separation of, 48, 61–62, 173, 251n26
family histories, use of, 143
family values, 162
fear of climate change, 235
Federal Energy Regulatory Commission (FERC), 19, 129, 132, 154
femininity, Karuk, 168, 172, 173, 175; and colonialism, 168, 170–171, 174–175, 188, 195–197; as resistance to colonialism, 175, 193–194. *See also* women, Karuk
feminist sociology: and gender constructions, 168–169; and Indigenous studies, 195–197; nature-social dualism in, 18, 175–177, 193–195
Fenelon, James: on civilized-savage dualism, 47; on food sovereignty, 137–138; on frontier genocide, 46; on holistic approach to climate change, 234; on infrastructure of empire, 95; on race formation and colonization, 27, 87; on race formation and land dispossession, 34, 36; on race formation of Indigenous people, 43–44, 73

Fiege, Mark, 7, 30
fire ceremonies, 40, 213
fire exclusion. *See* fire suppression
firefighting activities: damage to cultural resources by, 103–104, 120, 121, 228–229; denial of access to public lands by, 121; long-term implications of, 120–121, 124; and tribal sovereignty, 119–125, 228
Fire in California's Ecosystems (Stephens and Sugihara), 94
fire-line construction, 120, 124, 228–229, 252n18
fire management periods, 106–107
fire management policies, Karuk: communication of to U.S. Forest Service, 123–124; criminalization of, 97; ecological benefits of, 40–41, 89–91; and gender identity, 93; return of, 9, 104, 106, 121; as selfish, 98–99; and tribal sovereignty, 104–105, 117, 118, 119–125, 228; use of ridge systems in, 103–104
fire permits, 100–101
fire retardants, 120, 122–123, 123f, 228–229
fires: as cultural resource, 90; denial of as cultural genocide, 109–110; as destructive to forests, 59, 72, 94, 96; enhancement of forest through, 40–41, 90, 109; enhancement of traditional food sources through, 2, 12, 41, 74, 89, 90–91, 97, 109; expanding season of, 119; high-intensity, 32, 104, 111–112, 125–126, 227–228; low-intensity, 100; return of Karuk fire regimes, 9, 104, 106, 121; as tool for shaping ecology, 12–13, 40–41, 88–93, 98, 104–105, 110–111, 125; use of by Indigenous people, 40, 73–75, 89, 98, 213
fire suppression: accumulation of fuel through, 74, 77f, 100, 104, 121, 124; and climate change, 125–126; as colonial violence, 94–102, 103–105, 125; damage to historic trails, 122; destruction of cultural artifacts through, 96–97, 103–104, 120, 121, 228–229; establishment of, 75–76, 95–96; forced assimilation through, 109, 115–117; increase in high-severity fires, 76, 227–228; and Indigenous erasure, 76, 104–105; and land dispossession, 78, 97–98, 103–106, 106–110; maximizing of tree growth through, 102; military tactics in, 94–95, 97, 107, 120; and political sovereignty, 118; reduction in fires by Indigenous people, 98; and responsibilities to animate nature, 12–13, 105, 108–109, 110–111, 114; use of herbicides in, 107–108; use of scientific rhetoric in, 99. *See also* land management policies, Karuk; land management policies, nontribal; U.S. Forest Service
firewood, 65
First People, Indian potatoes as, 171
"first peoples," settlers as, 53, 101
First Salmon Ceremony, 10, 207
fish. *See* salmon
fish crews, 202, 220
fisheries programs, 158, 191–192
fishermen: anger of, 187; gender identity of, 181, 183, 184; as providers, 179, 180, 180f, 188, 204
fishing, 5f; and ceremonies, 10, 184, 207; criminalization of, 115; gender constructions of, 165–166, 179, 180, 186–189; and grief, 203; levels of, 145–146, 146f; moratorium on, 116; as responsibility to fish, 179, 189, 191–193
fishing regulations: anger at, 210–211; as forced assimilation, 12, 62, 66, 115
fishing rights, denial of, 60
food, store-bought, 65, 66, 69, 153
food assistance programs, 69, 113, 143, 153
food justice, 163
food preparation, attitude and, 149–150, 182
foods, traditional: affected by dams, 12, 130, 132, 144; barriers to, 65, 115, 132, 146, 211; enhanced by fire, 2, 74, 90–91; and fire retardants, 122–123; and food security, 152–153; and health, 72, 113, 130, 132, 148–149; mismanagement of, 12–13; as relations, 21, 161–162; shame and lack of, 151; as social glue, 151. *See also* acorns; deer; diet shifts; elk; lamprey; mushrooms; potatoes, Indian; salmon; sturgeon
food security, 113, 132, 152–153
food sovereignty movements, 21, 137–138, 162–164, 225
food studies, 20–21, 161
forced assimilation: alteration of gender structures as, 172–173; and blood quantum, 36, 38, 44, 48; and boarding schools, 6, 61, 173, 251n26; denial of use of fire as, 109, 115–117; as elimination of Indigenous category, 44, 61; and emotion norms, 218;

gendered resistance to, 194; illegality of Karuk cultural practices as, 61–67; land management actions through, 115–117; as ongoing, 46; and private land ownership, 58–59; resistance to, 220; and settler-colonialism, 114–115; through marriage, 38, 44, 174; and traditional foods, 12, 208
forced relocation of Indigenous people, 46, 48, 52–53
forest management. *See* fire management policies, Karuk; fire suppression; land management policies, Karuk; land management policies, nontribal
forest reserves, creation of, 96
forestry, Indigenous-Western collaborations in, 158
Fothergil, Alice, 194
frontier genocide, 38, 46. *See also* genocide
fry bread as "traditional food," 69
fuel, accumulation of, 74, 77f, 100, 104, 121, 124

garlon, 107–108
Garroutte, Eva Marie, 36, 38, 43, 44
gender, 165–197; and environment, 18, 179–183, 189–191, 194–195; Indigenous notions of, 165–168, 170–172, 177–178, 179–185, 188; and nature, 168–172, 175–176, 195–197; as social, 169, 176
gender arrangements, 170–171
gender binaries, 167, 172, 183
gender constructions: alteration of as erasure of Indigenous people, 172–175; and colonialism, 167–168; in feminist sociology, 168–169; and natural world, 175, 194–195; and responsibilities, 165–167, 182–183
gender expectations: and grief, 204–205; as less rigid among Karuk people, 178, 183; and management activities, 93
gender identity: and environmental degradation, 184; and fishing, 165–166, 179, 180, 186–189; and natural environment, 171–172, 180–182; and power relations, 188; and responsibilities to relations, 183–184; and sugar pine nuts, 205–206
General Integrating Inspection Report (Six Rivers National Forest, 1950), 99–100, 212
"General Recapitulation of the Expenditures incurred by the State of California For the Subsistence and Pay of the Troops," 50–51

genetic engineering, 230
genocide: at Captain Jack's Stronghold, 57; end of legal, 45; environmental degradation as, 17, 93, 108–110, 156, 160, 203, 207, 208–209, 214, 217, 225; gendered colonial violence as, 172; and grief, 207–209, 217; and Indigenous people as savages, 36; loss of salmon as, 150–151, 166, 189–191, 207; obliterate history of through erasure of Indigenous presence, 44; resistance to through knowledge of land, 52–53, 57; and shame, 214; state-sponsored, 6, 45, 50–51, 57; use of herbicides as, 108. *See also* cultural genocide; frontier genocide; structural genocide
geographic names: erasure of Indigenous presence through, 42–43, 250n15; "squaw" in, 175
geography, 87, 194, 254n3 (chap. 5)
Glaze, Laverne, 3f
Glenn, Evelyn Nakano, 86, 196
gluten, 157–158
Go, Julian: on endurance of colonialism, 8; on exclusion of groups in sociology, 110, 128; on imprint of empire on sociology, 7, 78, 127; on militarization of police, 95; on white male elites and sociology, 81, 176
Goffman, Erving, 185
gold mining. *See* mining
Goldstein, Alyosha, 87
Goldtooth, Tom, 118
Goodwin, David, 181, 202
Goodwin, Jeff, 219–220
Goodwin, Jennifer, 165, 179, 184, 187, 198
Goodwin, Jesse Coon, 63–64, 66, 188
Goodwin, Robert, 9, 41, 171, 207
Grande, Sandy, 84
grasslands, disappearance of, 76
Graves, Harvey, 96
"Great Idaho Fire" (1910), 96
Grey, Sam, 163, 228
grief, 113, 202–209, 204t, 217, 219t
"grounded normativity," 41, 111
group identity, 91, 196
guilt, responsibilities and, 186, 235

Happy Camp, California, 206
harassment by law enforcement, 60, 64–66, 65f, 115

Harley, F. W., 98
Harrington, John P., 57, 158
Harte, Bret, 52
harvest gatherings, 92, 115, 179–180, 205, 207, 211
Hawaii, 172
healing, 178
health: emotional dimensions of, 149, 201; and food insecurity, 153; linked to environment, 19–20, 147–149; and loss of salmon, 130, 132, 140–141, 143, 147–149, 150–151, 157; and traditional foods, 72, 113, 130, 132, 148–149. *See also* "The Effects of Altered Diet on the Health of the Karuk People"; foods, traditional; Karuk Health and Fish Consumption Survey
health care costs, 153–154
health researchers in settler-colonial structure, 146–147
health studies: benefits of TEK to, 137, 140–143, 157–159; Indigenous collaboration in, 17, 135, 225; pathologizing of Indigenous people in, 135–136, 145
heart disease, 130, 147
Henderson, H., 95
herbicides in fire prevention programs, 107–108
heteropatriarchies, 197
Highway 96, widening of, 59–60
Hillman, Leaf: on allotment system, 59; on attitude while preparing food, 182; on collective continuance, 185–186, 207; on communicating Karuk TEK to U.S. Forest Service, 155; on criminalization of traditional lifestyle, 62–63; on dominance in Karuk culture, 167, 183; on environmental degradation as genocide, 150; on extermination of nature as extermination of Indigenous people, 51; on fire as cultural resource, 88, 90; on hopelessness, 215; on impacts to Karuk sovereignty, 119–120, 121; on Karuk femininity, 168; on "kincentricity," 91, 213; on loss of cultural identity, 205–206; on providing for the community, 179–180; on single species management, 42, 76; on traditional land management, 249n1; on use of fire, 74–75; on U.S. Forest Service policies as genocide, 110, 160
Hillman, Lisa: on allotment system, 59; on climate change and TEK, 223; on collective continuance, 185; on female gender roles, 177–178; on Indigenous people as part of nature, 51; on Karuk femininity, 168; on loss of seed-grasses, 166; on loss of women's status in Karuk community, 174–175; on single species management, 42; on women working together, 165
History of Humboldt County (Elliott), 53
History of the Six Rivers National Forest (Conners), 212
Hochschild, Arlie, 200
Homestead Act (1862), 33–34, 47, 58
homosexuality, 183
Hoopa Valley Indian Reservation, 52, 238
Hoover, Elizabeth, 87, 135, 152
hopelessness, 215–216, 216t, 219t
HoSang, Daniel, 45
House, Freeman, 53, 54, 55, 57
Houston, Marge, 113, 122–123
Howard, E. H., 47
huckleberries, 74t, 179, 180–181
human-nature divide, myth of, 231
humans and nature work together, 21
Humboldt Bay, 47, 53
Humboldt County, California, 40, 54, 101
Humboldt Times, 52, 100
hunger, 55, 153
hunting and gathering: harassment by law enforcement, 64–66, 65f; levels of, 145, 146f; physical benefits of, 132, 149–150, 151. *See also* fishing; regulations
Huntsinger, Lynn, 102, 109
Hurtado, Albert L., 49
Huynh, Megan, 233, 237
Huyser, Kimberly, 135, 214

identity: and emotional experiences, 216; emotions and, 200, 204, 204t, 205–206, 209, 210, 210t, 211, 212–214, 215t, 216–217, 216t, 218–219, 219t; of fishermen as providers, 179, 180, 188; and traditional foods, 21, 151–152; and traditional knowledge, 110–111. *See also* gender identity
identity, group, 91, 196
identity, pan-Indian, 38
Íhuk maidens, 170, 214
imagination, collective, 105
imagination, ecological, 239
imagination, sociological, 156, 234, 238–239

imprisonment, 65, 101, 115, 121, 210
"Incendiary Problem," 94–95, 98–100
"An Increasingly Rare Sight in California Mid-Elevation Mixed Conifer Forests" (Sweitzer), 42
indentured servitude, 173–174
Indian Sovereignty Movement, 117
Indigenous anti-capitalism, 110
Indigenous colonial resistance, land and, 52–53, 57, 97–98, 110
An Indigenous History of the United States (Dunbar-Ortiz), 1, 7, 78
"Indigenous Knowledges Are Not Just for All Humanity" (Whyte, et al), 227
Indigenous people: and anti-capitalism, 83–84; and climate change ideas, 226–227; and climate resistance, 236; as "close to nature," 38; collaboration in academic research, 135–137, 154–156, 158–159; court victories for, 238; "disappearance" of, 6; ecological damage as disruption to cultural management, 55–56; elimination of for legitimation of land claims, 84–85; gender roles as less rigid among, 178; as knowledge holders, 71, 92, 106, 110–111, 134; lack of section in ASA, 6, 79, 128, 224; notions of apocalypse, 229–230; as outside sociological inquiry, 127–128; pathologizing of, 128, 135–136, 145, 214; political leadership of, 4–5, 6, 13–15, 117–118, 125–126, 161, 232–234; racial formation of, 43–45; and racism, 14; resistance to fire suppression policies, 97–98; right to vote for, 48; as savages, 13, 36, 37, 46, 50, 51–52; teaching agriculture to, 58; use of fire, 73–75, 89. *See also* Karuk people
Indigenous perspectives: on capitalism, 31; in health studies, 17, 135, 225; omission of in sociology, 6, 117, 127–128, 134, 136–137, 154, 224
Indigenous research, 129, 156
Indigenous scholars as underrepresented, 15–16
Indigenous sociologists, 69, 136
Indigenous studies, 7, 20, 79, 195–197
individualism, myth of, 234
inequality: importance of natural world for, 170; and land, 32; and local food movements, 163
inevitability, sense of, 53
"information-deficit," 235

Integrated Wildland Fire Management Program, 112
intellectual property, 14, 242
interconnections: and grief, 203; within Indigenous cosmologies, 154–156; in native-social dualism, 12–13, 139, 230–232, 237–238. *See also* cosmologies, Indigenous; "kincentricity"; responsibilities, cross-species
intergenerational transfer of knowledge, 92–93
intermarriage, 38, 44. *See also* marriage
"internal colonialism," 14, 79, 80
"interspecies ethnography," 194
"intervening factors" in Karuk Health and Fish Consumption Survey, 140, 149–152
interviews, 142, 143, 148, 149, 150, 201, 241
Iron Gate dam, 130
Ishi Pishi Falls, 153, 181, 184, 206

Jacob, Michelle M., 127
Jacobs, Finn, 209–210
Jagger, Alison, 232
Jasper, James, 201, 209, 219–220
Johnston, Fay H., 112
Johnston-Dodds, Kimberly, 94
joy, 202
Jump Dance, 186

Karuk Aboriginal Territory, 26, 60, 119, 130
Karuk Constitution, 119
Karuk Department of Natural Resources, 4–5, 14f, 104–105, 106, 132, 139, 155, 161
Karuk Draft Eco-Cultural Resource Management Plan (Karuk Tribe, 2010), 88–89, 90
Karuk Health and Fish Consumption Survey (Norgaard and Reed): absence of comparable studies, 140–142; commodity food consumption in, 113; findings of, 143–154; intervening factors in, 149–152; methodology of, 140–143; publicity of, 154; reasons for diet shifts, 64, 144–147; use of TEK in, 142
"Karuk K-12 Needs Assessment" (Karuk Tribe), 217
Karuk-owned lands: for agricultural purposes, 58–59; under management of U.S. Forest Service, 56, 58; widening of Highway 96 on, 59–60

Karuk people, 249n2; abundance of natural resources of, 1–2, 4, 25; and effective social action, 235–238; federal recognition of, 161, 238; gender arrangements among, 170–171; gender constructions of, 165–166; gender roles less binary among, 183; interactions with black settlers, 40; interactions with Chinese, 39; knowledge of land to resist genocide, 52–53, 57; loss of status of women among, 174–175; population of, 4–5, 50, 209, 251n19; relocation to Hoopa Valley Indian Reservation, 52; return to traditional lands, 57; treaties with, 49. *See also* Indigenous people; men, Karuk; women, Karuk

Karuk Public Domain Indian Allotments, 59–60

Karuk social management, 2–3, 74

Karuk Tribe Eco-Cultural Management Plan, 40–41

Katimin (Karuk center of the world), 9

Kemple, Thomas M., 224

"kincentricity," 47, 91–93, 105, 109, 113, 160, 171, 198, 215, 218

Klamath Basin Food System Assessment (KBFSA), 65, 113, 115, 211

Klamath Hydroelectric Project, 129–131

Klamath National Forest, 47, 75–76, 98

Klamath people, 238

Klamath River: emotions and decline of, 189–190, 202, 205, 207, 208, 210, 212, 215, 215t, 220; as orienting point, 202; salmon production in, 4

Klamath River Basin, 1, 5

"Klamath River Jack," 97–98, 208

Klein, Naomi, 28, 231

Klopotek, Brian, 53, 87

knowledge, traditional: and cultural reproduction, 110–111; extraction of, 28, 241. *See also* traditional ecological knowledge (TEK)

Kosek, Jake, 32, 33, 34, 36, 67

Kotok, E. I., 72

Krieger, Nancy, 158

LaDuke, Winona, 117, 232

Lake, Frank: on alteration of vegetation by fire suppression, 105; on criminalization of Indigenous fire use, 97, 100–101; on destruction of cultural artifacts by fire suppression, 96, 103–104, 121, 122; effects of fire suppression policies, 93; on enhancing forest health through burning, 90, 109; on fire ecology and TEK, 89; on grief and inability to provide, 203–204; on incendiary problem, 94–95, 98; on loss of social capital, 205; on political sovereignty of fire, 91; on powerlessness, 215; on shame and identity, 212; on social ties to salmon runs, 92

lamprey, 42; abundance of, 4; in creation stories, 151; and dams, 130; and erasure of Indigenous ecologies, 87; extractive technologies and decline of, 55; and harassment, 65f; households fishing for, 146f; intimate connections with, 160, 202; in Karuk diet, 144; low yields of, 145–146, 184; and "primitive accumulation," 31, 84; providing of, 186, 203, 204; as rarely seen, 42; socialization while processing, 182. *See also* foods, traditional

land: alteration of relationships with, 67–71, 108–110; American sense of entitlement to, 82; as animate, 108–109, 114; and elimination of Indigenous people, 84–85; and food sovereignty, 163–164; and Indigenous colonial resistance, 110; Karuk title to, 117; qualitative dimension of, 109; and racial inequality, 32; racial meanings of, 17, 30, 32–39; and resistance to genocide, 52–53, 57; as savage, 87; settler access to, 173–174; settler-colonialism and relationships to, 87; as unoccupied, 53; as wealth, 69

land access: and firefighting activities, 121; denial through race definitions, 48; through elimination of Indigenous category, 44; and treaty negotiations, 49–50; and whiteness, 20, 30, 33–34, 69–70. *See also* allotment system

land appropriation, 30

land dispossession: and fire suppression, 78, 103; generation of wealth through, 14, 33–36; as ongoing process, 103–104, 127; and primitive accumulation, 57, 67; and U.S. imperialism, 83; and white supremacy, 20, 30, 46, 69–70; and willful ignorance, 82

land management policies, Karuk: criminalization of, 25, 61–67, 97, 115; differences to nontribal management systems, 62–63, 63t; historically, 3–4; restoration of, 9–11;

shaping of ecology through, 88; and sovereignty, 104–105, 117, 118, 119–125, 228; and World Renewal Ceremony, 213. *See also* fire management policies, Karuk

land management policies, nontribal: anger toward, 210–211; benefits of TEK in, 17, 71–72, 89, 97, 137, 155; and colonialism, 32, 94–102, 107–110; differences to Karuk land management, 62–63, 63t; disregard of Indigenous land occupancy and title, 66–67; forced assimilation through, 115–117; Indigenous involvement in, 13, 154–156, 237; long-term implications of, 120–121; and military science, 94–95, 97, 107, 120; racialization of, 224–225; resistance to, 97–98, 110; scientific documentation of, 140; settler-colonialism and, 32, 103–106; and transfer of wealth, 114; U.S. Forest Service use of, 59, 94–102. *See also* fire management policies, Karuk; fire suppression; U.S. Forest Service

land occupancy, failure to recognize, 45–46, 57–60, 66–67

land ownership, private, 47, 58–59, 60

landscape productivity, 47

language, Karuk, 57, 220

Large, Judith, 191

LaRocque, Emma, 172

Latour, B., 169

licenses, hunting, 62. *See also* regulations

lightning, 119

Living in Denial (Norgaard), 229, 235, 236

local food movements, 162–163

Lockie, Steward, 231

logging. *See* timber industry

Long, Jonathan W., 90

lookout towers, 96, 122

loss, responsibilities and sense of, 187–188

Maddux, Phoebe, 151, 158

Magubane, Zine, 80–81

Manifest Destiny, 66

manzanita, 103

maps, 42–43

Mares, T. M., 163

marijuana cultivation, 60, 115

marriage, 38, 44, 170, 174

Marshall Doctrine, 46–47

masculinity: and domination, 189; and natural environment, 171–172; and providers, 183; restructuring of, 191; as socially constructed, 177

masculinity, Karuk: and activism, 191–193; and colonialism, 167–168, 172–175, 177, 188, 189–193, 197; and environmental degradation, 193; forced alteration of, 172, 173; as not static, 193; as resistance to colonialism, 177, 190, 193–194. *See also* fishermen; men, Karuk

matriarchies, 172

Mawani, Renisa, 224

McCaffrey, Sarah, 102, 109

McCovey, Mavis, 108

McPhillips, Kathleen, 194–195

medical plants from Chinese miners, 39

medical records, 142, 147, 148

medicine families, responsibilities of, 178

Melody, Shannon, 112

men, Karuk: gender constructions of, 165–166; gender identity of, 180, 181. *See also* fishermen; masculinity

mental health, 19, 200, 214, 217, 218

Messerschmidt, James W., 176–177

methodologies, Indigenous, 154

Middleton, Beth Rose, 106, 223

Mid-Klamath Watershed Council, 237

military science in nontribal forest management, 94–95, 97, 107, 120

militias, 50, 57

Miller, Robert, 46–47

mills, 101

Mills, C. W., 156, 234, 238

mining: and Chinese people, 39, 40; destruction from, 42, 55; restructuring of wealth and ecology through, 47–48, 54; and treaties, 49

Mirowsky, J., 187

miscarriages, 107, 108

missions, forced assimilation and, 172

Mojola, Sanyu, 194

Moore, Donald S., 32, 33, 34, 36, 67

moral accountability, 111

"moral batteries," 220

moral teachings: in face of climate change, 226–238; supported by traditional management, 13–14, 91–93, 111, 137–139, 162

Morehead, Janet (Wilder), 115, 175, 198

Moreton-Robinson, Aileen, 7, 20, 25, 69

Morris, Aldon, 7, 78, 136
Mortsolf, J. B., 96
"motherwork," 196
moving forward, 106
mushrooms: and anger, 211; impact of fire retardants on, 122–123; regulations on, 64; yields of, 146. *See also* foods, traditional
mussels, 25, 144

Nakata, Martin, 7
National Environmental Policy Act (NEPA, 1970), 120–121
National Forest System, 56, 58, 60, 119. *See also* Klamath National Forest; U.S. Forest Service
National Resources Conservation Service (NRCS), 119
Native American Cultural and Subsistence Beneficial Uses, 132
Native people. *See* Indigenous people
natural resource policies. *See* land management policies, Karuk; land management policies, nontribal
The Natural Wealth of California (Cronise), 25
nature: association of Indigenous people with, 38; community interactions in, 206–207; and emotions, 199–200, 201, 216, 225, 254n3 (chap. 5); and gender constructions, 168, 169, 175, 184; and gender identities, 182–183; and gender relations, 194–195; humans as part of, 91–93, 113, 226–227, 230–231; and masculinity, 171–172, 193; notions of, 139; as pure, 37; and race, 34–35, 36–37, 67–69, 68t; and racial formation, 29, 30–31, 33–34, 35t, 66–70, 68t; and resistance to colonialism, 233; role of in sociology, 30, 221, 224; as source of power, 170, 227
nature as animate. *See* "kincentricity"
nature-social dualism, 167, 168–170, 175–177, 195–197, 221, 231–232
neoliberalism, lack of connections in, 233–234
Neuner, Sophie, 181–182, 198
newspapers, 19, 52, 100, 208
Nishnaabeg people, 83, 172
Nixon, Rob, 43

No Dakota Access Pipeline movement, 238
Norton-Smith, Kathryn, 228

oak trees: and fire suppression, 107, 121; sudden oak death, 150, 185–186, 203, 209. *See also* acorns
O'Brien, Jean, 53
Omi, Michael: on internal colonialism, 80; on race as dynamic, 37; on racial formation, 16, 26, 34; on racial projects, 45; on sociohistoric context of racism, 30; on whiteness, 33
The Omnivore's Dilemma (Pollan), 161
"one eleven" tattoo, 175
oneness, sense of, 202
Orleans, California, 96, 98, 112f
Owens, Patricia, 81

PacifiCorp, 129–130, 131, 131f, 153–154
Pandian, Anand, 32, 33, 34, 36, 67
parenting, 194
Park, Lisa Sun-Hee, 17, 30, 34, 37, 46
Park, Robert, 80–81
Patel, Raj, 163, 228
pathologizing of Indigenous people, 128, 135–136, 145, 214
patriarchies, 172
Peek, Lori, 194
Pellow, David: on "critical environmental justice," 29; on environmental discrimination, 17; on environmental racism, 37, 46; on institutional racism, 30; on lack of Indigenous perspectives in sociology, 69–70; on racial formation, 34; on racialization "traveling," 51; on racial profiling, 66; on resistance to genocide, 53; on toxins and racism, 70–71
physical activity of hunting and gathering, 132, 149–150, 151
Pierce, Ronnie, 131f
Pikyávish. *See* World Renewal Ceremony
Pinchot, Gifford, 95–96
plantation lands, redistribution of, 33–34
"poaching," 62–64, 211
"poison food," white food as, 151, 158, 208
police activity in colonial framework, 95
political action in social movement scholarship, 233
political ecology, 117

political movements, Indigenous, 21, 133, 137–138, 161, 227, 232–233, 238
political sovereignty, 13–14, 46–47, 76, 91, 104–105, 117–118
Pollan, Michael, 161
Polletta, Francesca, 219
Polmateer, Mike, 183, 188, 220
porcupines, 42
postcolonial theory, 8, 79, 80–82; Indigenous critiques of, 7–8, 10, 28, 83–87, 196; and settler-colonialism, 84, 85–86, 102, 232
Postcolonial Thought and Social Theory (Go), 7
potatoes, Indian: disappearance of, 76, 152, 206; as First People, 171; and gender constructions, 166, 170; responsibility to, 179; in stories, 93, 185
poverty, 26, 114, 153
power: cultural dimensions of, 237; and gender identities, 188; importance of natural world for, 170, 227; Indigenous conceptions of, 232–234; Indigenous-state, 83–84; and masculinity, 177
powerlessness, 234, 235. *See also* hopelessness
Powers, Stephen, 57
power structures, emotions of environmental decline and, 200, 219–220, 219t
"Practicing *Pikyav:* A Guiding Policy for Research Collaborations with the Karuk Tribe," 9, 242
Pratt, Richard, 61
prayers, 179, 213, 220
Preston, Vikki, 209
pride, 180
"primitive accumulation," 31, 41–42, 47, 54, 57, 67
Princes Right to Conquest, 56
privilege, environmental, 229–230, 235, 236
providers: and elders, 11, 180f, 183, 189; fishermen as, 179, 180, 188, 204; Karuk women as, 170–171; and masculinity, 183
publicity: of "The Effects of Altered Diet on the Health of the Karuk People," 19; of Karuk Health and Fish Consumption Survey, 154
Pulido, Laura, 30–31, 70, 87, 159, 160
"pyrodiversity," 89
"pyro-kincentricity," 92

queer theory, 176
Quinn, Scott, 58

Rabbit, 11, 93, 202, 203, 212
"race making," 26–27, 29, 67–69, 68t
race/racism: and climate change, 70; families as resistance to, 196; for Indigenous people, 14; legal definitions of, 48; and natural environment, 67–69, 68t; as not static, 30; perpetuation through nature-society dualism, 232; settler-colonialism within, 86–87; as social construction, 26
racial capitalism, 33, 57, 70–71
racial categories: and access to natural environment, 34–35; as construct of colonizer, 44; as dynamic, 37–38; elimination of through destruction of nature, 38; forced assimilation as elimination of, 61; transfer of wealth across, 54–56
"racial-colonial formation," 22, 23, 27, 32, 250n12
"racial-colonial project," 30
racial formation: of colonial ecological violence, 190; as distinctive, 43; and ecological domination, 34; environmental exposure as, 17; importance of natural environment to, 29, 30–31, 33–34, 35t, 66–70, 68t; of Indigenous people, 43–45; land of recognition of Karuk land title and occupancy as, 60; and natural environment, 27, 30–31, 33–34, 35t; of the state, 47, 159; through comparisons with other oppressed peoples, 51; through environmental practices, 35t, 36, 67–71, 68t, 224–225; and wealth transfer, 26
racial inequality, importance of land in, 32
racial profiling, harassment by law enforcement as, 66
"racial projects," 26, 45, 53–55. *See also* forced assimilation; genocide; land occupancy
racism, anti-Chinese, 40
racism, institutional, 26, 30, 32–33
railroads, 40
Raphael, Ray, 53, 54, 55, 57
real estate prices, 60
reciprocal responsibilities, 111
recreation, regulations on, 62–65
Redding, Pierson B., 54
Red Power movement, 232–233

Reed, Ron: on anger as loss of salmon, 210; on "being Indian," 11, 93; on climate change opportunity, 223, 234; collaboration with nontribal agencies, 155; on cross-species responsibilities, 179; on cultural use species, 12, 116; effect of state policies on environmental health, 134; on enhancing what is used, 62; and environmental justice, 159; on exclusion of fires as exclusion of Karuk people, 102; fishing, 5f, 180f; on fishing as rite of passage, 181; on fixing the world, 9; on guilt and responsibility, 186; and health study proposal, 140–141, 157; on importance of burning forest for river health, 2; on inability of salmon subsistence, 153; on "Indigenous social management," 74, 92, 206; as insider, 241–242; and Klamath River Hydroelectric Project, 130, 131; in methodology of health study, 142; on shame and responsibility, 212–213, 214; on speaking out on behalf of fish, 192–193; on traditional foods and health, 72, 105

reflex emotions, 219

regulations: anger at, 210–211; as barrier to Indigenous foods, 62, 65, 115, 211; on fire suppression, 100–101; on fishing as forced assimilation, 12; on hunting, 62–64; and inability for fishermen to provide, 188; on mushrooms, 64. See also harassment by law enforcement

replanting programs, 124, 125

Republic of Nature (Fiege), 7, 30

reservation system, 50, 52

resistance: to assimilation, 13, 35t, 38, 175, 193–194, 220; collective continuance as, 191; to colonialism, 18–19, 110, 116–118, 127–128, 190–196, 220, 223; to denial of access, 116; direct action, 220, 235, 236–238; and emotions, 219–220; gendered dimension of, 175, 177, 189–191, 190, 193–194, 196; to land-altering policies, 110; "one eleven" tattoos as, 175; and relationship to the natural environment, 38, 127. See also activism; environmental justice

responsibilities: and activism, 192–193; to act on climate change, 226–227, 234–238; and anger, 187; and gender constructions, 165–167, 170–171, 182–183; and grief, 205; and guilt, 186; Indigenous ethics of, 41, 111, 138, 161–162, 170–172; of Karuk women, 178; and sense of loss, 187–188; and shame, 212–213

responsibilities, cross-species, 12–13, 14, 91–93, 137–138, 173, 179, 191. See also "kincentricity"

responsibilities, family, 41

responsibilities, sacred, 162

restoration ecology, 158, 237

Richards, R. T., 39

ridge systems, 103–104, 121

riparian areas, damage to, 120, 121

rites of passage, 181

riverine species, record-low harvest yields of, 145–146, 146f, 184. See also lamprey; salmon; sturgeon

road building, 120, 121

Robertson, Dwanna L., 129, 154, 156, 159

Robinson, C. J., 70, 194–195

role strain, 186–187

role stress, 186–187

Ross, L., 187

sacred sites, destruction of, 96–97

Safe Drinking Water Act (1974), 107

salmon: abundance of, 2, 4, 25, 42; in ceremonies, 144, 207, 213; cultural significance of, 4, 11, 92, 151–152, 186, 207–208; denial of access to, 12, 63t, 66, 132, 145–146, 210–211; and gender identity, 179–180, 181, 183, 191–193, 205, 210; as ideal food, 141, 149, 157; impact of loss on health, 130, 132, 141, 147–149, 148f; in Karuk diets, 4, 130, 143–144, 146, 152–153, 158; loss of as emotional burden, 19, 150, 151, 202–203, 206, 210, 215, 217; loss of as genocide, 56, 150–151, 190, 207, 209, 217; loss of as recent, 140, 184; as managed, 10, 41; record-low harvest yields of, 4, 145–146; as relatives, 160, 171, 182, 198, 215; responsibility to, 179, 192–193, 213; restoration of, 191–193, 220, 237. See also fishermen; fishing

Salmon River Restoration Council, 237

Schaefer, B. P., 95

Scheff, Thomas, 212

The Scholar Denied (Morris), 7

Schrock, D., 189

Schwalbe, Michael, 189

science, Western: collaboration with traditional knowledge, 132–135, 140–143,

145, 154–159; and cosmologies of colonialism, 15, 63t, 154–155, 196–197, 229; replacement of traditional knowledge by, 25–26, 39, 42, 229. *See also* traditional ecological knowledge (TEK)
seed-grasses, 166
self-reported data, 147, 148, 153
settler-colonialism: alteration of environment in, 221–222; and capitalism, 28; and climate change, 126; and climate injustice, 228–229; as dynamic, 87; emergence of, 6, 14–15; and forced assimilation, 114–115; health research in, 146–147; Indigenous erasure in, 73, 81–82, 87–88, 103, 133; key dynamics of, 77–78; legitimation of land claims in, 84–85; operation with other theories, 85–87; use of within sociology, 8, 85
Settler Colonial Studies, 102
settler environmentalism, 230
sex as biological, 176
shame, 212–214, 215t, 219t
shame, collective, 151, 186
Sherman, Jennifer, 191
Show, S. B., 72
Shriver, Thomas, 216
signal functions: of anger, 209, 210; of emotions, 218, 225; of grief, 204, 217
Simpson, Leanne: on alteration of gender structures, 172, 173; on capital, 31, 83, 84; on extraction of Indigenous, 28; on private property, 33
single species management, 42
Siskiyou County, California, 39
Siskiyou Volunteer Rangers, 51
Six Rivers General Inspection Report (1950), 99–100, 212
Skinner, Carl N., 89
Smith-Lovin, Lynn, 200, 204, 206, 211, 216
smoke, 111–112, 112f
Smokey Bear, 75f, 94, 96
Snipp, C. Matthew, 127
social, notion of: and gender, 169, 176; and whiteness, 81
social action, collective, 234
social action, Karuk experience and, 235–238
social capital, 205
social interactions: and grief, 204; and natural world, 206–207

social movements scholars, failure to understand Indigenous movements by, 232–233
social-nature duality. *See* nature-social dualism
social networks, 41, 84, 92, 152, 187, 206–207
social resources, disproportionate access to, 32–33
socio-economic stress, food insecurity and, 153
sociological imagination, 156, 234, 238–239
sociology: and activism, 238; and climate change, 238–239; colonial amnesia in, 82; emergence of, 7, 80–81, 176; emotional experience of environmental decline in, 217–218, 222; and epidemiology, 158; erasure of colonialism from, 80–82, 224; erasure of Indigenous perspectives in, 6, 117, 127–128, 134, 136–137, 154, 224; exclusion of social groups in, 128; imprint of socially dominant on, 127; natural environment in, 16–18, 29–31, 69–71, 88, 221, 224, 231–232; nature-social dualism in, 175–176, 195–197, 221; pathologizing of Indigenous people in, 128, 214; reproduction of white supremacy in, 136; role of nature in changing gender relations, 194–195; use of settler-colonialism within, 8, 85. *See also* feminist sociology
sociology of emotion, 206–207, 221, 225
sociology of gender, 175
Solnit, Rebecca, 229
sovereignty, biocultural, 118, 233
sovereignty, cultural, 118
sovereignty, food, 21, 137–138, 162–164, 225
sovereignty, Indigenous notions of, 137
sovereignty, Karuk, 117, 119–125, 228, 253n22
sovereignty, political, 13–14, 46–47, 76, 91, 104–105, 117–118
Special Committee on the Disposal of Public Land, 49
"squaw" in geographic names, 175
stand conversions, 76
Standing Rock reservation, 227, 238
starvation, 55, 153

the state: alteration of relationships with land, 43, 54, 58–60, 62–71, 68t, 78–88, 94–102, 108–110; appeals for protection to, 160; as colonial, 31–32, 77–78, 84, 107–108, 159; as not an ally, 160; as racialized, 26, 31, 33, 70, 159, 193–194; racial projects by, 45, 46–56, 56–60, 61–67
Stauffer, Renee, 208–209
Steinman, Erich W., 116, 117, 232–233, 238
Steinmetz, George, 81, 95
Stephens, Scott L., 94, 95–96, 97, 113–114
stigma of tradition, 178
Stoermer, Eugene F., 230
stories, Karuk, 91, 93, 151, 170–171, 185
stress, chronic community, 217
structural genocide, hopelessness and, 215–216
structural violence, environmental justice and, 221
sturgeon: abundance of, 4; and dams, 130; extractive technologies and decline of, 55; households fishing for, 146f; intimate connections with, 202; in Karuk diet, 144; low yields of, 145–146, 184. *See also* foods, traditional
Styker, Sheldon, 204
substance abuse, 153, 217, 218. *See also* alcohol use
sudden oak death, 150, 185–186, 203, 209
sugar pines, 103, 205–206
Sugihara, Neil G., 94, 95–96, 97, 113–114
suicide, 69
Suman, Seth, 99
Sunderland, Rose, 61
Supreme Court, Doctrine of Discovery and, 46
surveys. *See* "The Effects of Altered Diet on the Health of the Karuk People"; Karuk Health and Fish Consumption Survey
survival, ethics of, 235
Swillup Creek (California), 43
symbolic interactions, emotions and, 200
symbolic violence, 190

Talley, Geena, 207
tattoos, 175
Taylor, Dorceta E., 33
TEK. *See* traditional ecological knowledge (TEK)

"10 A.M. Policy," 96
Tengan, Ty P. Kāwika, 172
This Changes Everything: Capitalism vs. the Climate (Klein), 231
timber industry: damage to ecosystem through, 55; fire damage to, 59, 72; in Humboldt County, 54; production in, 100, 101, 194; protection of through fire suppression, 75–76; salvage sales, 47, 124; wildfire as threat to, 59
toxins: and Anthropocene, 230; environmental justice focus on, 18; exposure to, 36–37, 160, 194–195; fire retardants, 120, 122–123, 123f; placement of, 30, 216; and racial capitalism, 70–71; use in fire suppression, 107–108. *See also* herbicides in fire prevention programs
traditional ecological knowledge (TEK): as adaptable, 89, 108, 140, 155–156, 223; benefits of, 135, 137, 140, 156, 158–159; in collaborative research, 132–135, 140–143, 145, 154–159; and environmental health, 40, 137, 157–159; and fire, 74–75, 89–92, 97–98, 103–104, 106; and Karuk identity, 110–111; and "Klamath River Jack," 97; and "pyro-kincentricity," 92; replacement of by settler practices, 25–26, 39, 42, 101–102; and sovereignty, 14, 91, 104–105, 117–121; and traditional foods, 10–11, 25, 151–152; use in Karuk Health and Fish Consumption Survey, 140–143, 145; and U.S. Forest Service, 13, 154–156, 237. *See also* fire management policies, Karuk; land management policies, Karuk
traditions. *See* cultural production
Trafzer, Clifford E., 34, 35, 43–44, 46, 47, 73
trails, 116, 122
transgender, 176, 183
trauma, intergenerational, 114
treaties, 49–50, 56, 117
tribal identity, 44–45, 186
tribes: as "domestic dependent nations," 117–118; as leaders in climate change, 104–105, 106, 226–238; as unique, 44–45
Triplicate, Del Norte, 208
Tripp, Harold, 96
Tripp, William, 40, 106, 121, 123–124
trout, 4, 25, 144, 202. *See also* foods, traditional

Trump, Donald, 227, 233–234
Tsosie, Rebecca, 118
Tuana, Nancy, 232
Tuck, Eve: on decolonization, 79, 82–83, 138, 249n4; on importance of land, 85; on settler-colonialism, 84; on settlers and race, 86, 87

Udel, Lisa, 196
unemployment, 187, 191, 205
United States: as an empire, 80, 81–82; transfer of wealth to non-Indigenous settlers, 114
U.S. Army directing murder of Indigenous males, 173
U.S. Department of Agriculture, 113
U.S. Fish and Wildlife Service (USFWS), 62, 119
U.S. Forest Service (USFS): claiming of Karuk lands, 58; communication of Karuk perspectives to, 123–124; damage through firefighting activities by, 228–229; and elderberries, 198; fire suppression by, 12, 75–76, 88, 95–96; genocidal practices of, 110, 160; impact on acorns, 42, 185; and Karuk TEK, 13, 154–156, 237; land management policies of, 59, 94–102; laws on fire suppression, 62, 100–101; long-term consequences of policies, 121; management of Karuk-owned lands, 56; non-Indigenous visions of forest management, 94–102; refusal to acknowledge Karuk boundaries, 119–120; regulations on mushroom harvesting, 64; shaping of ecology, 42, 75–76, 88, 185; use of fire retardants by, 122–123; use of scientific rhetoric by, 99; and value of timber, 42. *See also* fire suppression; regulations
U.S. Indian Commission, 49
U.S. Secretary of Agriculture, 58
U.S. Secretary of the Interior, 58
U.S. Senate, treaty negotiations and, 50

Van Willigen, Marieke, 216
Varese, S., 118
Vasquez, J. M., 178
Veracini, L., 102, 116
Via Campesina movement, 21, 137, 163
Vietnam War, 107
villages, hydraulic mining and Karuk, 42, 55, 57
villages, intergenerational transfer of knowledge in, 92
violent masculinity, 191
voting rights for Indigenous people, 48

war zones, 120, 191
Washington Post, 19
water quality standards, 67, 107–108, 123, 124, 132
water systems: and fire retardants, 122; impacts of fire on, 90
Watts, Vanessa, 129
wealth: generation of, 10, 26, 36, 41–42, 67, 76; land as, 32–33, 69; and race, 29, 33–34, 36–38, 41–42, 54, 61, 69, 82; role of environment in production of, 25, 28, 66–67
wealth, transfer of: and environmental degradation, 86, 132, 190–191; and land management policies, 11–12, 108, 113–114; and racial categories, 54–56, 72–73. *See also* "primitive accumulation"
Webb, Gary, 216
Weber, Max, 168
Web of Science (WoS) Social Citation Index, 136
Weeks Act (1911), 96
West, Candice, 177, 189
West, Thomas, 96
Western Klamath Restoration Partnership, 106, 237
West Oakland Farmers Market, 163
Wetzel, C., 178
Whitbeck, Les, 208, 217
whiteness: as civilized, 50; and land access, 20, 30, 46, 69–70; and manipulation of natural environment, 33–34; and notion of social, 81; as pure, 37
The White Possessive: Property, Power, and Indigenous Sovereignty (Moreton-Robinson), 7
white supremacy: and colonialism, 20, 33–34, 36–37, 46–47, 50–55, 57, 86–87, 177; and environmental degradation, 52, 55–56, 67–70, 95, 98, 208–211, 224–225; as racial project, 53–55; reproduced in sociology, 136. *See also* "racial projects"

Whyte, Kyle Powys: on "ancestors' dystopia," 229, 236; on climate injustice, 125, 228; on "collective continuance," 177, 190–191; on "containment," 109; on "covenant of responsibility," 92–93, 111; on ecological restoration, 237–238; on Indigenous ecologies, 43, 73, 76, 104; on "kincentricity," 139, 162; on land dispossession, 83, 84–85; on settler ecologies, 78, 83, 95; on settler-Indigenous collaboration, 227; on U.S. imperialism, 86
Widick, Richard, 47, 52, 54, 55
Wildcat, Daniel, 226
wilderness, 10, 43, 53, 250n16
Wilderness Act (1964), 43, 250n16
wilderness in environmental justice, 18
wild game, 55
Wiley, Austin, 52
Wilkes, Rima, 127, 156
willful ignorance, 53, 82
Wilson, Paul G., 101
Winant, Howard: on internal colonialism, 80; on race as dynamic, 37; on racial formation, 16, 26, 34; on racial projects, 45; on sociohistoric context of racism, 30; on whiteness, 33
Wing, S., 144, 146
Witz, Anne, 168, 176
Wiyot people, 38
Wolfe, Patrick, 81–82, 84, 108
women, Karuk: gender constructions of, 166; gender identity of, 180–181; loss of status in community, 174–175; and "one eleven" tattoos, 175; as providers, 170–171; responsibilities of, 178; socialization of, 165, 181–182; as "taken," 174; in traditional stories, 170–171
World Renewal Ceremony, 9, 40, 59, 181–182, 210, 213

Yang, K. Wayne: on decolonization, 79, 82–83, 138, 249n4; on importance of land, 85; on settler-colonialism, 84; on settlers and race, 86, 87
Yes Magazine, 106
Yurok people, 1, 102, 108, 109, 130, 238

Zimmerman, Don, 177, 189

ABOUT THE AUTHOR

DR. KARI MARIE NORGAARD is a non-Native professor of sociology and environmental studies at the University of Oregon and author of *Living in Denial: Climate Change, Emotions, and Everyday Life* and other publications on the intersections of gender, race, climate change, and sociology of emotions. She has engaged in environmental justice policy work with the Karuk Tribe since 2003.

Printed in the United States
By Bookmasters